U.S. Marine Corps Physical Readiness Training for Combat

Plus

Rappelling

Department of Defense

Pentagon
Publishing

UNITED STATES MARINE CORPS
Marine Corps Combat Development Command
Quantico, Virginia 22134-5001

29 January 1988

FLEET MARINE FORCE REFERENCE PUBLICATION............0-1B

MARINE PHYSICAL READINESS TRAINING FOR COMBAT

1. PURPOSE

Fleet Marine Force Reference Publication (FMFRP) 0-1Marine Physical Readiness Training for Combat is intended for use by all Marines. It provides the information and references necessary to establish and conduct physical conditioning programs to prepare Marines for the physical demands of combat.

2. APPLICABILITY

Although not directive, this handbook is intended for use as a reference by all Marines in developing unit physical readiness training for combat. This FMFRP does not establish policy for the conduct of physical fitness training. Refer to Marine Corps Order 6300.3_ for applicable Marine Corps policy on physical fitness. It may also be used as a basis for instruction conducted by unit leaders or as a reference for instruction or training.

3. RECOMMENDATIONS

To improve this manual, the MAGTF Warfighting Center seeks comments and recommendations from the user. Officers and noncommissioned officers with constructive comments are encouraged to respond. Lack of response will be considered as concurrence. The User Suggestion Form which follows can be detached and forwarded to - -

Commanding General
Marine Corps Combat Development Command (WF 06)
Quantico, Virginia 22134-5001

4. DISTRIBUTION

Initial distribution of this FMFRP 0-1B is shown below. Addressees on this list should ensure that appropriate subordinate organizations are provided copies.

JOHN I. HOPKINS
Brigadier General, U.S. Marine Corps
Assistant Chief of Staff, Warfighting
Marine Corps Combat Development Command
Quantico, Virginia

DISTRIBUTION: TDJ

From:

To: Commanding General, Marine Corps Combat Development Command
(WF 06), Quantico, Virginia 22134-5001

Subj: FMFRP 0-1B, MARINE PHYSICAL READINESS TRAINING FOR COMBAT, RECOMMENDATIONS CONCERNING

1. In accordance with the Promulgation Page to FMFRP 0-1B, which invites individuals to submit suggestions concerning this FMFRP directly to the above addressee, the following unclassified recommendation is forwarded:

Page________Article/Para No.________Line No.________Figure/Table No.________

Nature of Change:_____Add_____Delete_____Change_____Correct

2. Proposed New Verbatim Text: (Verbatim, double spaced; continue on additional pages as necessary.)

3. Justification/Source: (Need not be double spaced.)

NOTE: Only one recommendation per page.

PREFACE

MARINE PHYSICAL READINESS TRAINING FOR COMBAT

Chapter 1
PHYSICAL READINESS LEADERSHIP
Section I. Role of Physical Fitness in Combat

1101. PURPOSE OF PHYSICAL FITNESS TRAINING

Physical fitness training in the Marine Corps has one purpose: to prepare Marines to physically withstand the rigors of combat. All other goals of physical fitness training are subordinate to and must support attainment of this goal. The idea that only infantry or reconnaissance units and their attachments normally face physically demanding combat is wrong. This error must not influence the priority commanders of combat support, combat service support, aviation, and headquarters units give to physical fitness for combat. Physical fitness for combat has a high priority for all Marines. A sound, effective unit program, requiring limited time and material, offers a greater payoff in combat than many more expensive and time-consuming training programs. This handbook describes the concept and provides the ingredients of a program for commanders to use in carrying out one of their most serious responsibilities: ensuring that their Marines are physically ready for combat.

1102. SCOPE OF THE MANUAL

This handbook provides guidance for all leaders, trainers, and planners of physical training programs. It describes unit physical fitness for combat training in the following chapters:

a. Physical Readiness Leadership. Chapter 1 provides guidance to leaders in the conduct of physical combat readiness training.

b. Physical Readiness Training Programs. Chapter 2 discusses how to structure programs to meet physical readiness goals in a variety of combat training situations.

c. Physical Conditioning Activities. Chapter 3 discusses the primary physical conditioning activities which commanders may use to attain readiness for combat goals. Sections in this chapter address foot marches under load, strength-building activities, and activities which build confidence and the aggressive spirit needed in combat.

d. Combat Water Survival. As an amphibious force, Marines cannot ignore the importance of combat water survival training. Chapter 4 describes a program for commanders to conduct this training.

e. Competitive Conditioning Activities. Chapter 5 describes competitive activities which stress the relationship to combat actions, featuring team-building types of competition.

f. Evaluation of Performance During Training. Chapter 6 describes tests which measure individual and unit physical fitness for combat.

g. The Human Body. Chapter 7 covers the structure and functioning of the body.

1103. COMMANDER'S ROLE

Major General Commandant John A. Lejeune in the 1921 edition of Marine Corps Manual expressed our philosophy of leadership. Among other things, General Lejeune stated that "it will be necessary for officers ... to endeavor to enlist the interest of their men in building up and maintaining their bodies in the finest physical condition." It is significant that the Commandant prefaced these writings in 1921 with the statement that his thoughts were greatly influenced by the experience of World War I. We have never varied as a Corps from this belief that one of a commander's most serious responsibilities is to have Marines physically prepared for combat. Senior commanders and their staffs carry out this role primarily by providing command emphasis, including planning, support, and supervision. Commanders at the battalion/squadron and company/battery level execute a program of physical fitness for combat by close supervision and, most importantly, leadership by example.

1104. HISTORY OF MILITARY PHYSICAL READINESS

Every war has revealed our military physical deficiencies during the initial periods of mobilization. This realization followed the Civil War and has recurred regularly with each national emergency.

a. Training programs in each war were geared to the physical need of the era. Success was dependent upon the amount of time available during training to physically prepare Marines for battle conditions. Frequently, casualties in initial engagements were attributed to the inability of our Marines to physically withstand the rigors of combat over rugged terrain and under unfavorable climatic conditions. With adequate preparation, Marines have always handled the test of battle.

b. During World War II, the first physical conditioning doctrine that could be scientifically justified by testing was introduced. As the war progressed, this program was effective in the physical conditioning of millions of Marines for combat.

c. Postwar periods have traditionally been a time of consolidation. Unfortunately, some leaders considered the conditioning phase of training to be a wartime tool. With such a philosophy prevalent between wars, physical readiness was relegated to secondary importance resulting in a serious lowering of combat effectiveness. The initial commitment of Marines in Korea dramatically displayed this failure to recognize the extreme physical demands of warfare. Call-up of large numbers of reserve Marines and pressing them into the war within weeks after call-up mainly caused the loss in effectiveness.

d. Costly lessons learned from our military experiences over a period of years and the course of several wars led to an increasing interest in the physical conditioning of the individual Marine. No longer can we afford to emphasize physical fitness during wartime and de-emphasize it during peacetime. It is evident that, in spite of increased mechanization and modern weapons, physical readiness retains a vital place in the life of each individual Marine and in every unit within the Marine Corps regardless of mission.

e. Commanders are well aware of the need for rugged and well conditioned Marines. The daily demands of housekeeping, maintenance, support, training, operations, and other time-consuming tasks make it necessary for commanders to set aside time for frequent, regular, and vigorous exercise periods.

1105. PHYSICAL DEMANDS OF COMBAT

There are three primary elements of effective physical fitness for combat: lower body strength and stamina; upper body strength and stamina; and a competitive, combative spirit. Cardiovascular functioning is not treated as a separate element since it is a necessary foundation for the other three. Exercises and activities designed to improve Marines' condition in the three primary elements will foster good cardiovascular and respiratory condition.

a. Lower Body Strength and Stamina. Some common demands that combat places on lower body strength and endurance are marching long distances under load and functioning effectively at the destination; moving quickly and evasively under fire; and carrying wounded Marines to safety.

b. Upper Body Strength and Stamina. Some common demands that combat places on upper body strength and stamina are rapidly emplacing crew-served weapons; handling large-caliber ammunition for extended periods; climbing walls, cliffs, and other high obstacles; and performing field maintenance on air craft or heavy machinery.

c. Competitive, Combative Spirit. While part of this critical aspect of combat readiness is mental and emotional, robust physical condition and a training program which promotes physical aggressiveness greatly enhance a competitive and combative spirit. Activities which force Marines to overcome natural physical fear, which directly leads to fatigue, are particularly valuable.

1106. PHYSICAL EFFECTS OF COMBAT STRESS

It has been thoroughly documented that the added uncertainty and stress of combat have a major physical effect on Marines as well as the acknowledged psychological effect. In other words, in combat, fear equals fatigue. In training, we cannot easily reproduce this combat stress which reduces the effectiveness of individuals and units. However, we can produce fatigue and psychological doubt in training situations by developing a challenging physical training program which stresses the participants. This compels Marines to exercise their ability to continue to function under difficult and trying conditions. Such activities as long foot marches under load and difficult confidence courses train Marines to overcome their own fear and fatigue. Rugged competitive activities such as martial arts and pugil stick fighting prepare Marines psychologically to overcome an opponent as well as their own fear and fatigue. This sort of physically demanding training builds self-confidence and unit morale. The adage that "the unit that sweats more in peacetime will bleed less in battle" certainly applies to its physical training program. Athletic coaches have long followed this approach by such practices as having teams shoot foul shots after wind sprints in order to strengthen their mental conditioning and toughness. Although pushing Marines to their limit is necessary, commanders often set lower standards initially than ultimately desired. The program should then call for steadily increasing the difficulty. A good guide for a starting point is a standard that will be physically and psychologically demanding for no less than three-fourths of the unit. Expressed another way, no more than one-fourth of the unit should be working at less than their limit. The program should progress quickly, augmented, if needed, by remedial training for those unable to meet the unit standards, until the training is challenging for all but the most exceptional Marines in the command. Commanders must never forget that Marines expect to be challenged and thrive on being pushed to their limit.

1107. LIMITS OF PHYSICAL READINESS

A well-conditioned Marine unit has a significant advantage in combat, but this does not mean that physical conditioning can substitute for technical knowledge or good planning. For instance, units must, as discussed in chapter 3, pursue a continuing and challenging program of marches under loads. This does not disprove the axiom that fighting men, if required to carry over one-third of body weight, will generally become ineffective in battle. Leaders and planners must decide what is essential to be carried and must use an effective logistic distribution system, such as unit trains, to make remaining items available when they are needed. In a similar vein, the valid requirement to foster a competitive and combative spirit in Marines does not excuse leaders from the responsibility to practice tactics which will not needlessly risk lives. Physical readiness for combat is a vital goal, but it is only one part of a unit's overall readiness for the lest of combat.

Section II. Fundamentals of Physical Fitness

1201. FITNESS FOR MARINES

Total combat readiness includes technical proficiency and mental, emotional, and physical fitness. If any one of these attributes is lacking, combat effectiveness suffers proportionately. Without technical fitness, Marines lack the knowledge and skill to fight; without mental and emotional fitness, they lack the incentive and will to fight; and without physical fitness, they lack the physical ability and confidence to fight. Physical fitness in a Marine means a whole and healthy body, the capacity for skillful and sustained performance, the ability to recover from exertion rapidly, the desire to complete a designated task, and the confidence to face any eventuality.

1202. COMPONENTS OF PHYSICAL FITNESS

A sound body, free of disease and defect, does not in itself constitute physical fitness. Before an untrained Marine can be considered physically fit for combat, the following traits must be developed:

a. Strength. Every Marine needs enough strength to perform the heaviest task encountered in routine and emergency activities. The basic areas where strength is required are in the arm and shoulder girdle, abdomen, back, and legs. Muscles increase in size, strength, and firmness with regular and strenuous exercises. Without work, they grow flabby and weak.

b. Endurance. Each Marine needs sustaining power to maintain the maximum ability without undue fatigue. There are two types of endurance:

(1) Muscular Endurance. The Marine needs muscular endurance to fight the enemy under the most tiring combat conditions. Muscular endurance is the ability to perform continuous work over long periods of time. Endurance depends on the bloodstream's ability to deliver large amounts of oxygen and nutrition to the muscle masses and then carry away the waste products quickly.

(2) Cardiovascular-Respiratory Endurance. The development of cardiovascular-respiratory endurance ("wind") is necessary to maintain muscular endurance. Cardiovascular-respiratory endurance depends on the efficiency of the blood vessels, lungs, and heart. The maximum effort a Marine can exert over a period of time is limited by the capacity to absorb oxygen and expel carbon dioxide. The average Marine's cardiovascular-respiratory capacity can be greatly increased by exercise.

c. Agility. A Marine must be able to change direction quickly and as faultlessly as possible. The ability to react instantly and to maintain orientation during rapid changes of body position is important to survival. This important ability may be developed by conditioning exercises that require varied and rapid changes of body position on the ground and in the air, such as obstacle courses.

d. Coordination. Coordination is the ability to move all parts of the body in a smooth, efficient, and concerted effort (commonly called timing). A well-coordinated individual does not make useless movements. An individual who moves with precision and accuracy saves energy. Coordination is best developed by practicing diversified muscular activities and skills affecting all body parts.

1203. TYPES OF EXERCISE

Basically, there are two forms of exercise: isotonic and isometric. Both forms are contained in the activities found in this manual.

a. Isotonic. Isotonic exercises are those in which the expenditure of energy is regulated and released during concentrated efforts. The regulated expenditure of energy is controlled by both the mode of exercise and the individual's effort. This type of effort is common to most exercise and sports. To develop endurance, coordination, and flexibility, isotonic exercise must be used. Strength can also be developed or increased through this type of exercise.

b. Isometric. Isometric exercises are those in which maximum effort is applied and held until the engaged muscle energy is depleted during a single contraction effort. The individual exerts full force against an immovable object for a relatively short period of time (6 to 10 seconds) and then repeats the contraction several times with a short period of rest between each contraction. This type of exercise develops only strength; therefore, it has limited application.

1204. PRINCIPLES OF PHYSICAL CONDITIONING

Men and women vary in their physical makeup. Physiological (body) function and reaction also vary in proportion to heavy demands placed upon the body. To attain the maximum benefit without sacrificing Marines' welfare, fitness must grow with a careful program of conditioning. To allow for adjustments in body functioning as the conditioning program progresses and to ensure attainment of objectives, the following principles of physical conditioning must be applied:

a. Overload. As strength and endurance increase, the physical load must be increased until the desired level of fitness is reached.

b. Progression. In beginning stages, the load must be moderate. Gradual progression from this low state of fitness to a higher state is possible through application of a progressive program.

c. Balance. An effective program utilizes various types of activities and provides for the concurrent development of strength, endurance, and coordination as well as basic physical skills.

d. Variety. Some programs fail because the routine becomes boring. The most successful programs always include conditioning activities, competitive events, and military physical skill development.

e. Regularity. There is no easy or occasional way to develop physical fitness. Regularity of exercise is a must, with daily exercise preferred.

1205. THREE STAGES OF PHYSICAL CONDITIONING

Unconditioned or poorly conditioned Marines pass through the following stages in reaching the desired state of physical condition.

a. Toughening Stage. This stage is approximately 2 weeks long and is usually characterized by muscular stiffness and soreness followed by recovery.

b. Slow Improvement Stage. This stage is approximately 6 to 10 weeks long and is characterized by slow and steady improvement until the desired level, or a high level, of fitness is attained.

c. Sustaining Stage. This stage goes on indefinitely in order to maintain the level of conditioning achieved by passage through the previous stages.

1206. EFFECTS OF CLIMATIC CONDITIONS

Temperature, both atmospheric and body, affects the physical performance of Marines. Proper maintenance of body temperature through warm up exercise, proper dress in cold weather, and removal or adjustment of clothing in hot weather is necessary for effective performance and health. Climatic factors to be considered are the following:

a. Exercise In High Temperatures and High Humidity. Marines can endure strenuous physical activity in extremely hot temperatures if they are given an opportunity to become acclimated and if they take enough salt and water. It is essential to continue physical training programs in hot climates. Marines can better withstand high temperatures when they are well-conditioned. High humidity combined with high temperatures presents a serious danger. These conditions prevent the natural cooling of the body by the evaporation of perspiration. Training schedules should conform to the provisions of the base commander in respect to wet bulb conditions. Those who conduct training under these conditions should monitor weight loss and be careful to make fluid replacement adjustments.

b. Exercise at High Altitudes. Certain problems are encountered in conditioning Marines stationed in high altitudes because the heart undergoes greater exertion during exercise. It is particularly important that only light exercise be given initially at such altitudes. A Marine's body gradually adjusts to high altitudes within a few weeks. After this adjustment, progressively greater amounts of exercise are possible.

c. Exercise in Arctic Regions. Military duty in the arctic is so arduous that a high level of physical conditioning is essential. Because of the difficulties of carrying on physical conditioning exercises in extreme cold, Marines should be conditioned to the highest level possible before they arrive. A sustaining program will then maintain that level. When exercising in cold weather, Marines should be required to remove excess clothing to prevent them from becoming damp with perspiration.

1207. WARMING UP AND COOLING OFF

It is a fundamental physiological principle that Marines should warm up gradually before taking strenuous exercise. Such action speeds up the circulation to prepare the body to take an overload and helps to prevent injury to muscles and joints after exercising. Marines should be kept mildly active, walking, stretching, or performing some other mild muscular activity until breathing and temperature have returned to normal. Marines should never be allowed to cool off too rapidly; in cool or cold weather, they should put on additional clothing during the cooling-off period.

1208. PHYSICAL ACTIVITY AS AGE INCREASES

Combat makes severe physical demands on Marines. All Marines, regardless of age, must be physically ready to meet these demands. There is no physiological reason to cease exercise or exertion with age. Increased age usually brings increased responsibility which, in many instances, leads to a routine that can become almost devoid of physical activity. The key to fitness with increased age is to continue exercising at a reasonable level and to include exercise of a vigorous type in the daily routine. Older persons who have not regularly maintained a reasonable state of physical fitness require more time than younger persons to become fit. Such individuals usually require a longer period of time to recover from physical effort than younger Marines. If general health is good, evidence shows that older Marines can develop and maintain a degree of fitness which permits vigorous activity and proper performance of duties. It is both the individual's and the commander's responsibility that all Marines continue a daily sustaining exercise program.

Section III. Goals of Physical Readiness Training

1301. NECESSITY FOR PHYSICAL READINESS TRAINING

A very important objective of training is attainment and maintenance of operational readiness. Marines must be physically ready for operations at any time and under all conditions of climate and environment. A combination of training to develop proficiency in physical skills and conditioning to improve strength and endurance results in physical readiness for combat. The degree of physical fitness required of Marines can be acquired through physical exercises in a normal well-coordinated and closely monitored training routine. The performance of purely military duties, such as drills and marching, is not enough to build all the desired areas of fitness. Few recruits are physically fit for the arduous duties ahead of them. The softening influences of our mechanized civilization and the difficulties of conditioning Marines make physical fitness more important than ever before. If Marines are to be fully developed to and maintained at the desired standard of physical fitness, a well-conceived plan of physical readiness training must be a basic part of every training program. Marines cannot be adequately prepared in any other way for the hard work and arduous demands required on the battlefield.

1302. OBJECTIVE OF PHYSICAL READINESS TRAINING

The overall objective of the physical readiness training program is to develop individuals and units who are physically able and ready to perform their duty assignments or missions during training and in combat. To attain the objective of physical readiness, exercise activities must develop the following areas:

a. Strength and Endurance. Developing adequate strength to perform required duties and adequate endurance to sustain activity over a long period of time.

b. Muscle Tone. Developing muscle tone adequate to maintain proper posture and reasonable weight control.

c. Skills. Developing military physical skills which are essential to personal safety and effective combat performance. As skill is developed, agility and coordination are attained. The essential skills are-

(1) Marching Under Load. Marching with individual and unit weapons and equipment.

(2) Running. Distance and sprint running on roads and cross-country.

(3) Jumping. Broad jumping and vertical jumping downward from a height.

(4) Dodging. Changing body direction rapidly while running.

(5) Climbing and Traversing. Vertical climbing of ropes, poles, walls, and cargo nets. Traversing horizontal objects such as ropes, pipes, and ladders.

(6) Crawling. High crawling and low crawling for speed and stealth.

(7) Throwing. Propelling objects, such as grenades, for a distance with accuracy.

(8) Vaulting. Surmounting low objects, such as fences and barriers, by use of hand assists.

(9) Carrying. Carrying objects and employing man-carries.

(10) Balancing. Maintaining proper body balance on narrow walkways and at heights above normal.

(11) Falling. Contacting the ground from standing, running, and jumping postures.

(12) Surviving in Water (Or Other Specialized Situations) Using water survival techniques.

d. Character Traits. Jnstilling character traits which help accomplish military missions to include-

(1) Confidence. Confidence develops through achieving progressively more difficult tasks as physical ability grows.

(2) Aggressiveness. Participation in combative activities and contests develops desire and willingness to overcome an opponent.

(3) Reaction Under Pressure. Competitive contests and game situations train Marines to think and to act quickly while under pressure.

(4) Teamwork. Teamwork develops through competitive events in which a number of Marines must coordinate their efforts to accomplish a physical task.

1303. BENEFITS OF EXERCISE

The benefits of exercise are not always understood. Some of the more important results of exercise are listed below:

a. Improved Muscle Tone. Muscular tone improves and, at the same time, muscular strength and endurance are built up.

b. Cardiovascular-respiratory Endurance. Cardiovascular-respiratory endurance, or wind, improves through a process of opening up dormant lung capacity to absorb greater amounts of oxygen.

c. Circulation. Circulation of the blood speeds up and extends to a greater portion of the body as exercise forces the blood to service all parts of the body. The efficiency and effectiveness of the heart, lungs, and blood vessels improve.

d. Flexibility. A wider range of muscular movement is possible and rapidity in physical skills grows.

e. Elimination of Body Waste. Bending and twisting the body and the general speedup of body processes caused by exercise regulate and help eliminate body wastes.

f. Tension. Working off excess nervous energy and relief from daily worries and cares relieve tension. Participation in exercises leaves little time for worry.

g. Sleep. Sleep improves because muscles are healthfully tired after about of exercise. A by-product of sound sleep is relief of tension.

h. Obesity Control.. Control of obesity (fat) is made possible by using up excessive amounts of fat-producing food elements.

i. Injury Susceptibility. Susceptibility to injury is reduced through exercise. Muscles, tendons, and joints are strengthened. Injuries such as hernia, back strain, and joint sprains are less likely to occur if muscles are maintained in proper tone.

Section IV. Leadership Roles

1401. PSYCHOLOGICAL LEADERSHIP

The full development of a Marine's resources is not all physical. To be effective in developing physical readiness, leaders must realize that mind and attitude are also important to success. The more important psychological considerations are to—

a. Promote Understanding of the Value of Physical Readiness. A desire to be physically ready should be created in all Marines. Motivation is increased and Marines take greater interest in their individual physical fitness if they understand the value and benefits of vigorous exercise. When Marines realize their efforts are an investment in their own personal welfare, it should not be difficult to obtain their cooperation. Marines should understand the objectives, the benefits, and the value of each type of exercise activity in their program. They should also understand the relation of physical readiness to survival in combat.

b. Maintain a Positive Approach. Physical readiness training for combat is strenuous and demanding. It is a responsibility of leadership to create an atmosphere where all desire to participate fully. This attitude should be fostered. A negative approach must not be identified with physical readiness training, even with those having difficulty. Only in unusual cases should fear of punishment be the motivating factor behind good performance. For those few who cannot keep up or attempt to malinger, an effective remedial program is essential.

c. Seek Cooperation and Develop Morale. In a program placing maximum physical stress upon individuals, it is necessary to gain their cooperation. Favorable reaction is enhanced by proper planning and organization, challenging requirements, use of competition, and application of a progressive program resulting in physical fitness. As physical fitness grows, morale also grows.

1402. COMMAND AND SUPERVISORY FUNCTIONS

a. Command Functions. Commanders should take the following actions to support physical readiness training:

(1) Lead by personal example.

(2) Instill command interest and indicate to subordinate personnel the importance of this training to the welfare of the organization.

(3) Allot sufficient time for the achievement of objectives and monitor the use of such allotted time. The substitution of other training or routine duties for scheduled physical readiness training is unsound and unwise.

(4) Assign and properly utilize qualified personnel to supervise and conduct physical readiness training. If leaders are not competent, take action to ensure they become competent quickly. Reassign those who do not meet this standard to other duties.

(5) Make necessary facilities and funds available to support a program to develop physical readiness within all personnel.

(6) Measure the physical fitness of individuals and units in order to evaluate progress and to determine if the program is successful.

b. Supervisory Functions. Leaders responsible for planning, conducting, and supervising physical readiness training should take the following actions:

(1) Prepare physical readiness training schedules which apply the principles of physical conditioning and which aim for a particular type of program plan.

(2) Provide for wide participation of as many Marines as possible. All Marines, regardless of position or age, will benefit from regular exercise. In some instances, special efforts are necessary to overcome obstacles to regular and frequent training. Special effort is also necessary to ensure remedial conditioning. Such conditioning should occur for those who are physically substandard and after extended absence due to leave, sickness, injury, and travel.

(3) Prevent waste or unwise use of time allotted for physical readiness training. Time-wasters include unprepared instructors; assignment of one instructor to a group larger than a platoon; progression which does not keep pace with the physical development of the Marines; extreme formality; inadequate equipment or facilities which require waiting turns to exercise; and lengthy rest periods between exercises which interfere with the application of overload.

(4) Ensure that the program contains vigorous physical activity. Such activity places progressively greater demands upon the body during each exercise session and also over the duration of the training program. To be of benefit, exercise must tire the muscles and cause the heart to increase its rate of beat.

(5) Set an overall objective for each physical fitness program.Observe the training as necessary to ensure that the established objectives are being achieved.

(6) Observe physical readiness training to insure the use of a positive approach. To implement a positive attitude, small-unit leaders and instructors should personally set the example; have an understanding, fair, and sympathetic attitude; recognize individual differences; and motivate Marines toward their best effort.

(7) Guide and inform small-unit leaders and instructors concerning approved techniques, directives, and literature. As necessary, arrange for local training of instructors to include clinics, conferences, schools, and demonstrations.

(8) Determine the effectiveness of physical readiness training by personal participation in and observation of training, analysis of field inspection reports, and analysis of individual physical fitness test scores. Scores may be combined to reflect the fitness of the unit.

1403. SMALL-UNIT LEADERS AND INSTRUCTORS

a. Responsibility. The instruction and conduct of physical readiness activities are the function of company/battery commanders, platoonleaders and persons assigned as instructors. Experience has proved the effectiveness of physical fitness development when conducted in company-and platoon-size units under direct control of the leader with overall supervision by the parent-unit commander. For example, all the platoons of a company may exercise at the same time under the general supervision of the company commander, with each platoon conducting the assignment separately and under its own leadership.

b. Leader's Assignment. A small-unit leader or an instructor in a school or training activity is assigned to a combat unit or to a support unit. In this assignment, the leader is responsible for all training to include physical readiness training. In a different situation, the noncommissioned officer (NCO) or officer may beassigned as a full-time physical readiness instructor. This contrasts to a unit leader assignment where only part of the time is devoted to such training. In either case, the leader will hold an important and vital position for the physical fitness of Marines.

c. Leader's Training. Leaders may come to the assignment either fully or partially trained. It may be their first responsibility for the development of physical fitness. If they have had previous training through experience, make certain that their information is supplemented with study of this manual. If they have had professional training in physical education during civilian life, but no military experience, they should also use this manual to learn the methods used by the Marine Corps. A new leader should take advantage of various ways to learn including attendance at leader training courses, self-study, practice, and discussion with more experienced leaders.

d.Leader's Objective. As a physical readiness training instructor, the leader has two general objectives. The first is to motivate Marines to want to be physically fit. The second is to conduct a program that will develop a high degree of physical fitnss. Motivated Marines will react enthusiastically to such a program. It aids greatly in achieving local program objectives.

e. Leader's Personal Fitnes. A unit leader who must instruct and demonstrate physical activities must be in physical condition to do the job without undue physical stress.The leader should be able to do those things that must be demonstrated. The leader's strength, endurance, posture, and skill should set the example. This does not mean that the leader must excel, as other Marines do not expect championship performance. However, they do expect, and deserve, a creditable showing of fitness for the job.

f. Leader's Knowledge. The leader must have three types of knowledge to properly administer physical readiness training. Theyare- -

(1) Knowledge of Marines. The leader must understand Marines, know how to lead and motivate them, understand how they learn, and apply this knowledge wisely in the day-to-day training situation.

(2) Understanding of Body Functioning. A more intelligent exercise program results from understanding and applying the principles which govern physical conditioning of the body. The leader with such knowledge can better prescribe, adjust, and regulate exercise types, amounts, and progression to attain fitness.

(3) Understanding Exercise Activities. The leader needs to understand the contribution each type of physical activity makes to physical fitness, and how to use each activity to develop fitness. Skill to demonstrate and lead the various activities is a necessary part of technique and is invaluable to the instructor or small-unit leader.

Chapter 2
PHYSICAL READINESS TRAINING PROGRAMS
Section I. Development of a Program

2101.GUIDANCE FOR PLANNERS

This chapter instructs planners on physical readiness training procedures. It contains program planning guidance including factors to consider when developing programs; steps in assembling a program; definitions of activity packages and systems of exercise; selection of activity packages; and selection of systems used in implementing those packages.

2102. MARINE CORPS MISSION

The mission of the Marine Corps is to seize and defend advanced naval bases and to perform other missions as the President of the United States may direct. Traditionally, to perform this mission, Marines have been projected into the area of operations by Navy vessels and have attacked hostile beaches over the shore in landing craft, assault amphibian vehicles, and helicopters. Recently, performance of this task has been somewhat complicated by the introduction of the concepts of maritime prepositioning and the air-landed Marine expeditionary-brigade. Under these concepts, Marines, landed at airfields in foreign countries, will join their equipment at commercial ports. The distance between these airfields and ports is often 10 or more miles. In time of international crisis, it can be expected that transportation assets will not be adequate. Marines from all elements of a Marine Air-Ground Task Force may have to march from the airfield to the site of their equipment. All Marines must be physically conditioned to rigors of conducting foot marches with individual weapons and equipment.

2103. EVALUATION OF FITNESS

The Marine Corps mission is such that all Marines must constantly and consistently achieve a high level of fitness that prepares them for the demands of combat. While Marines have many different MOS's, all Marines must be prepared for the demands of marching under load and performing basic infantry tasks, such as rear security and patrolling. While the physical fitness test remains our universal measure of individual fitness, it is the commander's responsibility to observe and evaluate the unit's ability to perform effectively in combat. Often the commander's most effective evaluation tool is weekly physical training sessions. The commander should participate in these sessions and see that they are fully integrated into the unit physical conditioning program.

2104. INTEGRATION OF TRAINING

Training time must be used efficiently and wisely. Every opportunity to integrate physical training into other training activities should be seized. For example, foot marches under load can be integrated into the normal daily activities as a way to move from one training or work site to another. This integration reflects realism in training and should be used to maintain overall proficiency. To accomplish this requires imagination and enthusiasm on the part of the commander. The area of operations in which the training is conducted determines what can be integrated and how. If possible, training should also consider the terrain and climate of the area in which the unit will subsequently conduct training or operations. Training should include familiarization with special equipment and the application of specialized techniques to tactical principles. Activities such as gun drills and command post emplacement drills also have physical conditioning value, particularly if supervised and done competitively or measured against time standards.

2105. ACTIVITIES AT THE UNIT LEVEL

Physical readiness training is a command responsibility and is generally conducted in Marine units at the company and platoon levels. Battalion/squadron commanders must constantly supervise, making sure that all Marines are physically prepared for combat.

a. Physical readiness training should usually be conducted weekly by each battalion/squadron-level unit, using events designed for all personnel (e.g., marches under load, battalion/squadron runs, military field days, etc.).These events enable the commanders to demonstrate personal leadership, and observe and evaluate the physical condition and combat physical readiness of their Marines. This exercise period should be designed to elevate the unit's morale and emphasize unit identity through the wearing of uniforms and the carrying of battalion and company guidons. The battalion/squadron commander should make every effort to be a visible participant in the unit's physical readiness program on a weekly basis. These activities could be conditioning marches under load or a weekly calisthenics/unit run.The suggested times for these activities are early on Monday morning or late on Friday afternoon after the training week has been completed. The uniform may be physical training gear orutilities and boots.

b. Commanders should be cautious of long runs in boots and utilities. This is not to suggest that units should not run in utilities, but to remind commanders that boots are designed for marching, not running. If commanders choose boots and utilities, they should select off-road routes which will provide a cushioning effect.

c. Units should be made well aware of the commander's intent before the unit falls out for exercise. If the exercise period is conducted early in the morning, it need not necessitate an extensive shower/clean-up period afterward, so long as all Marines from the commander on down remain dirty. Commanders should inform their Marines not to fall out for these sessions in clean, pressed uniforms.

d. The physical readiness training of headquarters units is often difficult to manage. Often physical training is left for individuals or sections to develop and manage on their own. This approach can be effective, but if not supervised, it can result in a "paper" program. Headquarters unit commanders should endeavor to conduct weekly integrated training which will enable them to evaluate the physical readiness of their personnel. Often this training has to be scheduled before or after normal working hours to maximize participation of the entire unit.

2106. STEPS IN PLANNING

To implement workable and effective programs (as directed by CMC ALMAR 261/87) the planner must-

a. Determine the Type of Program Needed. Marine units are inherently different in organization and mission. The physical readiness program must be tailored to the mission and to the current physical condition of most unit personnel. Programs to meet this need are of the following types:

(1) **Developmental Programs.** Marines in a beginning or poor state of physical readiness need a program which will develop strength, endurance, physical skills, and character traits which are beneficial to successful accomplishment of military missions. Such programs should be applied progressively to rise gradually to a peak of fitness and skill.

(2) Maintenance Programs. Once Marines reach the sustaining stage of conditioning, their goal is then to maintain this level by participation in a maintenance program.

(3) Remedial Programs. The term "remedial" is usually applied to those individuals or groups who possess substandard physical fitness. For example, a remedial physical conditioning program could be applied to persons who are overweight, who fail to reach physical fitness test standards, or who have missed extended periods of conditioning due to illness, injury, extended hospitalization, or other absence. Saturday training sessions are designed as remedial sessions for all individuals who either miss the daily training or have failed to meet adequate standards. Note that personnel in light duty or no-duty status should be expectedto attend all training sessions, observing, supervising, or assisting as needed.

b. Determine the Time Required. The amount of time for training operations varies considerably. However, every unit can find time to conduct physical readiness training. Frequent (i.e.,daily) physical training of short duration-30 to 60 minutes-is preferred over occasional longer periods. Other demands for training time are so urgent that every minute of time allotted for physical readinesstraining should be used. Determine the time required per week anddivide it into daily blocks.

c. Organize for Various Group Size. It is essential to stress exercise rather than formality. Marines must complete the program where they are- -on the training field, in the motor pool, on the range, next to the classroom, in the office area, in the shop, aboard ship, or elsewhere. Although desirable, it is not always possible to assemble company-size units for physical training. Platoon-size groups are appropriate for the proper conduct of physical conditioning activities. Certain situations may require exercise programs for section- or squad-size units. This manual outlines programs for all situations and types of organizations.

d. Allow for Weather and Exercise Area. In programing and scheduling, the climate and terrain often govern the selection of activities.

(1) Weather changes cause differences in temperature, rainfall, wind chill, and snow. These changes should be anticipated as they dictate the type of program. Alternate plans should be part of the schedule. Nevertheless, normal weather changes and rain should not drive training inside.

(2) Local terrain and available exercise areas may also influence the selection of activities and the type of program which it is possible to support. Some activities can be completed in nearly any area.

e. Plan for Seasonal Change. As most physical readiness training is conducted outdoors, it is necessary to recognize seasonal change. A program should be divided into fall, winter, spring, and summer parts. Seasonal change also causes change in light. For example, an early morning program started in the summer will have ideal lightconditions, yet in the fall or winter, darkness will occur at that same hour and interfere with the conduct of the program. Develop programs in seasonal blocks, and make provisions for anticipated changes.

f. Consider Needed Facilities. An excellent program can be conducted with practically no facilities since there are exercises which require no equipment. However, a better program can be developed when supported by certain facilities and items of equipment. Proper command support, plus ingenuity, will solve this problem. (Items of equipment, when necessary to support the recommended exercises, are included in the chapters on exercise activities.)

g. Specify Appropriate Uniform. The uniform worn for exercising depends upon the season of the year, the state of the weather, and local regulations. Whenever possible, Marines should be dressed alike. Undershirts are preferred as the upper garment when the weather permits. A uniform that restricts the free movement of the body should not be worn when exercising.

h. Consider Availability of Instructors. Leaders who can lead and direct the scheduled activity must be available. Organizational units should train junior officers and noncommissioned officers down to squad or section leaders to instruct and lead the various activities.

i. Select Activity and System. With the type of program needed and the objectives in mind, select an activity package or an integrated training package for each day's scheduled physical training. At this same time, the system to be used in employment of the selected activity or activities must be determined. This selection affects equipment, areas, instructors, transportation, and other support requirements.

j. Secure Command Participation and Support. Prepare and brief the commander to assure full understanding of the objectives and administration of the program. The full participation and support of the commander will greatly improve the level of success.

k. Supervise Execution. Determine needs, publish the program, and supervise its execution as a necessary part of developing the unit's training schedule.

2107. SELECTION OF ACTIVITIES AND SYSTEMS

a. Exercise Activities. The planner must determine exercise activities which will be appropriate to include in the program. An exercise activity is a single means of exercise usually identified by the name applied; for example, running, log exercises, and obstacle course. Many exercise activi ties can become part of activity packages.

b. Activity Packages. Many of the physical activities described in this manual are arranged in prescribed sequences and are known as activity packages. An activity package is a number of exercises of the same type, assembled as a group or a set, and arranged in a specific sequence. Exercise packages are organized in such manner that not more than 15 minutes will be required to complete the execution of any package. Each type of activity is explained in laterchapters. The number of available drills, tables, or circuits; the manner of organization; and the contribution each makes to the total program are covered. Full understanding of this information will greatly assist in developing effective programs. Various designations are used to identify exercise packages; for example, conditioning exercises when arranged in a set order are known as drills, and other packaged activities are designated as tables or circuits. The following are activity packages:

(1) Activity packages for groups:

o Conditioning Drill Two
o Conditioning Drill Three
o Rifle Drill
o Log Drill
o Grass Drill
o Running Tables
o Guerrilla Tables
o Circuit-Interval Table
o Combatives Tables
o Relay Tables

(2) Strength Circuits:

o Fixed Circuit
o Movable Circuit
o Simplified Circuit (Circuit Interval Table)

(3) Activity packages for individuals:

o The 6-12 Plan
o Weight Training
o Isometric Exercise

c. Advantages of Using Packages. These of exercise packages simplifies scheduling and conducting of exercise and results in the following benefits:

(1) Schedule development is simplified as the planner assembles packages which will satisfy the training objective. There is no need to deal with selection of individual activities or to be concerned about the amount of time to be expended on each.

(2) Any 15-minute period, and in some cases less time, can be scheduled or used to perform an activity.

(3) If longer periods of time are available or if the objective demands, several packages can be assembled to provide a more complete period of activity.

(4) Marines are assured a balanced set of exercises or activities as each package is carefully arranged to reach all muscle groups.

(5) The instructor can concentrate on the conduct of a vigorous workout as the type and duration of the activity have already been determined.

d. Nonpackaged Activities. Several types of activity are not packaged. Activities in this category are conditioning marches, unit runs, obstacle courses, combat water survival swimming, teamcontests, and team sports. These activities can be scheduled in combination with packaged activities, or they may be scheduled separately. Often they require a longer period of time. Most nonpackaged activities require a 50-minute period to satisfactorily complete their objective. For conditioning marches, a few hours are required. The benefits of these activities should not be overlooked as some desirable objectives cannot be attained without them.

e. Systems of Exercise. Several methods or systems of organizing exercise and activity packages can be used. Each system is based upon a specific organization as follows:

(1) Single Activity System. The unit leader immediately assumes command of the unit at the beginning of the exercise period. The leader moves the unit to a predesignated exercise site at double time, forms the unit in a circle around the leader, groundsclothing and equipment as appropriate, and quickly moves into the exercise routine. There is usually no time to teach; therefore,the Marines must know the activity to be used. At the conclusion of a 5- to 15-minute period, the unit leader returns the unit to the instructional area at double time and releases the unit for the next scheduled activity.

(2) Progressive Activity System. All Marines (company or platoon) complete activities in the same order during the period. For example, Drill One is followed by dual combatives, and finally a 1-mile run. This system is usually progressive from a warm up activity, such as Drill One, to an activity which contributes in a major way to one of the objectives, such as aggressiveness development through combatives or cardiovascular development, such as running.

(3) Rotating Activity System. Set up the same number of activities or stations as there are platoons in the company. Each platoon rotates through each station in turn. With four platoons in a 50-minute period, about 10 minutes can be devoted to each station. With three platoons, approximately 15 minutes may be spent at each station. Activities must be of a type that can be covered in the time allotted. For example, with three platoons, Station 1 could be Conditioning Drill One; Station 2, running; and Station 3, a team contest.

(4) Circuit System. Set up a number of stations to provide various types of exercise equipment and/or items of apparatus. The idea is to keep all Marines busy and exercising vigorously for a short period of time at each station. The fixed strength circuits and the movable strength circuits are examples of this system. Station changes must be rapid, and the exercise must be started quickly after each change. Since the objective is to exercise at top speed, the motivation comes from frequent change of activity by moving to another station. Rotation by station groups continues until all Marines have covered all stations.

(5) Interval System. This system stresses the development of strength and endurance. It involves heavy work for a given distance within a specified time, alternated with lighter work and recovery, but never stopping during the workout. This procedure is repeated, and the intensity is increased gradually as physical condition improves, but always with adequate recovery. The important concept is stress, recover, stress, recover, and so on. This system is often applied through running, but other activities of a continuous nature may also be used. An example is the Circuit-Interval Table.

f. Steps in Selecting Activities and Systems. The planner should follow these steps.

(1) Evaluate the needs of one unit with emphasis upon unit mission, objective, and time available.

(2) For each day, select an activity package or a combination of activities which will contribute to the objective.

(3) Then determine the system to be used in implementing the selected activities. There are several possible choices and much opportunity for flexibility in program development. For example, the weekly program may contain various exercise packages, systems, and time periods. Additional guidance is contained in following chapters.

2108. SAMPLE WEEKLY PROGRAMS

The variety of weekly programs given shows available alternatives. Programs and schedules other than those illustrated may be assembled to provide schedules to fit any situation. Many other scheduling combinations are possible through use of the packaged activities, varied time periods, and different types of assigned duty.

a. Sample A. Figure 2-1 illustrates a 60-minute time allotment in which the single activity system is used with activity packages that vary from day to day. In this program, 5 hours of training are scheduled for the week. This program needs some type of integrated unit training to supplement it each week. A bimonthly conditioning march of 10 miles in 3 hours is suggested to round out this program. The Saturday session is designed for remedial physical training.

NOTE: A physical training activity must be conducted at a minimum of once every 72 hours. After 72 hours, muscles begin to atrophy.

Figure 2-1 is available in the illustration section at the end of the book

b. Sample B. A more comprehensive weekly program (fig. 2-2) illustrates the scheduling of various-length periods, a variety of activities, and the use of three systems during the week. In this schedule, 3.5 hours are included for the week. The following should be noted concerning this schedule:

Figure 2-2 is available in the illustration section at the end of the book

(1) On Monday, a 60-minute period is available for physical readiness training. Three 20-minute packages are scheduled. Each platoon, under its own leadership, will progress through each scheduled activity in turn.

(2) On Tuesday, Wednesday, and Thursday, only 30-minute periods are available. Here each platoon leader supervises each platoon in a single activity system.

(3) On Friday, the company commander directs training to exhibit personal leadership example and supervision. In this example, only one strength circuit and one cross-country course are available, and each will accommodate only one platoon. For this reason, Conditioning Drill One station is included. A platoon rotates to each station within the 60-minute period.

(4) The Saturday session once again is designed for remedial physical training.
(5) Bimonthly, a 10-mile/3-hour march under load will be necessary to round out this program.
NOTE: Cardiovascular/respiratory exercise to improve endurance must be of long enough duration (20 minutes or more) to induce and maintain oxygen debt.

c. Sample C. A third sample schedule (fig. 2-3) illustrates the use of the single activity system for 5.25 hours of training. This schedule includes sustaining-type activities for a unit that has passed through both the toughening and slow improvement stages of conditioning. The use of competitive activities is featured to holdinterest and provide self-motivation. A bimonthly 10-mile/3-hour march under load will be necessary to round out this program. The Saturday session once again is designed forremedial physical training.

Figure 2-3 is available in the illustration section at the end of the book

NOTE: Thc body will respond to hard work or physical training without soreness and stiffness provided the work or physical training is conducted regularly.
NOTE: Excrciscs of short duration increase strength and bulk but do little to effect endurance.
NOTE: Thc times shown in the charts arc participation times.

Section II. Remedial Physical Training

2201. NEED FOR ATTENTION

Remedial physical training is a process by which physically substandard individuals are conditioned to meet standard levels of performance. To achieve a full degree of operational readiness, it is necessary to bring all Marines up to the prescribed standard. Experience has demonstrated that some Marines have difficulty due to a poor state of fitness, obesity, or lack of motivation. Attention to these individual deficiencies will help improve unit combat readiness.

2202. IDENTIFICATION OF INDIVIDUALS

The company or battery commander identifies persons who cannot achieve the prescribed level of proficiency through the physical fitness test or as a result of observation during training. The commander notes particular weakness of body parts as indicated by failure of certain test events. These Marines are then placed in a special remedial program either at platoon, company/battery, or battalion/squadron level. Extra time is devoted to overcoming the weakness. These extra periods of conditioning may be during or after duty hours, as determined by the commanders. Other persons need reconditioning as result of hospitalization. Restoring physical fitness to damaged areas of the body is accomplished through progressively graded physical activities under professional supervision, not by the physical fitness trainers.

a. Types of Deficiencies. Physical deficiencies which can be corrected by exercise fall into several categories.

(1) Lack of Strength in One or More Body Parts. The major muscle areas concerned are the arms and shoulder girdle, back, abdomen, and legs.

(2) Lack of Overall Endurance. Usually, there is deficient muscular and cardiovascular endurance.

(3) Deficiency in Coordination and Agility. In these cases, physical skill is not developed to a satisfactory degree in activities such as crawling, running, jumping, climbing, traversing, vaulting, pushing, pulling, lifting, and carrying.

(4) Overweight or Underweight. Either condition may interfere with physical fitness and mission accomplishment. Lack of exercise is not always the cause. The cause may be malfunction of normal physiological functions or it may be poor health habits such as over- or under-eating, lack of adequate rest, or overconsumption of alcohol.

(5) Lack of Motivation. Not all persons are motivated to attain or maintain **a** desirable state of fitness. Some Marines do not understand the importance of physical fitness, some find proper exercise too difficult, and others find it inconvenient.

b. Causes of Deficiencies. There are several causes for Marines being deficient in physical fitness:

(1) Absence of exercise.

(2) Exercise which fails to develop all muscle groups and components of fitness.

(3) Exercise which is not vigorous enough or which lacks progression.
(4) Injury or illness which depletes fitness.
(5) Inadequate amounts of sleep or rest.

c. Methods of Detection. The commander has several means by which to detect physical deficiencies:
(1)
from those who pass. Make an analysis to determine the cause of failure as revealed by test scores.
(2) Observe Marines as they perform physical tasks-
o Marines who have difficulty during training or in physical types of work,
o Marines who have become obese and therefore experience difficulty.
(3) Be alert to those Marines who are often on sick call, returnees from hospitalization, or newly assigned.

2203. ADMINISTRATION OF REMEDIAL ACTION

a. Group. Attitude Marines who are singled out as being physically deficient are self-conscious and are not always convinced they need extra help. Within a deficient group, motivation may be low. These Marines must be convinced that a remedial program which is tailored to their needs will help them and will eliminate their deficiencies. Encouragement is often needed and desirable.

b. Leadership. The leader of this group must study individuals and know their deficiencies. The unit leader should counsel Marines individually, observe them closely as they progress through the remedial program, maintain records, and adjust the program as required.

c. Measurement.. Whenever improvement in performance is noted, Marines should be measured by the physical fitness test either individually or by group. When testing reveals an individual to be satisfactory, that individual should be released from the remedial group. There may be exceptions to this policy in the case of Marines who are overweight or in the execution of an exercise program prescribed for some special purpose.

d. Organization of a Remedial Group
(1) A remedial group is usually a small group of Marines within a company or battalion. In some situations, it may be a larger group numbering 50 to 75 Marines. In the case of a larger group, divide the Marines initially into subgroups according to ability. Prescribe exercise loads commensurate with their ability. General conditioning is sufficient in the early part of the program to qualify Marines who are on the borderline. Qualification will reduce the size of the group, permitting more individual and specialized attention to those who remain.
(2) As the program progresses, regroup individuals who have like deficiencies in order that they can concentrate on their weakness. For example—

(a) One group may be weak in the arms and shoulders as revealed by failure in the chin up event. This group, with an assistant instruction charge, can work on pull-ups, rope climb, pushups, rifle or logexercises, horizontal ladder, and similar types of exercise.

(b) Another group may be weak in general endurance as revealed by the3-mile run. This group could profit from participation in conditioning drills, running, grass drill, and strength circuit.

(3) It may be necessary to form some groups to overcome weaknesses in skills, such as an inability to throw, to quickly change direction while running, to crawl rapidly, or to carry a load. Lack of coordination or lack of practice may cause these deficiencies. In this instance, instructors must provide an opportunity to practice and correct poor form and other errors as they are noted.

Chapter 3
PHYSICAL CONDITIONING ACTIVITIES
Section I. Marching Under Load

3101. GENERAL

Few physical fitness activities are as directly related to readiness for combat as foot marches under load. In addition to the obvious physical conditioning and unit cohesiveness benefits, marches under load prepare most Marines for numerous foreseeable tasks in combat. The idea that only infantry and reconnaissance units actually require training marches to prepare them for combat is clearly a mistake. In addition to the very real possibility of Marines from all types of units being used to fill combat-depleted infantry units, it is likely in maritime prepositioning force operations that aviation, combat service support, and command element personnel will have to march with weapons and equipment, from the arrival airfield to the equipment marshalling site. In American coastal cities, the harbor is nearly always 10 or more miles from the airport. It is difficult to imagine any different condition in a lesser developed country .A further combat-related benefit is that foot marching under load is the most effective way to develop leg strength in the context of a unit training program. In his classic study of troop performance in World War II, Men Against Fire, S.L.A. Marshall noted that the greatest single weakness of replacements of all occupational fields was leg strength. Truck drivers must push trucks which are stuck in mud, and headquarters communicators must climb hills to emplace antennas. The demands on the lower body in combat apply to all Marines, and foot marching is one activity which can help build these muscles while being conveniently integrated with the rest of the training program.

3102.TRAINING GOALS

The standard for success of a foot march is very simple to measure: did the unit arrive at the destination at the prescribed time with Marines in condition and required equipment present to accomplish the mission?A progressive program can increase a unit's readiness. It can instill pride by increasing distances and rates of march and by selecting routes over increasingly challenging terrain as the Marines become better conditioned. However, the ability to execute the mission at the conclusion of the march must remain the standard of success.

3103.MOVEMENT PLANNING

The basic considerations in planning a foot march are the mission, tactical situation, terrain and weather, and the units to participate. The success of the march will depend largely upon the thoroughness with which it is planned. A successful march is characterized by adherence to prescribed routes and time schedules, the efficient employment of the means available, and the ability of the unit to accomplish its assigned mission upon arrival at the destination.

a. Movement Orders. Movement planning culminates in the preparation and issuance of an operation order prepared in the standard, five-paragraph format. Necessary annexes are attached to furnish detailed information required for the movement. Written movement orders are rarely prepared at company level. A discussion of the annexes is contained in this chapter.

b.March Planning. March planning, as discussed here, is the planning conducted at battalion level. March planning may be organized into the following steps:

(1) **Preparation and Issuance of the Warning Order.** In order to afford subordinate units the maximum possible time to prepare for a pending move, a warning order containing all available information about the march is issued. The amount of planning time available will determine the time of issuance and the content of the warning order.

(2) Estimate of the Situation. In his estimate, the commander considers the mission, terrain, weather, time, and space factors, available routes, available transportation for the movement of equipment and/or shuttling of Marines, enemy capabilities, disposition of own forces, physical condition and training of Marines, and courses of action available to the command.

(3) Organization and Dispatch of a Reconnaissance. Party very march plan is based on as thorough and complete a ground reconnaissance as time and the situation will permit. Map and aerial reconnaissance are valuable in formulating a plan, but are not a substitute for ground reconnaissance. Route reconnaissance is accomplished by a reconnaissance party which usually consists of a reconnaissance element, an engineer element from the attached or supporting engineer unit, and a traffic control element. Unit standing operating procedures generally establish the basic composition of the reconnaissance party. It is modified as necessary to meet the requirements of a particular march. A recommended method of reporting information obtained by the route reconnaissance party is shown in figure 3-1.The minimum information required from the reconnaissance party is—

(a) Available routes and conditions.(Routes may be specified by higher headquarters.)

(b) Recommended rate of march.

(c) Selection of start point and release point, or confirmation of the suitability of start point and release point previously selected by map reconnaissance.

(d) Confirmation of location of the assembly or bivouac area.

(e) Location of critical points on the route.

(f) Distance between critical points on the route and total distance from start point to release point.

Figure 3-1 is available in the illustration section at the end of the book

(g) Location of obstacles and estimation of necessary Marines and equip- ment needed to repair and maintain routes.

(h) Number of guides required and their location on the route.

(4) Development of Detailed Movement Plan. Some of the elements included in a detailed movement plan are-

(a) Organizing the Column. To facilitate control and scheduling, units will be organized into serials and march units and given an order of march. In determining the order of march, the march planner must consider the enemy situation and the desirable order of arrival of the units at the destination. Where dispersion is required, a unit may be organized into two or more columns, each assigned a different route.

(b) Using Reconnaissance Information. Results of the route reconnaissance will be used to select the route(s); determine the start point, critical points along the route(s), and the release point; and select the rate of march.

(c) Determining March Computations. March unit pass time is based on the strength, formation, and rate of march. The pass time of the marching columns, plus necessary time distance computations, will be used to determine the completion time of the march.

(d) Drafting of Road Movement Table. Using the completed march computations, a draft road movement tabic is compiled.

(e) Checking the Plan. Using the draft road movement table and a road movement graph, the movement plan is checked to ensure that it conforms to the directive of the higher headquarters and the battalion commander's instructions.

(5) Preparation and Issuance of the Road Movement Order. After the plan has been checked and approved by the commander, an operation order is prepared and issued. The operation order may be in written form or issued orally, and is accompanied by a road movement table, overlay and/or strip map, and appropriate administrative details.

(6) Road Movement Table. A road movement table, prepared as an annex to the operation order, provides serial commanders with arrival and clearance times at critical points along the route of march. It also provides the column commander with information as to the proposed location of elements of the column at various times.

(7) Overlay/Strip Map. An overlay serves the normal purpose and should show, as a minimum, the present location of units, route of march, critical points, and the new location of units at the destination. A strip map is a schematic diagram of the route of march and shows landmarks and critical points with the distances between them. A strip map may be issued as an annex to the road movement order, in addition to or in lieu of an overlay.

c. March Computations. Prior to issuance of the operation order, the S-3 must verify time and space computations as they provide him with the necessary data for the preparation of a road movement table.

(1) Time-Distance. Time distance (TD) is determined by dividing the distance to be traveled (D) by the rate of inarch (R):

TD = D/RWhere TD = hours D = distance in kilometers R = kilometers per hour

(2) Length of ColumnThe length of column (LC) is used to determine the pass time of a column. The sum total of the following two parts determines the length of column. (See pars, (a) and (b)).The two parts are the space occupied by the Marines alone (including the distance between Marines) and the sum of the distances between the units of the column (column gap).

(a) The length of column of Marines alone is determined by multiplying the number of Marines by the appropriate factor selected from the table below. The length of column does not include distances between units. LC Marines = No. of Marines x factor.

(b) The total distance in meters between units is determined as follows:

Step One. Determine the number of gaps between serials (total serials minus one).

Step Two. Multiply the number of serial gaps from Step One by the length (in meters) between respective units.

Step Three. Determine the number of gaps between march units (total march units minus one, minus the number of serial distances).

Step Four. Multiply the number of march unit gaps from Step Three by the length (in meters) between respective units.

Step Five. Add the totals from Step Two and Step Four to get the total meters for the column gap.

EXAMPLE:A battalion foot column is organized into 12 platoon-sized march units and 3 company-sized serials. REQUIRED: total column gap distances when there are 100 meters between serials and 50 meters between march units.

(3) Pass-Time.Pass-time (PT) is the time a unit takes to pass a specified point. For foot columns, the pass-time is determined by applying the following formula:PT (minutes) = LC x FACTOR (for appropriate rate of march).

PASS-TIME FACTORS - FOOT TROOPS

.0150 for 4.0 km/h

.0187 for 3.2 km/h

.0250 for 2.4 km/h

.0375 for 1.6 km/h

EXAMPLE: Determine the pass-time of a unit whose length of column is 1,500meters and is marching at a rate of 4 km/h.PT (min) = 1,500 x .0150 (the factor for 4.0 km/h) = 22.5 mm.

(4) Completion Time. Completion time is the time of day that a march will be completed. Completion time is determined by using the following formula: Completion Time = SP time + TD + PT + Scheduled Halts (other than normal hourly halts).

EXAMPLE: A column's starting point (SP) time is 0700.The time-distance is6 hours and 40 minutes. Pass-time of the column is 30 minutes. A 35-minutelunch halt has been scheduled. What is the completion time of the march?Employing the 24-hour clock system, the formula can be applied as follows for simplified addition of the times:

HrMin SP time................0700

TD......................0640

PT......................0030

Lunch halt..............0035

Completion time..........13 105or 1445

The march will be completed at1445.

(5) Experience Tables. Based on previous movements made by a unit, data is accumulated to facilitate march planning. Such data includes approximate pass-times lor various elements of the battalion. The S-3 can utilize these data rather than computing them each time a march is scheduled. Such experience tables serve to reduce the time required to complete the computation phase of march planning. Matter appropriate to the unit standing operating procedure should be integrated therein.

d. Road Movement Graph

(1) A road movement graph is a time-distance diagram used in planning, preparing, or checking road movement tables, and for controlling marches. The graph provides a visual representation of a march plan so that conflicts and discrepancies may be prevented in the planning stage, before congestion occurs on the route. It is not usually issued as a part of the order. Road movement graphs may be applied to small units, to movements of a single column, or to a large organization scheduling separate elements, moving by various means, with different rates of march, over one or more routes.

(2) To construct a road movement graph, use the following steps: (See fig. 3-2.)

Figure 3-2 is available in the illustration section at the end of the book

Step One. Determine the number of hours available for the march. Designate the lower left corner of a sheet of graph paper as the SP time or the earliest even hour before the march is to begin. Select a scale and plot the hours available in sequence from the left on the horizontal axis.

Step Two. Determine the distance to be moved in kilometers. Indicate the SP at the lower left corner of the graph sheet and, using an appropriate scale, plot the number of kilometers on the vertical scale from the SP to the release point. Indicate the location of critical points on the vertical scale.

Step Three. At the proper distance from the SP, draw a horizontal line indicating the location of the RP. Indicate by a vertical line the hour when the movement must be completed. Plot lines representing route restriction, if any, at the proper distances and times on the graph.

Step Four. Determine the pass-times of foot and motor elements of the column.

Step Five. Starting at the SP at the specified hour, plot the movement of the head of the leading element. If the rate of march is 3 kilometers per hour, the lead element will have moved 1.5 kilometers in 30 minutes, 3 kilometers in 1 hour, etc. Plot the trace of the lead man to the RP. The last Marine will cross the SP in pass-times after the lead man. Measure this time on the graph and plot the trace of the last Marine of the column. The lines describing the head and tail of a march serial are parallel. Indicate the time subsequent serials reach the SP and plot the trace of the head and tail. Check to see that the plan complies with all restrictions and orders. If there are any violations or conflicts, the plan may be made to conform by changing the order of march, the starting time of the serials, the planned rate of march, or the organization of the column.

e. Conversion to the Metric System. The adoption of the metric system for expressing linear distances requires a simple means of conversion.

(1) To accomplish this, use the following tables:

o Multiply inches x 2.54 to obtain centimeters. Multiply centimeters x .39 to obtain inches,

o Multiply yards x .91 to obtain meters. Multiply meters x 1.1 to obtain yards,

o Multiply miles x 1.6 (or 8/5) to obtain kilometers. Multiply kilometers x .62 (or 5/8) to obtain miles.

(2) Linear distances can be accurately measured on maps which have been converted to the metric system by using a straightedge graduated in centimeters (cm).The centimeter scale can be used in lieu of the graphic scale when the scale of the map is shown, as follows:

Map Scale:	One cm Equals:
1:500,000	5,000 meters
1:250,000	2,500 meters
1:100,000	1,000 meters
1:50,000	500 meters
1:25,000	250 meters

3104. INDIVIDUAL LOAD

Since the backbone of Marine ground combat forces is the infantry, tactical mobility is largely dependent on the marching ability of the individual Marine. The load that each Marine must carry is the greatest impediment to mobility. In general, the commander must make every effort to reduce the individual load of his Marines to the absolute minimum. While the load has the greatest effect on Marines marching on foot, the value of carrying the minimum essential supplies and equipment applies for all Marine forces.

a. Fighting Load. The fighting load consists of items of clothing, equipment, weapons, and ammunition that are carried by, and essential to, the effectiveness of the combat Marine and the accomplishment of the immediate mission. The fighting load should not weigh more than 40 pounds. The commander must ensure that his Marines' loads- -and his own- -be stripped to the bare minimum. The addition of small, light-weight items in excess of the limit should not be tolerated, since cumulatively, these items will take a toll in energy. Every extra pound a Marine must carry decreases his combat effectiveness.

b. Existence Load. The existence load consists of items other than those in the fighting load that are required to sustain or protect the combat Marine, that may be necessary for increased personal and environmental protection, and that are not normally carried by the individual. The existence load is normally carried by the Marine's unit. Marching, but not engaged in combat, a Marine may be expected to carry a maximum of 50 pounds of supplies and equipment.

c. Considerations

(1) The primary consideration is not how much a Marine can carry, but how much he can carry without impaired combat effectiveness-moral or physical. The combat strength of a unit is not counted simply in numbers of Marines, but in the numbers of willing and physically able Marines.

(2) It is generally better to risk temporary inconvenience from lack of health and comfort items than to exhaust Marines due to overloading.

(3) It is a fundamental truth that men become physically exhausted more quickly when under the stress of combat. Marines must be required to carry less into battle than they are conditioned to carry in training. Marines should be conditioned for carrying weight, but should be equipped in combat for fleetness of foot.

(4) A common mistake is to base the fighting load on the gear and supplies necessary to meet every contingency. The commander cannot reasonably expect the Marines to carry enough gear for every possible eventuality. The items to be carried must be based on reasonable expectations.

(5) It is the commander's responsibility to procure the transport to carry any additional gear. As a rule of thumb, a rifle company, or unit of similar size, requires one 5-ton truck and trailer in support to carry additional gear. In cold weather, or other conditions in which the necessary personal gear increases, this requirement will increase.

(6) The commander must ensure that the supply system provides, on a dependable and timely basis, the balance of essential supplies and equipment not carried by the unit. Marines must feel confident that they will be supported with the necessary supplies and equipment.

(7) In training, commanders must implant in their Marines the pride in operating under austere conditions. To effectively demonstrate the need for this spartan mentality, the commander must personally set the standard. Marines should be trained in field craft, foraging techniques, and the use of caches and field expedients. Maximum use should be made of captured stores.

3105. MARCH TECHNIQUES AND PROCEDURES

a. Organization. A command executing a march is organized into march units, march serials, and march columns, as necessary for control.

(1) March Units. A march unit is a unit of command which moves and halts at the command of a single commander. The march unit normally corresponds to one of the smaller Marine units such as a platoon or company.

(2) March Serials. A march serial consists of one or more march units organized under the senior officer and given a numerical or alphabetical designation to facilitate control. A serial is usually a battalion, but may be a company if the battalion is marching alone.

(3) March Columns. A march column is composed of elements of a command moving over the same route. It may be composed of one or more serials. To facilitate control, a column commander is designated. A column is normally a regiment or larger unit, but may be a battalion if marching alone.

b. Order ofMarch. The order of march is determined by the mission, terrain, and the commander's desired order of commitment of units into action.

c. Control and Coordination Measures. The commander establishes initial control of the march by designating control measures in his march order. The most frequently used measures include:

o Start point and release point.
o Time at which head or tail of column passes the start point.
o Rate of march.
o Order of march.
o Route of march.
o Location of command post.
o March communications.

d. The Pacesetter. The pace-setter is an experienced individual carrying the same load as the majority of the Marines and marching from 4 to 10 meters at the head of the column. (See fig. 3-3.)The pace-setter's primary duty is to maintain the rate of march ordered by the column commander.

(l) Setting the Pace. The pacesetter does this by establishing the pace (length of step) and cadence (steps per minute) to obtain the prescribed rate of march. He should be of medium height so normal strides will be taken. Overstriding or understriding tends to tire leg muscles quickly and affects the combat efficiency of marching Marines. The officer marching at the head of the column supervises the pacesetter to ensure that he takes normal strides and maintains a uniform cadence.

Figure 3-3 is available in the illustration section at the end of the book

(2) Reducing Accordion Effect. Accordion effect occurs when the column alternately lengthens and shortens, causing the rear elements of the column to double time to maintain distance. It is caused by changes in the rate of march by the lead elements, usually after passing a slowing obstacle. To reduce this effect as much as possible, lead elements should slow their rate of march for a sufficient time after crossing obstacles to allow the rear elements to maintain the prescribed distance without running. The rear elements can signal the head when the last man has cleared the obstacle and the rate can be resumed. Since some accordion effect is nearly inevitable, the order of march should be periodically rotated to prevent units in the rear from becoming physically exhausted at the conclusion of the march.

e. Length, Rate and Pace

(1) **Length of March.** The length of march varies depending on terrain and weather, enemy situation, and physical and mental condition of Marines. The normal length of march for a 24-hour period is from 20 to 32 kilometers (12 to 20 miles) marching from 5 to 8 hours at a rate of 4 kilometers (2.5 miles) per hour (km/hmi/h).A march in excess of 32 kilometers in a 24-hour period is considered a forced march. Well-trained units that have been progressively trained will be able to achieve a unit goal of 60 miles in a 3-day period.

(2) Rate of March. The same factors of terrain, weather, load to be carried, and condition of Marines affect the rate of march. The commander considers these and selects a rate which will place his unit at the destination in the shortest possible time in condition to accomplish the mission. Unit standing operating procedure should state the standard rate of march for that unit over normal terrain for both road and cross-country movement and in both day and night. The commander then modifies this rate if the situation requires. Normally prescribed standards are as follows:

Figure is available in the illustration section at the end of the book

(3) Pace and Cadence. The normal pace is 30 inches. A pace of 30 inches and a cadence of 106 steps per minute result in a speed of 4.8 kilometers per hour or 3 miles per hour and a rate of 4 kilometers per hour or 2.5 miles per hour if a 10-minute rest halt per hour is taken. Since the pace of each man may vary, the cadence may have to be adjusted to maintain the prescribed rate. Slope of ground and surface footing affect pace. A normal cadence is relatively simple to maintain on level or gently rolling terrain on firm footing. Mud, sand, loose gravel, and hills all greatly affect pace.

f. Halts. Halts during a day's march are taken at regular intervals to rest personnel and adjust loads.Halts are specified by standing operating procedures or march order. Such factors as midday heat or enemy action may force the adoption of long midday halts or night marches. At long halts, each unit disperses to a previously selected location near the route of march.

(1) Normal Time and Frequency. Under normal conditions, a 15-minute halt is made after the first 45 minutes of marching. Following the first halt, a 10-minute halt is made after each 50 minutes of marching. Variations of this schedule are made when a scheduled halt time occurs when passing through a built-up area or when cover and concealment are required by the tactical situation and none is available. Observation posts may be established, if required for security of the unit during halts.

(2) Actions at Halts. All units in the column should be halted at the same time. At the halt signal, Marines should move to the side of the road, staying within the immediate vicinity of their unit. Marines should remove or loosen gear and sit or lie with feet elevated. Commanders inspect Marines and equipment, and corpsmen administer medical treatment, as required.

g. Communications. The four primary means of communications used in controlling foot marches are radio, visual, sound, and messenger.

(1) Radio. Radios are normally used for communications between platoons and higher headquarters in the march column. Radio transmissions should be held to the absolute minimum necessary for control and should be at minimum power required.

(2) Visual.. Visual communications may include flashlights, luminous markers, panel sets, pyrotechnics, and hand and arm signals. When contact with the enemy is not expected, the loss of light discipline from some of these signals may be a lesser security risk than the risk of disclosure from radio transmissions.

(3) Sound. Sound communications include voice and such signaling devices as horns and whistles. Some of their uses include assembling Marines at the conclusions of halts and providing warnings of air or chemical attacks.

(4) Messengers. Messengers are particularly effective during periods of radio silence and reduced visibility. Messages should be simple and brief to preclude mistakes by the messenger.

h. Security. Both active and passive measures are used against attack by enemy aircraft and indirect fire weapons during movement. Active measures include the use of air sentries and organic and attached weapons in accordance with the unit air defense standing operating procedure. Passive measures include the use of concealed routes and assembly areas, night marches, and extended interval between elements of the column. With imagination and planning, all of these measures can be practiced in training without detracting from the physical conditioning benefit of the march. In fact, such drills as rapidly taking cover during simulated attacks and then quickly resuming the march can prove physically challenging.

i. Reconnaissance. Reconnaissance determines in advance critical points along the march route such as bridges, fords, and obstacles in order that congestion or delay may be held to a minimum and local security provided. Paragraph 3103.C describes organization of the reconnaissance party.

3106.TACTICAL MARCHES

These marches are movements of Marines and equipment not in direct ground contact with the enemy but expecting early ground contact either en route or upon arrival at the objective. Some characteristics of this type of movement include security elements to the front, flanks, and, if necessary, to the rear of the march column. Additional considerations include the selection of concealed areas on defendable terrain for start and release points and rest stops. In training, infantry units of regimental size or smaller will normally derive the greatest benefit from tactical marches. The security requirements usually dictate a slower rate of march and, therefore, reduce the physical conditioning value for other types of units which would seldom move in combat with the expectation of early ground contact with the enemy. Two key points should be remembered regarding tactical marches in a training program.

a. Security. Although security requirements often slow the rate of march, it is a valid and important training objective for infantry units to work toward reducing march times while still practicing sound tactical security. March standing operating procedures, good land navigation skills, and simple but complete march orders contribute as much to this goal as physical conditioning. The idea that proceeding slowly enhances security is seldom true. Keeping the enemy off balance by rapid movement frequently is the best guarantee of security.

b. Training. In peacetime training, it is critical that support personnel who normally are located with infantry units in combat, participate completely with infantry units in forced march training. Not only must they not slow the infantry rate of march because of inferior conditioning, but they must be thoroughly familiar with the infantry unit's standing operating procedure for security on the march. Personnel from within the ground combat element such as artillery forward observer teams and combat engineer units, can be easily identified for this training and should be habitually associated with the same infantry units. Personnel from other elements, such as low-altitude air defense teams and radio battalion detachments, are more difficult to identify and earmark for support of the same infantry unit. Their normal mode of employment is in vehicles and their units frequently provide general support vice direct support. This does not, however, lessen their requirement to be prepared to provide support to dismounted infantry over terrain or in a tactical situation not conducive to vehicles. Such training needs to take place, and planning and cooperation among the elements of the MAGTF can overcome the obstacles.

3107.ADMINISTRATIVE MARCHES

These marches are movements of Marines and equipment made when no enemy interference is expected except from aircraft or indirect fire. During these marches, units are administratively grouped for ease of control and speed of movement. The term administrative foot marches should not be taken to imply that tactical considerations are disregarded. There are simply less stringent security measures used than when ground contact with the enemy is a distinct possibility. The normal formation is the route column with one file moving on either side of the road and with negligible security. Air sentries are used and dispersion between individuals and units is practiced. Although roads usually provide the fastest route, there may be reasons to conduct these marches cross-country (particularly as part of a training program). Administrative marches are the type of foot marches most practical for training programs for non-infantry units. These marches may also play a key role in the physical conditioning program of infantry units, because of the greater speeds possible due to lessened security requirements.

3108. FORCED MARCHES

A forced march is a march which covers a greater distance than 32 kilometer (20 miles) in a 24-hour period. Normally the extra required distance should be achieved by increasing the number of hours marched in a day beyond the norm (8 hours) rather than increasing the rate of march beyond the norm (4 kilometers per hour/2.5 miles per hour). However, there will be occasions when the situation or mission demands an increase in the rate of march.

a. Maximum Recommended Distances. The maximum recommended distances for forced marches are-56 kilometers (35 miles) in 24 hours; 96 kilometers (60 miles) in 48 hours; or 128 kilometers (80 miles) in 72 hours.

b. Sample Time Breakdown. A sample time breakdown for a forced march of 52 kilometers, beginning at first light is as follows:

Figure is available in the illustration section at the end of the book

3109. NIGHT MARCHES

Night marches are characterized by closed formations, more difficult control and reconnaissance, and a slower rate of march than day marches.

a. Control. Control is increased by reducing the distance between individuals and units, and by using connecting files to maintain contact between platoons and companies. Connecting files are normally constituted from personnel from the rear most march unit.

b. Safety on Roads. Because of reduced visibility, night marches on roads used by vehicles require attention to safety procedures to prevent accidents. Commanders should not think these safety measures apply only to peacetime training. In combat, nearly all vehicles operate with blackout lights at night, making visibility even more difficult than in garrison driving. If consistent with the tactical situation, the following measures can reduce the possibility of marchers being struck by vehicles:

(1) Use off-road trails and routes as much as possible.

(2) Guards to the front and rear of march units should be marked with strips of luminous tape and carry red-filtered flashlights. These measures are generally consistent with light discipline, while still affording reasonable warning to approaching drivers.

Section II. Conditioning Drills (One, Two, and Three)

3201. GENERAL

a. Description and Objective. Conditioning drills are calisthenic exercises. Each drill contains seven exercises organized and numbered in a set pattern. Each drill takes 15 minutes to complete. The objective of conditioning drills is to exercise all major muscle areas in order to develop strength, endurance, coordination, and flexibility.

b. Area and Equipment. Any level area is satisfactory for conduct of the drills. Drills One and Two contain ground exercises. If ground conditions are unsatisfactory, Drill Three should be used as it contains no ground positions. Usually, no equipment is required;however, if the group exceeds a platoon in size, an instructor's stand is necessary.

c. Formation. The extended rectangular formation is used in this drill. (See app. A.)

d. Starting Level and Progression. The starting level is six repetitions of each exercise. An increase of one repetition for each three periods of exercise in which the drill is performed is an acceptable rate of progression. This rate is continued until 12 repetitions can be completed. To maintain, continue the drill at 12 repetitions. To progress, move to 6 repetitions at a more difficult drill which exercises the same muscle group. Progression can also be gained by moving from Drill One to Drill Two, as Drill Two is more demanding.

e. Starting Positions. Starting positions vary with the exercise and are explained in each exercise.

f. Leadership. A principal instructor demonstrates and leads the drills. The instructor must be familiar with leadership techniques peculiar to conditioning drills to include the exercises, commands, counting cadence, cumulative count, formation, method of teaching the exercises, and utilization of assistant leaders.

g. Use With Other Programs. Conditioning Drills One, Two, and Three reach all major muscles of the body. They are easy to learn and to perform, and they are simple to administer and supervise. These features, coupled with the short time required for completion, the fact that no equipment is necessary, and adaptability to most areas of execution, make these drills possible in any programs.

3202. CONDITIONING DRILL ONE

This conditioning drill is similar to the calisthenics drill commonly known throughout the Marine Corps as the "Daily Seven."The "Daily Seven" may be used as a substitute for Conditioning Drill One.

a. Exercise l:High Jumper

(1) Starting Position. Feet separated shoulder width, knees flexed, body bent forward at the waist, arms aligned with the trunk and hips, elbows locked, palms facing, fingers extended and joined, head and eyes to the front.(See fig. 3-4, A.)(Elbows remain locked throughout the exercise.)

Figure 3-4 is available in the illustration section at the end of the book.

(2) Cadence. Moderate.

(3) Movement. A four-count exercise: at the count of—

(a) ONE -- Take a slight jump into the air, swinging the arms forward and up to shoulder level.

(b) TWO -- Take a slight jump into the air and swing the arms downward and back, returning to the starting position.

(c) THREE -- Take a vigorous leap into the air, swinging the arms forward and up to an overhead position, momentarily looking skyward, on returning to the ground the knees are flexed, head and eyes return to the front.

(d) FOUR--Repeat the action of count two.

b. Exercise 2: Bend and Reach

(1) Starting Position. Feet spread more than shoulder width, arms overhead, elbows locked, palms facing, fingers extended and joined, head and eyes to the front.(See fig. 3-4, B.)

(2) Cadence. Moderate.

(3) Movement. A four-count exercise: at the count of—

(a) ONE-Bend at the knees and waist, swing the arms straight downward and reach between the legs. Touch the ground as far to the rear as possible and look to the rear. (Elbows remain locked throughout the exercise).

(b) TWO-- Recover sharply to the starting position.

(c) THREE--Repeat the action of count ONE.

(d) FOUR-- Repeat the action of count TWO.

c. Exercise 3: Pushup

(1) Starting Position. Front leaning rest position. To assume this position there is a silent one-two count:on the silent count of one, assume the squatting position, heels together, elbows locked inside the knees, hands flat on the ground directly beneath the shoulders; on the silent count of two, thrust the legs to the rear, toes and heels together, body straight from head to heels.(Seefig. 3-4, C.)

(2) Cadence. Moderate.

(3) Movement. A four-count exercise: at the count of—

(a) ONE -- Flex the elbows lowering the body until the thick portion of chest touches the ground.

(b) TWO -- Raise the body until elbows are straight and locked.

(c) THREE--Repeat the action of count ONE.

(d) FOUR -- Repeat the action of count TWO. (On returning to position of attention, the silent one-two count is used in reverse).

d. Exercise 4: Trunk Twister

(1) Starting Position. Feet are spread more than shoulder width apart, fingers placed behind neck, thumbs pointing downward, elbows back. (See fig. 3-4, D.)(Elbows remain well back throughout the exercise).

(2) Cadence. Slow.

(3) Movement. A four-count exercise: at the count of-

(a) ONE-- Keeping the knees locked and the back straight, bend forward at the waist sharply, with a slight recovery.

(b) TWO-- Twist the trunk to the left vigorously at the waist, keeping the elbow back. The left elbow is higher than the right.

(c) THREE-- Twist vigorously to the right, so the left elbow comes under the right.

(d) FOUR-- Straighten sharply to the starting position. NOTE: Do not attempt to touch the elbows to the knees on counts two and three.

e. Exercise 5: Squat Bender

(1) Starting Position. Feet are spread less than shoulder width apart, hands on hips, thumbs in small of back, elbows back. (See fig. 3-4, E.)

(2) Cadence. Moderate.

(3) Movement. A four-count exercise: at the count of—

(a) ONE-- Assume the squatting position, maintain balance on the balls of the feet, with trunk erect thrust arms forward to shoulder level, elbows locked, palms down.

(b) TWO-- Recover to starting position. Elbows are well back.

(c) THREE-- Keeping the knees locked, bend forward at the waist, touching the ground in front of the toes.

(d) FOUR-- Vigorously recover to the starting position.

f. Exercise 6: Body Twist

(1) Starting Position. On the back, arms extended sideward on the ground, palms down. The legs are raised to a near vertical position, feet together, knees locked.(See fig. 3-4, F.)

(2) Cadence. Slow-fast.

(3) Movement. A four-count exercise: at the count of-

(a) ONE-- Lower legs slowly to the left until they touch the ground near the left hand, keeping the knees straight and shoulders on the ground.

(b) TWO-- Recover the starting position by quickly raising the legs, keep knees straight and feet together.

(c) THREE-- Repeat movement of count ONE, except the movement is to the right side.

(d) FOUR-- Recover sharply to the starting position.

g. Exercise 7: Stationary Run

(1) Starting Position. Position of attention. (See fig, 3-4, G.)

(2) Cadence. Fast.

(3) Movement.

(a) At the command of execution, start running in place at double-time cadence, lifting the left foot first time cadence. Follow the instructor as he counts two repetitions of cadence;

e.g., 1, 2, 3, 4--1, 2, 3, 4. The instructor then gives informal commands such as FOLLOW ME, running on the toes and balls of the feet, keeping the back straight, speeding up the cadence to a sprint, raising the knees high, leaning forward at the waist, and pumping the arms vigorously.

(b) To halt the exercise, the instructor will count two repetitions of cadence as the left foot strikes the ground: 1, 2, 3, 4 - 1, 2, 3, HALT.

NOTE: When counting cadence, the instructor counts only as the left foot strikes the ground. The duration of the exercise is approximately 1 1/2 minutes.

3203. CONDITIONING DRILL TWO

a. Exercise 1: Jumping Jack

(1) Starting Position. Feet separated more than shoulder width, arms overhead. (See fig. 3-5, A.)

(2) Cadence. Moderate.

(3) Movement. A four-count exercise: at the count of~

(a) ONE-- Jump to position with the feet together and assume the squatting position, swinging the arms sideward and downward, placing the hands palms down on the ground, elbows locked inside the knees.

(b) TWO-- Recover to the starting position by jumping to the side straddle and swinging the arms sideward overhead.

(c) THREE-- Repeat the action of count ONE.

(d) FOUR-- Recover to the starting position.

b. Exercise 2: Turn and Bend

(1) Starting Position. Side straddle, arms overhead. (See fig. 3-5, B.)

(2) Cadence. Moderate.

(3) Movement. A four-count exercise: at the count of-

(a) ONE-- Turn the trunk to the left and bend forward over the left thigh, attempting to touch the fingertips to the ground outside the left foot. Keep the left knee straight. On successive repetitions attempt to touch farther and farther to the side.

(b) TWO-- Recover to the starting position.

(c) THREE-- Turn the trunk to the right and bend forward over the right thigh, trying to touch the hands to the ground outside the right foot. Keep the right knee straight.

Figure 3-5 is available in the illustration section at the end of the book.

(d) FOUR-- Recover to the starting position.

c. Exercise 3: Eight-Count Pushup

(1) Starting Position. Position of attention. (See fig. 3-5, C.)

(2) Cadence. Moderate.

(3) Movement. An eight-count exercise: at the count of—

(a) ONE-- Assume the squatting position, palms on the ground directly beneath the shoulders, elbows locked inside the knees.

(b) TWO-- Thrust the legs to the rear assuming the front leaning rest position.

(c) THREE-- Flex the elbows until the thick portion of the chest touches the ground.

(d) FOUR-- Raise the body on a straight plane until the elbows are locked.

(e) FIVE-- Repeat the action of count THREE.

(f) SIX-- Repeat the action of count FOUR.

(g) SEVEN-- Recover to the squatting position as in count ONE (elbows locked inside the knees).

(h) EIGHT—Return sharply to the position of attention.

d. Exercise 4: Turn and Bounce

(1) Starting Position. Feet spread more than shoulder width apart, arms sideward at shoulder level, palms up. (See fig. 3-5, D.)

(2) Cadence. Slow.

(3) Movement. An eight-count exercise: at the count of—

(a) ONE-- Turn sharply to the left as far as possible, then recover slightly.

(b) TWO-- Again turn to the left as far as possible and recover as in ONE.

(c) THREE-- Repeat the action of count TWO.

(d) FOUR-- Recover sharply to the starting position.

(e) FIVE—Turn sharply to the right as far as possible, then recover slightly.

(f) SIX—Again turn to the right as far as possible and recover as in FIVE.

(g) SEVEN-Repeat the action of count SIX.

(h) EIGHT-Return to the starting position.

NOTE: The head and hips remain to the front throughout the exercise and the knees and elbows are locked at all times.

e. Exercise 5: Squat Stretch

(1) Starting Position. Attention. (See fig. 3-5, E.)

(2) Cadence. Moderate.

(3) Movement. A four-count exercise: at the count of—

(a) ONE-- Squat, placing the hands on the ground about 12 inches in front of the feet.

(b) TWO— Keeping the fingertips on the ground, straighten the knees completely and raise the hips.

(c) THREE-- Recover to position ONE.

(d) FOUR-- Recover to the starting position.

f. Exercise 6: Leg Circular

(1) Starting Position. On the back, arms stretched sideward, palms down, feet raised foot from ground, knees straight. (See fig 3-5,F.)

(2) Cadence. Slow.

(3) Movement A four-count exercise: at the count of—

(a) ONE-- Swing the legs as far as possible to the left, keeping the knees straight and the legs together.

(b) TWO-- Swing the extended legs over head with the thighs as close as possible to the trunk.

(c) THREE-- Swing the legs as far as possible to the right.

(d) FOUR-- Recover to the starting position.

g. Exercise 7: Stationary Run

(1) Starting Position. Position of attention. (See fig. 3-5, G.)

(2) Cadence. Fast.

(3) Movement

(a) At the command of execution, start running in place at double-time cadence, lifting the left foot first. Follow the instructor as he counts two repetitions of cadence; for example: 1, 2, 3, 4--1, 2, 3, 4. The instructor then gives informal commands such as FOLLOW ME. Running on the toes and balls of the feet, keeping the back straight, speeding up the cadence to a sprint, raising the knees high, leaning forward at the waist, and pumping the arm vigorously.

(b) To halt the exercise, the instructor will count two repetitions of cadence as the foot strikes the ground: 1, 2, 3, 4-1,2, 3, HALT.

NOTE: When counting cadence, the instructor counts only as the left foot strikes the ground. The duration of the exercise is approximately 1 1/2 minutes.

3204. CONDITIONING DRILL THREE

a. Exercise 1: Side Straddle Hop

(1) Starting Position. Position of attention. (See fig. 3-6, A.)

(2) Cadence. Moderate.

(3) Movement. A four-count exercise: at count of-

(a) ONE-- Take a slight jump into the air, moving the legs sideward (more than shoulder width apart); at the same time, swing the arms overhead (to an overhead position) clapping the palms together.

Figure 3-6 is available in the illustration section at the end of the book.

(b) TWO-- Take a slight jump into the air, swing the arms sideward and downward returning to the starting position.

(c) THREE-- Repeat the action of count ONE.

(d) FOUR-- Repeat the action of count TWO.

b. Exercise 2: Back Bender

(1) Starting Position. Standing, feet 12 inches apart, fingers placed behind the head. (See fig. 3-6, B.)

(2)Cadence. Slow.

(3) Movement. A four-count exercise: at the count of-

(a) ONE-- Bend the upper trunk backward, raising the chest high, pulling the elbows back, and looking upward. Keep the knees straight.

(b) TWO-- Recover to the starting position.

(c) THREE-- Repeat the action of count ONE.

(d) FOUR-- Recover to the starting position.

c. Exercise 3: Squat Thrust

(1) Starting Position. Position of attention. (See fig. 3-6, C.)

(2) Cadence Moderate.

(3) Movement. A four-count exercise: at the count of—

(a) ONE-- Assume the squatting position; heels together, placing the hands flat on the ground, shoulder width apart, elbows locked and inside the knees.

(b) TWO-- Thrust the legs to the rear, assuming the front leaning rest position, body in line from head to toe, heels and toes together.

(c) THREE-- Return to the squatting position as in ONE.

(d) FOUR-- Return to position of attention.

d. Exercise 4: Side Bender

(1) Starting Position. Feet are spread more than shoulder width apart, arms are raised sideward and overhead, thumbs interlocked palms to front, fingers extended and joined, elbows locked. (See fig. 3-6, D.)

(2) Cadence. Slow.

(3) Movement. An eight-count exercise: at the count of—

(a) ONE-- Bend to left as far as possible, then recover slightly.

(b) TWO-- Again bend to the left as far possible, then recover slightly.

(c) THREE-Repeat the action of count TWO.

(d) FOUR-- Recover sharply to the starting position.

(e) FIVE-- Bend to the right as far as possible, then recover slightly.

(f) SIX-- Again bend to the right as far as possible, then recover slightly.

(g) SEVEN-- Repeat the action of count SIX.

(h) EIGHT-- Recover sharply to the starting position.

NOTE:Keep the elbows and knees locked throughout the exercise. The bend should occur to the side and not the front.

e. Exercise 5: Knee Bender

(1) Starting Position. Feet are spread less than shoulder-width apart, hands on hips, thumbs in small of back, elbows back. (See fig 3-6, E.)

(2) Cadence. Moderate.

(3) Movement. A four-count exercise: on the count of—

(a) ONE-Do a knee bend, lean trunk forward at the waist, thrust arms between legs until the extended fingers touch the ground palms to the ground, hands 6 inches apart.

(b) TWO~Recover sharply to the starting position.

(c) THREE-Repeat the action of count ONE.

(d) FOUR-Repeat the action of count TWO.

f. Exercise 6: Bottoms Up

(1) Starting Position. Front leaning rest position. A silent one-two count is used as in the pushups. (See fig. 3-6, F.)

(2) Cadence. Moderate.

(3) Movement A four-count exercise: at the count of-

(a) ONE-- With the weight on the hands, and knees locked, jump forward bringing the feet as close to the hands as possible; look to the rear.

(b) TWO--Keeping the knees locked, thrust the legs backward assuming the front leaning rest position.

(c) THREE-- Repeat the action of count ONE.

(d) FOUR-- Repeat the action of count TWO.

g. Exercise 7: Stationary Run

(1) Starting Position. Position of attention. (Sec fig. 3-6, G.)

(2) Cadence. Fast.

(3) Movement

(a) At the command of execution, start running in place at double time, lifting the left foot first. Follow the instructor as he counts two repetitions of cadence; for example: 1, 2, 3, 4--1, 2,3, 4. The instructor then gives informal commands such as FOLLOW ME, running on the toes and balls of the feet, keeping the back straight, speeding up the cadence to a sprint, raising the knees high, leaning forward at the waist, and pumping the arms vigorously.

(b) To halt the exercise, the instructor will count two repetitions of cadence as the left foot strikes the ground: 1, 2, 3, 4 - 1, 2, 3, HALT.

NOTE: When counting cadence, the instructor counts only as the left foot strikes the ground. The duration of the exercise is approximately 1 1/2 minutes.

Section III. Rifle and Log Drills

3301. RIFLE DRILL

a. Description and Objective. Rifle exercises are conditioning exercises performed with a rifle. Each drill contains six exercises (fig. 3-7) and they are numbered in a set pattern. The drill takes 15 minutes to complete. The objective of rifle drills is to exercise the arms, shoulders, and back muscles in order to develop strength and endurance, particularly in the upper body. In units without rifles, log drills may be substituted.

b. Area and Equipment. Any level area is satisfactory for conducting this drill. Each Marine completes these exercises from a standing position and no ground contact is required. Each Marine will need a rifle and, if the group exceeds a platoon in size, then the instructor will need an instructor's stand.

c. Formation. The extended rectangular formation is used in this drill. (See app. A.)
d. Starting Positions. Starting positions vary with the exercises and are explained in each exercise. As in all set conditioning drills, the command used to start the exercise is STARTING POSITION, MOVE. The following directions apply to rifle drill.
(1) In those exercises which start from the rifle downward position, on the command MOVE, execute port arms as prescribed in NAVMC 2691, Drill and Ceremonies Manual, and then assume the starting position. The command to return to the men to the position of attention at the conclusion of the exercise is POSITION OF ATTENTION, MOVE.
(2) In exercises which terminate in the rifle downward position, on the command MOVE, execute the position followed by order arms as prescribed in NAVMC 2691.
(3) In the exercises which terminate in a position other than the rifle downward position, Marines first assume the rifle downward position before executing port arms and order arms.
(4) These movements are executed without command. This procedure promotes uniformity, but precision is not expected. To be effective, rifle exercises must be strenuous enough to tire the arms, but not to the point where the arms cannot move with precision.

e. Leadership. A principal instructor demonstrates and leads the drill. He must be familiar with leadership techniques for conditioning exercises and the peculiar techniques for rifle drill.

3302. EXERCISES PERFORMED WITH RIFLES

The exercises of rifle drill are outlined in the following paragraphs,
a. Exercise 1: Foreup, Behind Back

(1) Starting Position. Rifle downward, feet together.(See fig. 3-7, A.)

Figure 3-7 is available in the illustration section at the end of the book.

(3) Movement. A four-count exercise: at the count of—
(a) ONE-- Swing the arms forward and upward to the overhead position. Inhale.

(b) **TWO**-- Lower the rifle to the back of the shoulders. Exhale.

(c) **THREE**-- Recover to position ONE and inhale.

(d) **FOUR**-- Recover to the starting position and exhale.

b. Exercise 2: Lunge Side, Turn and Bend

(1) Starting Position. Rifle downward, feet together.(See fig. 3-7, B.)

(2) Cadence. Moderate.

(3) Movement. An eight-count exercise:at the count of—

(a) **ONE**-- Lunge sidewards to the left, swing the rifle forward and upward to the overhead position.

(b) **TWO**-- Turn the trunk to the left and bend forward over the left hip. At the same time, swing the rifle to a **low** horizontal in front of the left ankle.

(c) **THREE**-- Recover to position ONE.

(d) **FOUR**-- Recover to the starting position.

(e) **FIVE, SIX, SEVEN, and EIGHT**-- Repeat on the right side.

c. Exercise 3: Foreup, Back Bend

(1) Starting Position. Rifle downward, feet together.(See fig. 3-7, C.)

(2) Cadence. Moderate.

(3) Movement. A four-count exercise:at the count of—

(a) **ONE--** Swing the arms forward and upward to the overhead position.

(b) **TWO**-- Bend backward, emphasizing the bend in the upper back. The face is up.Keep the knees straight.

(c) **THREE**-- Recover to position ONE.

(d) **FOUR**-- Recover to the starting position.

d. Exercise 4:Up and Forward

(1) Starting PositionRifle downward, feet together.(See fig. 3-7, D.)

(2) Cadence. Fast.

(3) Movement. A four-count exercise:at the count of—

(a) **ONE**-Swing the arms forward and upward to the overhead position.

(b) **TWO-** Swing the arms forward to shoulder level.

(c) **THREE**-Recover to position ONE.

(d) **FOUR**-Recover to the starting position.

e. Exercise 5: Foreup, Full Squat

(1) Starting Position. Rifle downward, feet in narrow stance. (See **fig.** 3-7, E.)

(2) Cadence. Moderate.

(3) Movement. A four-count exercise: at the count **of**—

(a) **ONE**-- Swing the arms forward and upward to the overhead position.

(b) **TWO**-- Swing the arms down to shoulder level and assume the squatting position.

(c) **THREE**-- Recover to position ONE.

(d) **FOUR**-- Recover to the starting position.

f. Exercise 6: Arms Forward, Side Bend

(1) Starting Position. Sidestraddle, regular stance, rifle forward.(See fig. 3-7, F.)
(2) Cadence. Moderate.
(3) Movement. A four-count exercise:at the count of-

(a) ONE-- Bend the trunk to the left. Keep the knees straight.
(b) TWO-- Recover to the starting position.
(c) THREE-- Bend the trunk to the right. Keep the knees straight.
(d) FOUR-- Recover to the starting position.

NOTE:Keep the rifle on the same level as the shoulders throughout the exercise.

3303. LOG DRILL

a. Description and Objective. Log exercises are conditioning exercises performed with a log. Each drill contains six exercises and they are numbered in a set pattern. The drill takes 15 minutes to complete. The objective of log drills is to develop strength and muscular endurance and, in this instance, under maximum loads. Log exercises also develop teamwork. Log may be used in lieu of conditioning drills after the Marines have become somewhat conditioned.

b. Area and Equipment. Any level area is satisfactory for conducting this drill. Each Marine completes these exercises from a standing position and no ground contact is required. If the group exceeds a platoon in size, then the instructor will need instructor's stand. Each six-person group or eight-person group will need a log. The logs should be from 6 to 8 inches in diameter. They may vary in length from 14 feet (for 6 people) to 18 feet (for 8 people).The logs should be skinned, smoother, and dried. The 14-foot logs should weigh approximately 300 pounds and the 18-foot logs, approximately 400 pounds. Rings should be painted on the logs to indicate each person's position. When not in use, the logs should be stored on a rack to keep them off the ground.

c. Formation. All the Marines assigned to the same log team should be about the same height at the shoulders. The recommended method of dividing the platoon is to have the Marines form a single file or column with short people to the front and tall people to the rear. Have the Marine assume their positions in the column according to shoulder height, not head height. When they are in position, they are given the command COUNT OFF BY SIXES (OR EIGHTS), COUNT OFF, to divide them into six- or eight-person log teams. Each team in turn, can then proceed to the log rack, shoulder a log, and carry it to the designated exercise area. The log teams form in columns in front of the instructor. With the Marine holding the log in the chest position, have them face the instructor and ground the log at least 10 yards between columns and 10 yards between log teams within the columns.

d. Starting Positions. The Marines fall in, facing the log, with their toes about 4 inches from it. The basic starting positions and commands are as follows:(See fig. 3-8.)
(1) **RIGHT-HAND STARTING POSITION, MOVE.** It the command MOVE, move the left foot 12 inches to the left, and lower the body into a flat foot squat. Keep the back straight, head up, and arms between the legs. Encircle the far side of the log with the left hand. Place the right hand underneath the log.(See fig. 3-8, A.)

(2) LEFT-HANDED STARTING POSITION, MOVE. These commands are executed in the same manner as in paragraph a. except that the left hand is underneath the log and the right hand encircles its far side. (See fig. 3-8, B.)

Figure 3-8 is available in the illustration section at the end of the book.

(3) RIGHT SHOULDER POSITION, MOVE. At the command MOVE, pull the log upward in one continuous motion to the right shoulder. At the same time, move the left foot to the rear and stand up, facing left. Balance the log on the right shoulder with both hands.(See fig. 3-8, C.)This movement cannot be performed from the left-hand starting position because of the position of the hands.

(4) LEFT SHOULDER POSITION, MOVE. These commands should be given from the left hand starting position. At the command MOVE, pull the log upward in one continuous motion, to the left shoulder. At the same time, move the right foot to the rear and stand facing right. Balance the log on the left shoulder with both hands.(See fig. 3-8, D.)This movement cannot be performed from the right-hand starting position.

(5) WAIST POSITION, MOVE. From the right hand position pull the log waist high. Keep the arms straight and fingers laced underneath the log. The body is inclined slightly to the rear and the chest is lifted and arched. (See fig. 3-8, E.)

(6) CHEST POSITION, MOVE. This command should be given after the waist position has been assumed .On the command MOVE, shift the log to a position high on the chest, bring the left arm under the log and hold the log in the bend of the arms.(See fig. 3-8, F.)Keep the upper arms parallel to the ground.

(7) To move the log from the right shoulder to the left shoulder, the command is: LEFT SHOULDER POSITION, MOVE. On the command MOVE, push the log overhead and lower it to the opposite shoulder.

(8) To return the log to the ground from any of the above positions, the command is: STARTING POSITION, MOVE. At the command MOVE, slowly lower the log to the ground. The hands and fingers must be kept from under the log.

e. Leadership. A principal instructor demonstrates and leads the drill. He must be familiar with the leadership techniques for conditioning exercises and the peculiar techniques for log drill.

3304. EXERCISE PERFORMED WITH LOGS

The exercises of log drill are outlined in the following paragraphs. Figure 3-9 graphically explains log drill.

a. Exercise l:Two-Arm Pushup

(1) Starting Position. Right or left shoulder position. Regular stance. (See fig. 3-9, A.)

(2) Cadence. Moderate.

(3) Movement. A four-count exercise:at the count of—

- **(a) ONE--** Push the log overhead until the elbows lock.
- **(b) TWO--** Lower the log to the opposite shoulder.
- **(c) THREE--** Repeat the action of count ONE.
- **(d) FOUR--** Recover to the starting position.

b. Exercise 2: Forward Bender

(1) Starting Position. Chest position. Regular stance.(See fig. 3-9, B.)

(2) Cadence. Moderate.

(3) Movement. A four-count exercise:at the count of-

(a) **ONE**-- Bend forward at the waist, keeping the back and legs straight.

(b) **TWO**-- Recover to the starting position.

(c) **THREE**-- Repeat the action of count ONE.

(d) **FOUR**-- Recover to the starting position.

c. Exercise 3: Straddle Jump

(1) Starting Position. Right or left shoulder position, feet together, fingers interlaced on top of the log.(See fig. 3-9, **C.)**

(2) Cadence. Moderate.

(3) Movement. A four-count exercise:at the count of—

(a) **ONE**-- Jump to a side-straddle. Pull down on the log with both hands to keep it from bouncing on the shoulder.

(b) **TWO**-- Recover to the starting position.

(c) **THREE**-- Repeat the action of count ONE.

(d) **FOUR**-- Recover to the starting position.

Figure 3-9 is available in the illustration section at the end of the book.

d. Exercise 4: Side Bender

(1) Starting Position. Right shoulder position, feet regular stance.(See fig. 3-9, D.)

(2) Cadence. Moderate.

(3) Movement. A four-count exercise:at the count of—

(a) **ONE**-- Bend sideward to the left as far as possible, bending the left knee.

(b) **TWO**-- Recover to the starting position.

(c) **THREE**-- Repeat the action of count ONE.

(d) **FOUR**-- Recover to the starting position.

(4) After completing the required number of repetitions, change shoulders and execute an equal number of repetitions to the other side.

e. Exercise 5: Knee Bend

(1) Starting Position. Right or left shoulder position. Narrow stance. Fingers interlocked on top of the log.(See fig. 3-9, E.)

(2) Cadence. Slow.

(3) Movement. A four-count exercise:at the count of—

(a) **ONE**-- Flex the knees to a quarter-squat position.

(b) **TWO**-- Flex the knees to a half-squat position.

(c) THREE-- Lower the body to a three-quarter squat position.(Lean slightly forward.)

(d) **FOUR**-- Recover to the starting position.

NOTE:Pull forward and downward on the log throughout the exercise.

f. Exercise 6: Overhead Toss

(1) Starting Position. Right or left shoulder position, regular stance. The knees are bent to a quarter-squat.(See fig. 3-9, F.)

(2) Cadence. Moderate.

(3) Movement. A four-count exercisc:at the count of-

(a) ONE-Straighten the knees and toss the log into the air approximately 12 inches overhead. Catch the log with both hands and lower it toward the opposite shoulder. As the log is caught, lower the body into a quarter-squat.

(b) TWO-- Again toss the log into the air and when caught return it to the original shoulder.

(c) THREE-- Repeat the action of count ONE.

Section IV. Grass Drills

3401. GENERAL

a. Description and Objective. Grass drills are extremely strenuous exercises and are performed at top speed for only short periods of time. No cadence is counted but the Marines continue to execute the multiple repetitions of the command until the next command is given. The grass drill consists of two drills:Drill One and Drill Two. Each drill contains six exercises. The objective of the drills is to decrease reaction time, to develop cardiovascular endurance, and to provide a vigorous workout for all major muscles.

b. Area and Equipment. Any level area suitable for ground contact and of a size to accommodate the group is adequate. No equipment is needed.

c. Formation. All movements are executed in place. The extended, rectangular formation is recommended for a platoon- or company-size unit. The circle formation is suitable for groups of squad or section size. At the beginning of an exercise program, 2 to 3 minutes of grass drills will insure a good workout.

d. Progression. Progression is gained by gradually increasing the length of time devoted to the drills. As the physical condition of the Marines improves, the periods should be gradually lengthened to 5 minutes. As the second drill is more difficult than the first, some progression can be attained by initially executing Drill One; then as the program and the Marines progress, introduce Drill Two. To extend the duration of the drill, it may be necessary to repeat the drill.

e. Starting Position

(1) The drills are started from the GO position. Other basic positions are FRONT, BACK, and STOP. (See fig. 3-10, A.)

(a) GO. Running in place (top speed): on the toes and balls of feet, knees raised high, arms pumping, body bent forward at waist.

(b) FRONT. Prone position: elbows bent (along body), palms flat on ground directly under the shoulders, legs together and straight.

(c) BACK. Supine position (flat on back): arms extended near side on ground with palms down, legs together and straight, feet toward the stand or instructor.

(d) STOP. Football lineman stance: feet spread and staggered, left arm across left thigh, right arm straight, knuckles on ground, head up, back parallel with ground.

(2) To assume the FRONT or BACK position from the STANDING, GO, or STOP position, vigorously get into the prescribed position as quickly as possible. (See fig. 3-10, B.)

(3) To change from the FRONT to the BACK position, quickly do a pushup, move the feet several short steps to the right or left, lift the arm on the side toward which the feet move, and thrust the legs vigorously to the front. (See fig. 3-10, C.)

Figure 3-10 is available in the illustration section at the end of the book.

(4) To move from the BACK to the FRONT position, sit up quickly, place both hands on the ground to the right or the left of the legs. Move the feet several short steps to the rear on the side opposite the hands. When the feet are opposite the hands, thrust the legs vigorously to the rear and lower the body to the ground.(See fig. 3-10, D.)

f. Leadership. A warm up activity of lesser intensity should proceed grass drill. During the instructional phase and conduct of these drills, the following points should be applied.
(1) The instructor executes only GO and STOP with the Marines. This allows the instructor to supervise the drill.
(2) The commands for grass drills are given in rapid succession without the usual preparatory command.
(3) To prevent confusion, the instructor should give the commands sharply to distinguish them from comments or encouragement.
(4) As soon as the Marines know the drill, they should respond to the instructor's commands and perform all exercises vigorously and as rapidly as possible. All exercises are executed continuously until the next command is given. Insist on top speed performance; anything less is not effective.
(5) The commands peculiar to each exercise are identical to the name of the exercise.
(6) Marines are not to be required to assume the position of attention once the drills are started. To halt the drill for instructions or for rest, the command UP is used. At this command, the Marines assume a relaxed standing position. Do not demand formality. At the conclusion of a fast and vigorous 5-minute grassdrill, it is physically impossible for people to stand at attention.
(7) The sequence of commands for the execution of grass drills should occur as follows:Drill One. GO, FRONT, Bouncing Ball; GO, BACK, Bicycle; GO, Full Squatter; GO, BACK, Situps; GO, FRONT, Mountain Climber; GO, FRONT, Roll Left; GO, STOP, UP.

g. Use With Other Programs. Since grass drills can be executed in a short period of time, they may be executed where only a few minutes are available for exercise or in conjunction with another type of activity. Grass drills are an excellent substitute for running when time is a factor.

3402. GRASS DRILL ONE AND TWO

a.Grass Drill One
(1) Bouncing Ball. From the FRONT position, push up, supporting the body on the hands (shoulder-width apart) and feet. (See fig. 3-11.)Keep the back and legs in line and the knees straight. Bounce up and down by a series of short, upward springs from the hands, hips, and feet simultaneously,
(2) Bicycle. From the BACK position, raise the legs and hips. Keep the elbows on the ground and support the hips with the hands. Move the legs vigorously as if pedaling a bicycle.
(3) Full Squatter. From the STOP position, assume a full knee bend, the feet on line, hands on hips. Bounce up and down in place by short, bouncing jumps.

Figure 3-11 is available in the illustration section at the end of the book.

(4) Situps. From the BACK position and with arms stretched overhead, sit up, reach forward, and touch toes. Return to the supine position.

(5) Mountain Climber. From the STOP position, place both hands on the ground directly under the shoulders. Thrust the right leg to the rear, knee straight. The left foot should be close to the left hand, the left knee outside the left arm. Shift the weight to the hands, thrust off with the rear (right) foot and bring that foot up close to the right hand, the right knee outside the right arm. At the same time, thrust the left leg vigorously to the rear, knee straight. Continue at a fast cadence, alternating the legs.

(6) Roll Left. From the BACK or FRONT position, make one complete roll in the direction commanded. On completing the roll, return to the FRONT or BACK position.

b. Grass Drill Two

(1) Legs Over. .From the BACK position and with arms stretched overhead, palms up, raise the legs upward and then swing them backward over the head until the toes touch the ground behind the head. Return legs to the starting position. (See fig. 3-12.)

(2) V-Up and Touch Toes. .From the BACK position, raise the legs with the knees straight, sit up until the trunk and legs form a V, and touch the toes with the hands. Return to the BACK position.

(3) Rocker. .In the FRONT position, clasp the hands behind the back, arch the body, holding the head back. Start rocking, using the front part of the trunk as a rocker.

(4) Bounce and **Clap Hands**. The procedure is the same as for bouncing ball, but while in the air, clap the hands. This requires a more vigorous bounce or spring. (See fig. 3-12.)

(5) Leg Spreader. .From the BACK position, raise the legs so that the heels are 10 to 12 inches from the ground, spread them apart as far as possible, then close them together. Open and close legs as rapidly as possible.

(6) Forward Roll. .For forward roll from the STOP position, place both hands on the ground, tuck the head, and do one complete forward roll, keeping the legs tucked as you roll, and come back to the STOP position.

Figure 3-12 is available in the illustration section at the end of the book.

Section V.Guerrilla Exercise

3501.GENERAL

a. Description and Objective. Guerrilla exercises are individual exercises of various basic skills that are performed rapidly while moving forward in a circle formation. There are two tables of guerrilla exercises. Each table takes 15 minutes to complete. The objective of this exercise is to increase strength and endurance, aid flexibility, and develop coordination.

b. Area and Equipment. Any level area is suitable for conducting guerrilla exercises. No ground contact is required other than the hands. There is no equipment requirement.

c. Formation

(1) The circle formation (app. A) is used for guerrilla exercises. Each platoon forms its own circle and engages in guerrilla exercises under a platoon instructor. If the platoon exceeds 30 Marines, double or concentric circles may be used.

(2) When the circle is formed, the instructor steps into the center of the circle and moves clockwise in a small circle. He commands:QUICK TIME, MARCH, 1-2-3-4. (Rapid cadence of approximately 130 counts per minute. Cadence and step are maintained between exercises.)

(3) To reform the platoon after completing guerrilla exercises, the instructor halts the Marines and places the base man or platoon guide where he wishes and commands:

(a) BASE MAN (or platoon guide),POST.

(b) FALL OUT AND FALL IN ON THE BASE MAN (or platoon guide).

d. Progression. Progression may be attained by moving from table I to table II or by shortening the quick time marching periods between exercises and performing all exercises a second time.

e. Leadership

(1) To execute the exercises, the Marines continue at quick time while the instructor simultaneously explains and demonstrates the exercise to be performed, and then commands the Marines accordingly. In each instance, the preparatory command will be the name of the exercise and, in all instances, the command of execution will ***MARCH***.. *To* terminate each exercise, the command *is* ***QUICK TIME, MARCH.*** *The* Marines immediately pick up the step as the instructor counts cadence.

(2) Unless specified differently, each exercise should be continued for 20 to 40 seconds depending upon the vigor of the exercise. The leader can determine the duration of each exercise by observing its effect upon the Marines.

(3) To form for double guerrillas, the commands for pairing the Marines (who are in circle formation) are-

(a) PLATOON, HALT.

(b) FROM (designate an individual), BY TWO'S, COUNT OFF(.Examplel-2; 1-2; 1-2; etc.)

(c) EVEN NUMBERS MOVE UP BEHIND ODD NUMBERS(At this time, adjust pairs according to height and weight.)

(d) YOU ARE NOW PAIRED UP FOR DOUBLE GUERRILLAS. To change the Marines position, merely command CHANGE.)

(e) FORWARD, MARCH.

f. Place in the Program. Many Marines have not had the opportunity to perform the simple skills involved in guerrilla exercises. The conduct of these exercises is a simple matter since they can be performed easily and quickly in almost any situation. The tables of exercise are applicable to all personnel. The tables can constitute a station within a 1-hour period or be completed within a separate 15-minute period.

3502.GUERRILLA TABLES

a. Table I

(1) Double Time.(See fig. 3-13, A.)Hold arms in the thrust position. Execute a double-time run, maintaining the circle formation and the prescribed distance between your fellow Marines. Duration-1 minute.

(2) All Fours.(See fig. 3-13, B.) Face downward. Support the body with the hands and feet. Walk forward hands first.

(3) Crab Walk.(See fig. 3-13, C.) Get in the sitting position, face upward, and lift the hips. Support the body with the hands and feet. Walk forward feet first.

(4) Squat Walk.(See fig. 3-13, D.) Assume a full knee bend position. Grasp the ankles (left ankle with the left hand, right ankle with the right hand). Walk forward.

(5)Broad Jump.(See fig. 3-13, E.) Jump forward on both feet in a series of broad jumps. Swing the arms vigorously to assist the jumps.

(6)Toe-Touch Walk.(See fig. 3-13. F.) Walk forward, bending at the waist and touching one hand to the toe of the opposite foot while it is on the ground. Raise the trunk to the vertical position between steps. Keep the knees straight.

(7)Bottoms-Up Walk.(See fig. 3-13, G.) Assume the front leaning rest position and move the feet toward the hands in short steps, keeping the knees locked. When the feet are as close to the hands as possible, walk forward on the hands to the front leaning rest position.

(8) Straddle Run.(See fig. 3-13, H.) Run forward, leaping to the right from the left foot and to the left from the right foot.

(9) Fireman's Carry .(See fig. 3-13,1.)

(lO)Single-Shoulder Carry(See fig. 3-13, J.) Two men execute the carries as indicated by the diagram. No. 1 man executes one type; No. 2 man executes the other.

Figure 3-13 is available in the illustration section at the end of the book.

b. Table II

(1) Double Time.(See fig. 3-14, A.)Hold arms at the thrust position. Execute a double-time run, maintaining the circle formation and the prescribed distance between your fellow Marine.Duration-1 minute.

(2) Toe-Grasp Walk.(See fig. 3-14, B.) Bend forward and grasp toes. With knees slightly bent, walk forward.

(3) Hand-Kick Walk..(See fig. 3-14, C.) Walk forward, kicking the moving foot upward on every step. At the same time, lean forward and touch the elevated toe with the hand of the opposite arm.

(4) Pike Juinping.(See fig. 3-14, D.) Jump forward and upward from both feet, keeping the knees straight, and at the same time, swing the legs forward and touch the toes with the hands at the top of each jump.

(5) Squat Jump.(See fig. 3-14, E.) Leap forward from the squatting position, with the hands on the ground and the arms between the legs. Land on the ground with hands and legs extended. Bring up the legs to the squatting position.

(6) Steam Engine..(See fig. 3-14, F.) Lace the fingers behind the neck and walk forward in the following manner: as the left leg moves forward, raise the knee high, bend the trunk forward, and touch the outside of the right elbow to the outside of the knee. Then lower the left leg and step forward on the left foot and raise the right leg. Repeat with the right leg and left elbow.

(7) Knee-Touch Walk.(Scc fig. 3-14, G.) Walk forward, bending the knees and touching the ground on each step. The knees are bent and straightened on each step.

(8)Hobble Hopping.(See fig. 3-14, H.) Hold foot behind back with opposite hand and hop forward. On the command CHARGE, grasp the opposite foot with opposite hand and hop **forward.**

(9) Cross Carry.(See fig. 3-14,1.)(10)Saddle Back Carry. (See fig. 3-14, J.) Two Marines execute the carries as indicated in the diagram. No. 1 Marine executes onetype;No. 2 Marine executes the other.

Figure 3-14 is available in the illustration section at the end of the book.

Section VI. Running and Orienteering

3601. RUNNING

a. General

(l)Description. The general form and technique for all types of running are fairly constant. (See fig. 3-15.)The head is erect, body slightly forward without bending at the waist, and the arms are at a loose thrust position alternating from front to rear in straight planes. A cross-body arm movement wastes energy .The movement of the legs and feet will be discussed in subsequent paragraphs dealing with the different types of running. Of primary importance is the fact that in all types of running, the toes should be pointed straight ahead. Toeing out is a common error in both running and walking and should be an item of individual correction.

(2) Objective. The objective of running is to develop cardiovascular endurance. Despite the fact that Marines have developed their muscle structure and the strength of their muscle tissue, unless they have developed cardiovascular endurance to a satisfactory degree, they are not entirely physically fit or combat ready. Cardiovascular endurance (wind) depends on the efficiency of the lungs and heart. The efficiency of the lungs and heart depend on the amount of oxygen the lungs can absorb with each breath inhaled and the amount of carbon dioxide the lungs can expel. The process of absorbing oxygen and expelling carbon dioxide (cardiopulmonary process) is performed by the blood that circulates through the lungs. The condition of this process will determine the amount of effort a person can exert over a period of time. Running is one of the best activities to develop this vitally important endurance.

b. Running Skills. In the development of running skills, individuals may require instruction to improve their proficiency. Some important skills to consider are-

Figure 3-15 is available in the illustration section at the end of the book.

(1) Action of the Arms. Arm action is important. Check to see that arms are held loosely and that the action is relaxed. The faster the run, the more rapid the arm action.

(2) Breathing. Allow the individual to breathe through the mouth as the body demands a large supply of oxygen. Oxygen can be inhaled in greater quantities through the mouth.

3602. TYPES OF RUNNING

a. Double Time

(1) Description and Objective. Double timing is marching at the rate of 180 steps per minute, each step being 36 inches in length. It takes practice to double time with precision in formation. The Marines should keep in step, placing their feet flat on the ground. This, however, should not be a stamping motion, but should be done with as slight a jolt as possible. Double timing is like a jog, the difference being that in a jog the feet are lifted well off the ground and the running motion is bouncy. In double timing, the feet skim the ground and there is no bounce to the ran. Double timing is a vehicle for teaching proper running form and for the development of the cardiovascular system.

(2) Area and Equipment. This type of running can be completed over a variety of surfaces. Usually a Marine uses a field or road. There is no equipment requirement.

(3)Progression. There is no set standard for alternating quick time and double time in the early conditioning of Marines. A general rule is to begin with enough quick-time marching to ensure a thorough warming up, then double time about 100 paces. Change again to quick time until the individuals have made a reasonable recovery from the running, then double time another 100 paces. The amount of double time can be increased and the quick time decreased from week to week, until the individuals are double timing about 1800 yards. This type of training should be given at least twice a week, but by no means is it adequate as the sole means of conditioning.

(4) Leadership

(a) The instructor should be to one side of the column or group and toward the rear so there is a full view of all Marines. Inexperienced instructors have a tendency to supervise from a position too far forward.

(b) Select an individual who can maintain the proper cadence to act as the guide during double-time running.

(c) There are several ways for the instructor and group to count cadence while double timing. If not contrary to local policy, learn several methods and use them for variety.

b. Wind Sprints

(1) Description and Objective. This type of running involves a series of 30- or 40-yard dashes, usually conducted in successive waves of squads. Each squad is in line and the squad leader on the right flank. Wind sprints assist indeveloping speed and cardiovascular endurance. Any flat and level area may be used which will permit the squad to form a line and run the required distance.

(2) Progression. One or two 30-yard sprints will be adequate at the beginning. As time passes, sprints can be lengthened and up to six or seven sprints may be used.

(3) Leadership

(a) At the command READY (given by squad leader), each runner assumes the sprinter's starting position. At the command GO, the squad sprints approximately 30 yards, takes 10 yards to stop, and lines up immediately with the squad leader who repeats, READY, GO, and again the squad sprints. At the conclusion of the third sprint, the squad waits until all the squads of the platoon have made three sprints. Then they all line up and the squad leaders conduct three more wind sprints in the opposite direction.

(b) Valuable time is gained by having each squad ready to go when the preceding squad has moved off its second sprint mark.

c. Cross-Country Running

(1) Description and Objective. Cross-country running is a distance run conducted on a course laid out along roads, across fields, over hills, through woods, and on any irregular ground. The cross-country ran may be utilized as a conditioner or as a competitive event; the objective is to cover the distance in the shortest possible time. The course should be 2 to 2.5 miles in length and be laid out to avoid heavy vehicular traffic. The course should be marked by directional arrows until the runners know the course. These runs build leg muscles, increase lung capacity, and develop endurance. Any local area of varied terrain is suitable.

(2) Progression. In the mass training of a large group, leaders should be stationed at the head and the rear of the column and should make every effort to keep the individuals together .After determining the abilities of the individuals in cross-country running, it is advisable to divide the unit into three groups. The poorest conditioned group starts first, and the best conditioned group, last. The starting time of the groups should be staggered so that all of them finish about the same time. In preliminary training, the running is similar to ordinary road work in that it begins with rather slow jogging, alternating with walking. The speed and distance of the run is gradually increased. As the condition of the individual improves, occasional sprints may be introduced. At first the distance run is from 1/2 to 1 mile. It is gradually increased to 2 or 2.5 miles. Well-conditioned personnel can run 2 to 2.5 miles within a 15-minute period.

(3) Leadership. Marines should not be required to take part in distance running until they have been through a progressively scheduled training program which requires a considerable amount of running. Cross-country runs should be scheduled occasionally to provide variety in the program. Cross-country running has the advantage of allowing mass participation. Interest can be stimulated by putting the runs on a competitive basis. As a single activity, short cross-country runs can be scheduled once a week, gradually increasing the distance as the physical conditioning improves.

(4)Use With Other Programs. Cross-country running can be combined with other activities such as conditioning exercises.

d. Fartlek Training

(l)Description and Function. Fartlek training is various running exercises conducted along an intense combat-like course. The running is conducted to overwork the lungs, allowing only partial recovery which is followed by another intense period of overload. This sequence is repeated for the duration of the workout. Fartlek training is a useful combat training method and general cardiovascular conditioner. The entire unit must be in uniformly good to excellent condition to qualify for this training.

(2) Progression. The following is an example of one fartlek session:

(a) Warm up by stretching 3-5 minutes.

(b) Warm up running easily 5-10 minutes.

(c) Run at a fast, steady pace for 3/4 to 1 3/4 miles (dependent on terrain).

(d) Walk/jog at a moderate pace for 5 minutes (recovery).

(e) Run easily sprinting 15 to 20 meters occasionally.

(f) Run full speed uphill for 175 to 200 meters.

(g) Warm down by running easily for 1/2 to 1 mile.

(h) Warm down by stretching 3 to 5 minutes.

(3)Leadership. To add variety, try incorporating combat movements, terrain association, a series of exercises or carrying table of equipment weapons (hit and roll). This will take some initiative in setting the course, but it allows for variety. In this manner, the fartlek course can be used as a training session or part of one. It works extremely well with highly motivated, competitive Marines.

3603.ORIENTEERING

Orienteering is land navigation over a prescribed course as a timed, competitive event. It is an excellent way to integrate land navigation training with physical conditioning and requires Marines to be skilled in both areas. Commanders adjust the training as needed to change the degree of difficulty or to emphasize either navigation or conditioning. Taken to the extreme, an orienteering exercise is conducted in full combat gear over rugged terrain against aggressors. Orienteering requires the following physical skills: endurance, speed, strength, and all-around conditioning.

Section VII. Circuit Training

3701. GENERAL

a. Description and Objective. A strength circuit consists of a series of stations where individuals in small groups exercise vigorously for a short period of time and then move (on signal) to the next station where a different form of exercise is conducted. This rotation continues until all groups move through all stations. Strength circuits contain no set or specific types of exercise stations within the circuit. The objective of circuit training is to develop strength. There are three general types of circuits.

(1) Fixed Circuit. This is a circuit in which apparatus of an immovable type (fixed into the ground) is used.

(2) Movable Circuit. This circuit consists of individual exercise apparatus which is portable and can be moved to and from the training area.

(3) Simplified Circuit. This circuit requires no equipment or apparatus.

b. Formation. The exercises are done at will, but rapid, steady, and continuous work is required of all. Each Marine's nervous and muscular system reacts differently to timed vigorous exercises. His performance should be measured on how many movements per exercise he can complete as an individual. For example, one Marine may be able to complete 5 movements, while another may be able to complete 20, and yet each is receiving the maximum benefit. All three circuits contained in this chapter are designed for platoon-sized groups. Expansion beyond this capacity requires a large amount of equipment, as each Marine in the fixed and movable types of circuits must have an item of equipment available for exercise at each station. A group larger than a platoon could be exercised through use of the simplified type of circuit; however, the group would be unwieldy and control could be a problem.

c. Place in the Program. All circuits illustrated can be completed in a 15-minute period. This feature allows the exercise of a platoonor smaller group on the circuit for a single 15-minute period, or the scheduling of the circuit as a 15-minute period within a longer period. A circuit can thus be utilized within a rotating activity system of scheduling. Choice of a circuit by the unit depends upon area, facilities, and other local factors; however, there is a circuit for every need.

3702. FIXED STRENGTH CIRCUIT

a. Description and Objective. The strength circuit is an arrangement of various types of exercise apparatus which are fixed in position. (See fig. 3-16.)Seven basic exercises are used and each exercise requires an apparatus. All apparatus of one type are positioned together to constitute a station. Each station will accommodate 10 Marines. The circuit is designed to be accomplished in 15 minutes when conducted on a time-rotation basis, as normally executed for unit training. For individual use, a Marine may complete the circuit by executing a specific number of repetitions for each exercise. In this case, the time required for completion of the circuit would vary slightly depending on the number of repetitions the Marine completed. The objective of this circuit is to provide a series of exercises which will improve and maintain the strength of the body's major muscle groups.

Figure 3-16 is available in the illustration section at the end of the book.

b. Starting Level. Marines must be thoroughly warmed up prior to participating in the circuit system. If Marines have not engaged in vigorous exercise immediately prior to starting the circuit, then an instructor should execute the following warm up exercises. These exercises should be conducted in the normal formation for set drills. Seven repetitions of each exercise will normally provide sufficient warm up. The exercises are-

(1) High jumper, Conditioning Drill 1.
(2) Bend and reach, Conditioning Drill 1.
(3) Squat bender, Conditioning Drill 1.

c. Progression. The instructor can adjust the circuit system through several methods to accommodate participants of varying physical ability. One method of adjusting an exercise is to change the method of executing the leg lift or to select a heavier weight for the barbell curls. When it becomes apparent during a unit program that the overall fitness of the group has Improved, then the exercise may be made more strenuous in two additional ways. First, the instructor can have one or more of the supplementary stations added. Second, the time spent exercising at each station can be increased in 5-second increments to a maximum of 60 seconds. Also, as Marines learn the circuit, the 45-second period for movement and instruction between stations can be eliminated, thus allowing only enough time to change stations.

d. Starting Positions. The Marine starts on any station, exercises steadily for a certain period (45 seconds initially), then moves on command to the next type of apparatus where he again exercises steadily for an equal period. The Marine continues until the required exercise is completed at each station. If it is desired to expand the number of stations in the circuit to accommodate more participants at one time, the instructor can provide four additional supplementary exercises, requiring no equipment. This will increase the amount of time required to complete the circuit.

e. Leadership. Close control of all Marines is necessary to ensure that a minimum amount of time is spent in moving them to their initial stations and in moving them between stations. One instructor can control the activity on the strength circuit. A stop watch or wrist watch with a second hand is required. When Marines arrive at the strength circuit, the instructor will form them for exercise and conduct the warm up drill. The group is then reassembled and formed into a number of files equal to the number of stations being used in the circuit. Each file is then directed to a station. As soon as all participants have reached an exercise position at a station, the instructor gives the command READY, GO. After 45 seconds of exercise, the instructor gives the command STOP, CHANGE OVER. The instructor allows Marines 45 seconds for moving to the next station and for preparing for the next exercise before the command EXERCISE is again given. In lieu of verbal commands, a whistle may be used to stop and start the exercises. For large groups, a megaphone or loudspeaker is useful.

f. Individual Conditioning Program. For the Marine working alone on the strength circuit, it would be impractical to time the exercise periods. For individual exercise, the Marine should select a number of repetitions of each exercise to complete, then rotate to the next stations after completing these repetitions. The number of repetitions selected should be at or near the maximum that the Marine is capable of doing without halting for rest.

g. Fixed Circuit Stations

(1) Primary Stations (See fig. 3-16.)

(a) Pullups/Chinups. A horizontal bar placed 8 feet above the ground is necessary for each Marine at this station. The Marine will also need a space on the bar that is 45 inches wide. On the command**EXERCISE**,the Marine grasps the bar with both palms facing either forward or to the rear, arms fully extended, and feet free of the ground and executes the exercise as described in MCO 6100.3_.The exercise is repeated as many times as possible until the command **STOP** is given. Then the Marine moves to the next exercise. If a Marine has done his maximum number of pullups/chinups prior to the command **STOP**, he will remain in the "dead" handing position until the command **STOP** and move to the next station.

(b) Twist Grip. The apparatus is a horizontal bar, free to turn, held between uprights placed 30 inches apart. The bar is 52 inches above the ground. A weight of 20 pounds is attached to the center of the bar by a light rope long enough to permit the weight to rest on the ground. The Marine stands at arm's length from the bar and grasps it with his hands on either side of the rope, palms down, thumbs under the bar. On the command **EXERCISE,** the hands are rotated so that the backs of the hands are rotated away from the body, thus winding the rope on the bar. The elbows are kept straight to ensure that the exercise is performed by the hand and forearm. When the weight is drawn up to the bar, the bar is then rotated in the opposite direction to lower the weight to the ground. This exercise is continued until the command **STOP** is given. The Marine then moves to the next station.

(c) Pulley Weights. The apparatus is a T frame with a system of pulleys that suspends a weight of about 90 pounds. The weight is attached to a light steel cable which has a drawbar attached to the other end. The Marine grasps the drawbar and sits down directly under the bar, legs extended to the front and arms extended overhead. The exercise is executed by pulling the drawbar down behind the head, then extending the arms slowly again until they are fully extended overhead. The exercise is repeated as many times as possible until the command STOP is given. The Marine then will move to the next exercise. Upon completion of the exercise, the weight is lowered slowly to the ground.

(d) Barbell Curls. A barbell is necessary for each Marine at this station. The barbell is constructed of 1 1/4-inch pipe 5 feet long, and two concrete-filled No. 10 cans. Each barbell should weigh about 40 pounds. Variance in the weight of the barbells, up to about55 pounds, will allow appropriate overload to be applied to Marines who are above average in strength or weight. The Marine grasps the bar with the palms forward and assumes a standing position with the barbell held in front of the hips, hands approximately shoulder's width apart. On the command **EXERCISE,** the Marine flexes the elbows and draws the barbell up until it touches the upper chest. The elbows remain at the sides. Breath is inhaled with the upward movement and exhaled as the barbell is lowered to the starting position. The exercise is repeated as many times as possible until the command **STOP** is given. The Marine moves to the next apparatus.

(e) Step-Up. The apparatus is a platform or ledge 18 inches high and of such size. At accommodate 10 Marines. The Marine faces the platform and on the command **EXERCISE,** steps up onto the platform, bringing his trailing foot up beside the leading foot. He then steps back down to the original position, stepping down first with the same foot he initially used in stepping up. After 10 repetitions of the exercise, he changes the order of moving the feet to use the opposite leg for stepping up. Repeat this exercise until the command **STOP** is given. The Marine then moves to the next station.

(f) Leg Lift. The apparatus is a horizontal bar constructed as described in (a). To prevent the body from swaying, a horizontal back support is added 40 inches below the horizontal bar. The arms are kept fully extended. On the command **EXERCISE,** the Marine jumps up, grasps the bar with the palms forward and the back support behind him. The exercise is executed by raising the legs to a horizontal position then slowly lowering them to the vertical position. The Marine does not flex his knees. He does not swing the legs to the rear of a vertical position to gain momentum for raising them in the next repetition of the exercise. The movement is repeated until the command **STOP** is given. The Marine then moves to the next exercise. If unable to raise his legs to a horizontal position without flexing his knees, the Marine flexes his knees and draws them up to the chest, then lowers his legs to the vertical position.

(g) Rope Climb. The rope climb is 20 to 30 feet high with five ropes suspended from a horizontal bar which forms the uppermost part of the framework. To prevent the horizontal bar from sagging and to provide safety, only five ropes are attached to it. There are two frameworks per station. The ropes are 6 feet apart. Any method may be used to climb the rope, and the Marines climb as high as possible. Marines who are proficient should climb therope several times during the time allotted. Inexperienced Marines should be cautioned to take care during descent to avoid rope burns on their hands.

(2) Supplementary Stations. The following exercises are designed to expand the basic circuit by being inserted in specific places within the system. For each supplementary station used, there should be adequate room for 10 Marines to exercise.

(a) Bent Knee Situp or Bottoms-Up. These calisthenics are designed to strengthen the abdominal muscles. These exercises will be inserted between the pullup/chinup and twist grip stations. The primary stomach exercise is the situp. In case of inclement weather or other conditions that make ground contact undesirable, the bottoms up exercise is used. In situps on the command **EXERCISE,** the Marine lies on his back (supine position) with knees flexed and both feet flat on the ground and executes the exercise (minus an assistant) as described in MCO 6100.3_.The Marine then returns to the starting position, repeating the exercise until the command **STOP** is given. The Marine then moves to the next station. In bottoms up on the commandEXERCISE,the Marine assumes the front leaning rest and executes the bottoms-up exercise as described in Conditioning Drill 3. He continues this exercise at a moderate cadence until the command **STOP** is given. The Marine then moves to the next station.

(b) Pushup. This exercise is designed to strengthen the arm- and shoulder-girdle muscles. It should be included between the twist grip and pulley weight stations. Upon the command **EXERCISE,** the Marine executes the pushup as described in Conditioning Drill 1. He continues this exercise at a moderate cadence until the command **STOP** is given. The Marine then moves to the next station.

(c) Knee Bender. This exercise is designed to build leg muscles and is included between the pulley weight and barbell curl stations. On the command **EXERCISE,** the Marine executes the knee bender as described in Conditioning Drill 3.He continues this exercise at a moderate cadence until the command **STOP** is given. The Marine then moves to the next station.

(d) Trunk Twister. This exercise strengthens the major muscles of the trunk and is included between the step-up and pullup/chinup stations. On the command EXERCISE, the Marine executes the trunk twister as described in Conditioning Drill 1. He continues this exercise at a moderate cadence until the command **STOP** is given. The Marine then moves to the next station.

3703. MOVEABLE STRENGTH CIRCUIT

a. Description and Objective. The exercises in this circuit are progressive and the course is planned to gain and hold the interest of the participating groups. The circuit consists of a series of stations, with each station designed to develop a particular group of muscles. Along with muscular development, correct posture and deep rhythmic breathing should be stressed on this circuit at all times.(See figs. 3-17 and 3-18.)

Figure 3-17 is available in the illustration section at the end of the book.

b. Equipment. The equipment is set up in files. Six files of 8 stations will accommodate a platoon

c. Formation. The platoon marches to the area where the equipment is positioned and forms a file within each lane of stations, covering down on a piece of equipment. Movements are made on the double, the important factor being that no time is wasted in getting to work.

d. Progression. Initially 40 to 45 seconds per station is adequate. As individuals become stronger the time should be increased in 5-second increments until a minute to a minute and a half is reached.

e. Leadership

(1) The leader stands in front of the barbell station and controls the rotation from this position. The leader supervises the entire group, with the assistance of several instructors who move about in the platoon correcting and encouraging the Marines.

(2) The leader starts each group but does not count cadence nor lead them through the exercises. Each individual exercises rapidly but individually.

(3) As each Marine finishes his repetitions with the barbell, he places the barbells on the ground. The leader calls, READY, followed by the command, FALL OUT ONE. All will double-time to the station directly in front of them, while the individuals on the barbell stations will do an about face to the rear station in their lane.

f. Movable Circuit Stations. The Marine can obtain best results on the movable circuit if the exercises on the various stations are given in the following manner:

(1) Station 1—The Barbell. (See fig. 3-17, 1.)The exercises at this station stress proper posture; deep, rhythmic breathing; and development of the muscles of the arms, shoulders, and upper body. An instructor teaches the proper methods of lifting before the exercises begin. The methods are to lift with the legs, to keep the back straight, and to merely grip with the hands. Two recommended exercises are given below. The instructor will use only one exercise per period .He may use either exercise. At successive periods, the other exercise is used.

(a) Exercise 1--Two-Hands Military Press. (See fig. 3-18, 1.)Grasp the barbell with both hands, knuckles up at shoulder width, and lift to the chest. Steadily press to arm's length overhead; lower to the chest resisting weight all the way. Inhale as the weight is pressed up and exhale as the weight is brought down.

(b) Exercise 2--Two-Hands Regular Curl. (See fig. 3-18, 1.) Lift the weight to the waist, with the palms of the hands out, heels together, stomach in, chest lifted and arched, shoulders back, elbows in close to the sides; inhale deeply and curl the weight to the shoulders, using the arms only, at the same time keeping the elbows close to the sides; exhale Emphasize posture and the use of the arms only. This is a very valuable exercise for the development of the biceps and the grip and should be repeated from 8 to 16 times, depending on the ability of the Marine.

(2) Station 2--The Jump Rope. (See fig. 3-18, 2.) This exercise develops strength and agility in the legs and stamina of the whole body. It makes the Marine agile on his feet and increases his footwork efficiency and timing. Each Marine should progress until able to jump rope at least 3 minutes at top speed.

(3) Station 3--The Twist Grip. (See fig. 3-18, 3.) The twist grip is an excellent exercise for the hands and forearms, and adds greatly to the Marine's ability in hand-to-hand combat. The handle is gripped and twisted, winding the rope until the weighted can is level with the height of the hands, which are held horizontal. The weight is lowered in the same manner; the Marine resists the weight all the way, occasionally stopping the twisting motion and alternately removing first one hand, then the other, from the handle. A variation of the above exercise is to wind the handle with the palms up and the arms bent and with the elbows held close in to the sides. Each Marine should maintain a good posture and keep the stomach muscles taut throughout this exercise.

(4) Station 4--The Incline Plane. (See fig. 3-18, 4,)The use of the incline plane is a very strenuous exercise and well designed for the development of the abdomen. Although 6 to 10 repetitions are sufficient for the beginners, more can be added as ability increases. Marines with hernias or recent operations will be excused from participation at this station.

(5) Station 5--The War Club. (See fig. 3-18, 5.) The war club is a simple and effective means of exercising the principal muscle groups of the body, especially those of the trunk, back, and shoulders. To gain the maximum benefit from this exercise, the Marine must keep both feet flat on the ground at all times. Throughout the exercise period, the weight is swung from arm's length as follows:

(a) As in chopping wood, first on one side, then on the other.

(b) As a batter warming up with a number of bats.

(c) In large circles, first with one hand and then with the other.

(6) Station 6--The Bicycle Ride. (See fig. 3-18, 6.) The bicycle ride is well suited to exercising many of the muscle groups of the body, particularly those of the abdomen. Vary the speed of the exercise, but keep everyone "riding" the entire period. A variation exercise may be performed by placing the legs together, raising them slowly to a height about 2 feet from the ground, and then lowering them slowly to the ground.

(7) Station 7--The Step-up. (See fig. 3-18, 7.) The step-up exercises the legs. The step-up is performed by initially stepping up with the left foot, followed by the right, then stepping down with the left foot followed by the right. Continue for 20 seconds, then change to the right foot as the lead foot for 20 seconds.

(8) Station 8--The Isometric Pull. (See fig. 3-18, 8.) Two trainees work at this station with a cable pull and perform the following exercises:

(a) Initially start with one Marine in the supine position and one Marine sitting. The sitting Marine lowers the upper body to the ground and pulls the partner up to the sitting position. The partner then performs this same action and this is continued for 20 seconds at a rapid rate.

(b) During the last 20 seconds, the same action takes place but, in this case, the Marine in the supine position resists the pull of the partner for approximately 5 seconds before allowing to be pulled up into the sitting position.

3704. CIRCUIT-INTERVAL TABLE

a. Description and Objective. The circuit-interval table is designed to develop strength and endurance within a short period of time, with no equipment requirement, through a rapid and vigorous routine of exercise. Fifteen minutes is an adequate period to execute all exercises and to secure a vigorous workout with the circuit-interval principle.

Figure 3-18 is available in the illustration section at the end of the book.

b. Formation and Starting Positions

(1) A leader forms platoons or smaller groups in an oval or circular formation with 3- to 5-yard intervals between Marines. The Marines face to the right and move forward at quick time and then double time. (See fig. 3-19, A.)After running several platoon circle laps, the leader calls out the name of an arm and shoulder exercise from the list below, orders quick time and commands, for example, PUSHUPS. On this command, all Marines immediately hit the ground and individually and rapidly begin doing pushups. No cadence is counted. (See fig. 3-19, B.) After 30 seconds of exercise, the leader commands. **QN YOUR FEET, FORWARD, MARCH** The platoon resumes the quick time cadence and the leader, when ready, gives the necessary commands for double time. The double time is continued for one or more laps and the leader calls out the name of the next exercise and the process is repeated. This continues, with running between each exercise, until every body part has been exercised.

(2) The instructor controls the running and quick time to observe the effects of the exercise upon the Marines. Cadence, step, and precision are not important to the objective and the instructor should not use them. What is important is speed and the instructor should stress this. After the exercise period is started, the Marines do not stop. This circuit method emphasizes stress and recovery, the recovery occurring during the quick time periods.

c. Activities. The leader can use the following exercises. These exercises can be repeated if necessary, during a second round. The leader can use other calisthenic exercises in the circuit-interval table.

(1) Arms and shoulders-push-ups.

(2) Stomach situps.

(3) Back-squat thrusts.

(4) Legs-bicycle (on back).

d. Progression. lhe progress is controlled by the leader, who must pace the running, quick-time movement, and exercise in such a way that Marines will receive a vigorous workout yet be able to participate throughout the 15-minute period. Marines who are in the initial stages of physical condition will not be able to double time or exercise as long as those who are better conditioned. The idea is to set a pace which can be increased during each workout, thus progressing gradually to a higher level of physical fitness.

e. Leadership. The platoon leader, platoon sergeant, or section leader can lead the group. The leader must execute the exercise with the unit to feel the effects and thereby adjust the pace.

f. Place in the Program. This activity may be scheduled whenever a short period of time is available. The only requirement is that enough space, indoors or out, be available to form the circle.

Figure 3-19 is available in the illustration section at the end of the book.

Section VIII. Basic Physical Skills and Obstacle Courses

3801. GENERAL

The purpose of this section is to list the basic military skills and the methods for their development. Many of these skills are best developed by obstacle courses but other drills are also discussed. The section explains types of obstacle courses, details of construction, and methods of negotiating the various obstacles.

3802. BASIC PHYSICAL SKILLS

a. Objective. The objective is to develop proficiency in the various military physical skills which are essential to personal safety and effective combat operations. In travel by foot over rugged terrain and in the execution of combat duties, Marines must be trained to perform basic skills, such as running, jumping, climbing, and carrying. During training, Marines will develop agility and coordination in these skills .Fast and skillful execution of these skills may mean the difference between success and failure on the battlefield.

b. Place in the Program. These skills are practiced throughout the entire program in many different activities. Many of these skills are best practiced on obstacle courses.(See par. 3803.)

c. Description. These basic skills are the minimum skills required by all Marines. The basic skills are as follows:

(1) Running. Running is used to strengthen the legs and develop cardiovascular endurance. Marines should be exposed to running in various situations: on roads, over rough ground, up and down hills, cross-country, and running over low obstacles.

(2) Jumping. In broad jumping, the take-off foot is planted firmly and the spring comes from the extension of this leg as the other leg reaches for the far side of a ditch or similar obstacle. (See fig. 3-20.) The arms are forcibly raised forward and upward to assist in propelling the body up and forward. Landing may be on one or both feet depending upon the length of the jump. In vertical jumping downward from a height, the jumper should aim his feet at the desired landing spot and he should also jump with the knees slightly bent and feet together, with the trunk inclined slightly forward. As the feet touch the ground, the shock is absorbed by bending the knees into a full squatting position. If the height is too great or the ground too hard to absorb the shock, then the jumper should forward roll or side roll thus eliminating some of the momentum.

Figure 3-20 is available in the illustration section at the end of the book.

(3) Dodging In combat situations, it is often times necessary to change directions quickly. To execute this movement while running, a lead foot is firmly planted, left foot if the direction is to the right and right foot if the direction is to the left. The opposite foot is moved toward the new direction. The knees are slightly flexed during the movement and the center of gravity is low and balanced. (See fig. 3-21.) At the time of the change of direction, the head and trunk are turned quickly in the new direction.

Figure 3-21 is available in the illustration section at the end of the book.

(4) Climbing and Surmounting. All Marines should know how to climb and surmount various types of obstacles. (See fig. 3-22.)

Figure 3-22 is available in the illustration section at the end of the book.

(a) Vertical Climbing, as in Climbing a Rope or Pole. The hands grasp the rope or pole overhead with the palms toward the face. Gripping the object, the body is pulled upward with the arms and shoulders, assisted by the feet which grip the object and assist by pushing downward. If shoulder girdle strength and body coordination are not adequate to permit alternating the hands, the arms act together in pulling upward.

(b) Climbing as in Surmounting a Wall. Ingoing over a wall, the body should be kept as close to the top as possible, since in combat operations it is important to offer as small a target as possible to the enemy. If an individual climbs a wall while carrying a rifle, both hands should be freed by slinging the rifle over the back. There are two methods commonly used for surmounting a wall of moderate height, but only one for dropping from it.The methods are as follows:

1 Run, Jump, and Vault. Approach the wall at a run, jump forward and upward at the wall and place one foot against it as high up as possible. Use the foot in contact with the wall to help push the body upward while grasping the top of the wall with the hands. Pull the body up with the arms, assisted by pressure of the foot against the wall and swing the legs over, propelling the body weight over the wall.

2 Hook and Swing. Approach the wall at a run and jump forward and upward. Hook one elbow over the wall, locking the arm in place by pulling up until the top of the wall is underneath the armpit. Grasp the top of the wall with the other hand. Draw the leg which is closer to the wall up as far toward the top as possible. Then swing the other leg over the top of the wall.The body is then carried over with a rolling motion. A variation of this leg action can be used by Marines who are unable to draw up the leg as described. While hanging with both legs fully extended, start a swinging motion with the legs together. When the legs have enough momentum, swing the outside leg over the top of the wall with a vigorous kick, then follow with the body.

3 **Dropping**. All drops from a height are executed in the same manner regardless of the method used to gain the top. One hand is placed against the far side of the wall while the other hand grasps the top. From this position, the body is rolled over the wall and vaulted away from it with the legs swinging clear. As the body passes over the wall and drops, it should at all times face the wall. This will keep the rifle and other equipment clear. Break the fall by retaining a grasp on the top of the wall as long as possible.

(c) Climbing Ladders and Cargo Nets. Rope ladders, stationary vertical ladders, and cargo nets employ the same general technique. The important element is to grasp the side supports firmly in the hands about shoulder height and place the feet on a rung which would cause the body to be fully extended. In movement upward, one hand is moved upward and a new grasp is secured and, at the same time, the opposite leg moves up a rung. As the knee straightens, the body is elevated. This process is repeated using the opposite arm and leg. Alternation continues in this manner until the climber reaches the objective.

(5) Traversing Horizontal Objects. The traversing of horizontal objects puts heavy stress on the arms and shoulder girdle area as the feet are usually suspended in the air with all of the body weight on the arms and shoulders.(See fig. 3-23.)

(a) Traversing Horizontal Ropes or Pipes. The hands grasp the horizontal support overhead with the palms facing. To propel the body forward, one hand is released and moved forward to secure a new grasp. At the same time, the opposite side of the body is swung forward (some people are able to "walk" in the air, keeping the body to the front and moving the legs in time with the arms as in walking on the ground). The other hand is then released and moved forward; this alternation is continued until the objective is reached.

Figure 3-23 is available in the illustration section at the end of the book.

(b) Traversing Horizontal Ladders. In this situation, the movement is the same as used in traversing a rope or pipe. The hands, however, are placed on the rung with the palms away from the face. Other than this difference the technique is the same.

(6) Crawling. Crawling in combat situations is an often used skill. Crawling may be high or low.(See fig. 3-24.)

Figure 3-24 is available in the illustration section at the end of the book.

(a) High Crawl.

In the high crawl, the Marine moves on hands and knees, moving one hand and the opposite knee and then continuing to move the hands in alternation with the opposite knee following the companion hand.

(b) Low Crawl. The Marine is in the prone position usually with the forearms and palms of the hand on the ground. He propels forward by bending the knee of one leg and pushing with the inside edge of the shoe. At the same time, the opposite arm moves forward and pulls to the rear. The body remains low and movement is continued by bending the opposite knee and pushing, and at the same time sliding the opposite arm forward and pulling. Alternation of hands and legs continues until the objective is reached.

(7) Throwing. Throwing may be executed from the kneeling or standing positions. The object to be thrown is held in the hand, and the throwing arm is bent at the elbow; the hand is then moved to the rear until the hand is behind the ear. The body is turned so that the lead foot and balance arm on the side toward the target are pointing at the target. (See fig. 3-25.) The balance arm is used to sight over and align the throwing hand and the target. When properly aligned, the elbow is moved rapidly forward until it is at a point just in front of the body where the arm is straightened and the wrist snapped. This whip motion propels the object to the target. Underhand throws secure momentum by the thrower bending his knees and swinging the throwing arm to the rear. As the knees are straightened, the arm is forcefully swung forward from the shoulder and the object released.

Figure 3-25 is available in the illustration section at the end of the book.

(8) Vaulting. Vaulting is employed to overcome low barriers or fences. (See fig. 3-26.)The object to be surmounted is approached at an angle. The hand on the side next to the obstacle is placed on the top of the obstacle and, with a straight arm, the body weight is pushed upward. At the same time, the leg on the side next to the obstacle is thrown upward and over the top followed by the other leg (side approach). In landing, the weight comes down on the leading leg first followed by regaining the balance on both legs. The free arm serves as a balance. A direct (front) approach can be used at which time both legs go over the object together.

(9) Man Carrying. There are three basic individual means of carrying personnel in combat situations and one of these methods may be used in carrying objects.

Figure 3-26 is available in the illustration section at the end of the book.

(a) Fireman's Carry."A" stands sideways in front of "B", "A" bends his knees and leans forward, placing one arm through "B's" crotch, grasps the wrist of "B's" arm, which is hanging over the shoulder, and then "A" runs forward.

(b) Saddle-Back Carry. With his back toward "B", "A" stands in front of "B". "B" mounts "A's" hips and clasps his arms in front of "A's" chest. "A" grasps "B's" thighs.

(c) Single-Shoulder Carry. "A" stands facing "B"."A" assumes a semi-squatting position. "B" leans forward until "B" lies across "A's" right shoulder. "A" clasps his arms around "B's" legs and straightens up, lifting "B" from the ground. "A" then runs forward. This method may also be used to carry heavy objects.

(10) Balancing. Balancing the body while walking or running on a narrow object when crossing obstacles is a skill which requires practice and confidence. Balance is required in negotiating a log placed across a stream, in crossing a narrow beam or rail, and in similar situations.(See fig. 3-27.) To perform this skill, place the feet on the object to be crossed, hold the arms to the sides at shoulder level, and fix the eyes on the object approximately 5 yards in front of the feet. Generally, it is not a good practice to look down at the feet. Walk the beam by placing first one foot and then the other in the center of the beam, thereby moving forward, using the arms to aid in maintaining balance.

Figure 3-27 is available in the illustration section at the end of the book.

(11) Falling. Injury can be avoided if Marines are taught to fall properly. They should know how to use the body momentum to their advantage during a fall rather than to try resisting that force. (See fig. 3-28.) If enough force is present, such as occurs during a fall while running or in jumping downward from a height, individuals can extend their hands to catch the weight. At the same time, duck the head and roll forward onto the feet. The key to falling without injury from the standing position is relaxation and rolling on the outside of the leg, hip, and buttocks to take the brunt of the fall.

Figure 3-28 is available in the illustration section at the end of the book.

3803. OBSTACLE COURSES

a. General. Obstacle courses are a valuable part of physical readiness training. The challenge presented by the obstacles assists in developing and testing the basic physical skills. In many combat situations, success will depend upon the Marine's ability to perform one or more of these skills, often while carrying field equipment and when fatigued. In this section, two types of obstacle courses will be discussed-Conditioning Obstacle Course and Confidence Obstacle Course.

b. Course Safety Precautions. Commanders and course instructors should take certain precautions to prevent injury on obstacle courses. A few of the precautions are:
(1) Inspect the course for faulty construction of obstacles, protruding nails, rotten logs, condition of the landing pits, and other hazards to safety.
(2) Conduct warm up exercises before the unit runs the course.
(3) Explain and demonstrate the correct techniques for negotiating all the obstacles before allowing the Marines to try them.
(4) Give Marines at least two weeks of conditioning exercises before scheduling the obstacle and/or confidence courses.
(5) Ensure that negotiation of the higher and more difficult obstacles are under the supervision of an instructor.
(6) Do not permit individuals who have neither practiced the basic skills nor run the conditioning obstacle course to participate in the confidence obstacle course.
(7) Weather conditions may cause footing or handhold surfaces to be slippery. If such is the case, postpone training on the course.

3804.CONDITIONING OBSTACLE COURSES

a. Description and Objective. The Conditioning Obstacle Course is commonly known as the Obstacle Course. This course consists of fairly low obstacles which are designed to be negotiated quickly. The obstacles serve to test various basic skills, and running the course is a test of the Marine's physical condition. After receiving instruction and an opportunity to practice the skills, Marines run the course against time.

b. Area and Equipment
(1) Complete standardization of obstacle courses should not be attempted since topographical conditions always vary. Commanders should use ingenuity in constructing a course, making good use of streams, hills, trees, rocks, and other natural obstacles. Since the course is eventually run at high speed, it should not be dangerous.

(2) The course should be wide enough for six or eight men to run simultaneously, encouraging competition. The lanes for the first several obstacles should be wider and the obstacles themselves easier than those that follow. This avoids congestion until the contestants scatter out over the course. The last two or three obstacles should not be too difficult and should not involve high climbing. This prevents injuries and falls resulting from fatigue.

(3) The total distance of the course should range from 300 to 450 yards and include from 15 to 25 obstacles. Normally the obstacles should be 20 to 30 yards apart and arranged so that those which exercise the same groups of muscles are separated.

(4) The obstacles should be substantially built. Peeled logs, 6 to 8 inches in diameter, are ideal for many of the obstacles. Sharp points and corners should be eliminated. Landing pits for jumps or vaults should be filled with sand or sawdust to prevent injuries.

(5) The course should be constructed and marked so that it is not possible to sidestep or detour obstacles. However, it is desirable to provide alternate obstacles of varying degrees of difficulty.

(6) The course should be in the shape of a horseshoe or figure eight so that the finish is close to the start and signs should be placed to indicate the course route.

c. Leadership. Before Marines run an obstacle course, they should be instructed in the proper technique of negotiating each obstacle. In each case this technique should be explained and demonstrated in detail, with emphasis on avoiding injury. Every individual should be given an opportunity to practice on each obstacle until he becomes reasonably proficient at negotiating it. Before the course is run against time, it is advisable for individuals to make several runs at a slower pace. During such practice or trial runs, the instructor should observe the performances and make appropriate corrections. Marines should never be permitted to run the course for time until they have practiced on all obstacles. The best method of timing the runners is to have the timer stand at the finish and call out the minutes and seconds as each individual finishes. If several watches are available, each wave may be timed separately. If only one watch is available, the different waves should be started at regular intervals, such as every 30 seconds. If an individual fails to negotiate an obstacle, a previously determined penalty should be exacted.

d.Types of Obstacles

(1) Jumping-Type Obstacles. These obstacles may be ditches which are cleared with one leap, trenches which the individuals can jump into, heights which require jumping downward, or hurdles which an individual can jump over. (See fig. 3-29.)

Figure 3-29 is available in the illustration section at the end of the book.

(2) Dodging-Type Obstacles. Obstacles of this type are usually mazes consisting of posts set in the ground at irregular intervals. (See fig. 3-30.)The intervals between posts should be rather narrow so that the Marines must pick their way carefully through and around them. Lane guides may be constructed to guide the Marines to dodge and change direction. Obstacles may be put into a maze pattern to cause the Marines to change direction.

Figure 3-30 is available in the illustration section at the end of the book.

(3) Vertical Climbing and Surmounting Type Obstacles. These obstacles may be climbing ropes, either plain or knotted and 1-5 inches in diameter; cargo nets or walls 7 or 8 feet high; or vertical poles 6 to 8 inches in diameter and 38 feet high.(Seefig. 3-31.)

Figure 3-31 is available in the illustration section at the end of the book.

(4) Horizontal Traversing-Type Obstacles. Horizontal obstacles may be pipes, beams, ladders, or ropes. (See fig. 3-32.)

(5) Crawling Type Obstacles. Obstacles which require crawling may be constructed of large pipe sections, low rails, and wire. (See fig. 3-33.)

Figure 3-32 is available in the illustration section at the end of the book.

(6) Vaulting-Type Obstacles. Obstacles of 3 to 3.5 feet in height such as low walls or fences may be used as a vaulting obstacle.(See fig. 3-34.)

(7)Balancing-Type Obstacles. Beams, logs, and planks may be used as balancing-type obstacles.(See fig. 3-35.)These items may be used to span water obstacles and dry ditches, or raised off the ground somewhat to simulate natural depressions.

Figure 3-33 is available in the illustration section at the end of the book.

3805.CONFIDENCE OBSTACLE COURSES

a. Description and Objective. The Confidence Obstacle Course is commonly known as the Confidence Course. This course is composed of higher and more difficult obstacles than those used in the conditioning course. The confidence obstacle course is designed to give Marines confidence in their mental and physical capacities and to cultivate their spirit of daring. They are encouraged but not compelled to negotiate this course and the course is not run against time. The negotiation of a confidence course, however, is strenuous enough to be an excellent physical conditioner. Marines should NEVER attempt to take the obstacles at high speed and should not compete for speed. The obstacles vary from fairly easy to extremely difficult. Some are of considerable height to accustom Marines to climbing such heights without fear. Considerable emphasis is placed on obstacles that train and test an individual's balance.

Figure 3-34 is available in the illustration section at the end of the book.

b. Area and Equipment

(1) The confidence course accommodates four platoons, one platoon at each group of six obstacles. The course should be made up of about 24 obstacles, numbered and marked as follows:! to 6, white numbers on red background; 7 to 12, black numbers on a white background;

13 to 18, white numbers on a blue background; and 19 to 24, white numbers on a black background.

(2)A few simple pieces of equipment should be provided for individuals who do not have the strength, courage, or ability to negotiate the obstacles.

Figure 3-35 is available in the illustration section at the end of the book.

c. Formation. The obstacles should be divided into groups of six, and each group is designated by a different color. Each platoon starts at a different color. Individuals are separated into groups of 8 to 12 at each obstacle. At the starting signal from the company commander, they proceed numerically through. Anyone may skip an obstacle who is afraid to try. Individuals proceed from obstacle to obstacle until time is called, then assemble as ordered.

d. Leadership. If the Marines are new to the confidence course, an instructor will demonstrate or will give a brief orientation at each obstacle, including an explanation and demonstration of a method of negotiating it. Marines are encouraged to try the various obstacles, but they are not compelled to do so. No compulsion is to be used. The manner of negotiating any obstacle is left to the discretion of the Marine. However, the instructor assists anyone who experiences difficulty. Instructors must supervise closely at all times to prevent injuries, as some of the obstacles are quite high. Also, some of the obstacles should not be used when slippery or wet. The example of instructors and especially selected demonstrators will serve to inspire the individuals to greater effort.

e. Negotiating the Obstacles. Although personnel need not conform to any one method of negotiating the obstacles, there should be some uniformity in the approach to them. A general method of negotiating the obstacles is indicated below.

(1) Red Group This group contains the first six obstacles. (See fig. 3-36.)

(a) The Belly Buster Individuals may vault, jump, or climb over. Warn them that the log is not stationary.

(b) Reverse Climb. Climb the reverse incline and go down the other side to the ground.

(c) The Weaver. Move from one end of the obstacle to the other by weaving the body under one bar and over the next.

(d) Hip-Hip. Step over each bar, either alternating legs or using same lead leg each time.

(e) Balancing Logs. Step up on log, retaining the balance, walk or run along log.

(f) Island Hopper. Jump from one log to another until the obstacle is negotiated.

(2)White Group. This group is composed of the second six obstacles. (See fig. 3-37.)

(a) Tough Nut. Step over each X in the lane.

(b) Slide for Life. Climb the tower, grasp the rope firmly, and swing the legs upward. Hold the rope with the legs to distribute the weight between them and the arms. Braking the slide with the feet and legs, proceed down the rope. Warn Marines that there is danger of getting rope burns on their hands. When the rope is slippery or wet, this can be a dangerous obstacle.

(c) Low Belly Over. .Mount the low log and jump onto the high log, both arms grasping over the top of the log, the stomach area in contact with it. Swing the legs over the log and lower the body to the ground.

(d) Belly Crawl Move forward under the wire, belly down, to the end of the obstacle.

(e) The Dirty Name Mount the low log and jump to or reach the higher logs in succession, then jump or drop to the ground. Warn the Marines about the height of the final log.

(f) The Tarzan Mount the lower log and walk the length of it and each successive, higher log until reaching the horizontal ladder. Grasp two rungs of the ladder and swing the body into the air. Negotiate the length of the ladder by releasing one hand at a time and swing forward, grasping a more distant rung.

(3) Blue Group. This group is formed by the third group of six obstacles. (See fig. 3-38.)

(a) High Step over. Step over each log, alternating the lead foot or using the same lead foot.

(b) Swinger. Climb onto the swinging log and over to the ground on the opposite side.

(c) Low Wire. Move under the wire on the back, using the hands to raise the wire to clear the body.

(d) Swing, Stop and Jump. Gain momentum with a short run, grasp the rope, and swing the body forward to the top of the wall. Release the rope while standing on the wall and jump to the ground.

(e) Six Vaults. Vault over the logs, using one or both hands.

(f) Easy Balancer. Walk up one inclined log and down the one on the other side to the ground.

(4) Black Group. The last group is formed by the final six obstacles. (See fig. 3- 39.)

(a) Incline Wall. Approach the underside of the wall, jump up and grasp the top and pull the body up and over. Slide or jump down the incline to the ground.

(b) Skyscraper. Jump or climb to the first floor, climb up the corner posts or assist each other to any desired floor. Descend to the ground in any desired manner.

(c) Jump and Land. Climb up the ladder to the platform and jump to the ground.

(d) Confidence Climb. Climb the inclined ladder to the vertical ladder. Go to the top of the vertical ladder, then down the other side to the ground.

(e) Belly Robber. Step on the lower log and assume the prone position on the horizontal logs. Crawl over the logs to the opposite end of the obstacle.

(f) The Tough One Climb the rope or pole on the higher end of the obstacle, then go down the ladder and across the log platform. Climb over or between the logs at the end and go down the rope or pole to the ground. Vault over the final log.

Figure 3-36 is available in the illustration section at the end of the book.

Figure 3-37 is available in the illustration section at the end of the book.

Figure 3-38 is available in the illustration section at the end of the book.

Figure 3-39 is available in the illustration section at the end of the book.

Section IX Individual Exercise Programs

3901.GENERAL

Marines at times will **be** stationed on independent duty and consequently will be responsible for their own physical fitness program. This section will assist them in understanding the need for exercise and will aid in the planning and execution of an individual exercise program. Exercise activities included are the Bench Conditioner, 6-12 Plan, Weight Training, and Isometric Contraction. Keeping physically fit is a problem that faces every Marine. Even though we are frequently engaged in training that requires some physical effort, in many cases, it is not enough to prepare us to meet the intense physical demands of combat. Attaining a satisfactory level of physical readiness is not an insurmountable objective. Available time appears to be the most difficult obstacle to the development of physical readiness. In most cases, regular physical training programs are centralized, requiring the individual to temporarily leave the work area. The problems involved in setting an hour aside two or three times each are numerous. However, most of us can devote a few minutes each day to physical fitness with little, if any, impact on our daily work schedule, especially if it does not require us to leave our work area.

a. Type of Program. There are many good physical fitness programs available to the individual or group. Regardless of the type or duration, to be effective, the program must contain exercises that are strenuous and are challenging to the individual. Space will not permit the inclusion of all available means of individual exercise. The programs selected for this chapter have met the requirement of minimum space and minimum time.

b. Need to Augment Program. These programs are quite strenuous and will develop a satisfactory level of physical readiness. However, if the Marine desires additional development of endurance, it is recommended that he supplement these programs with a 15- to 30-minute period of wind sprints and double timing on an alternating daily basis.

c. Progressive Training. If Marines are performing duties which require little or no physical activity, they must plan a physical conditioning program that assures a moderate beginning, moderate but steady progression, and sufficient warmup before starting the vigorous exercise. To avoid organic of bodily harm, a Marine activity without adequate warmup. He should conduct conditioning programs on a daily basis over an extended period of time, never on an unduly accelerated or crash basis.

3902.THE BENCH CONDITIONER

a. Description and Objective. The bench conditioning program uses a modified bench to employ both isotonic (moving) and isometric (stationary) exercises are as the nucleus of the program. The exercises are designed to develop strength and endurance in all the major muscle groups of the body. The principles of progression, overload, and balance are employed when the exercises are performed properly.

b. Area and Equipment. The conditioning apparatus can be constructed in any unit motor pool with welding equipment found in most salvage yards.(See figs. 3-40, 3-41.)Additionally, there are available any number of commercially produced apparatus available. The important thing is that in utilizing this equipment the following exercise routine be adhered to.

c. Starting Level and Progression.. The program consists of two tables, each with 10 exercises. The Marine controls progression by required repetitions or, in some cases, by application of maximum effort. Each table can be completed within 15 minutes.

d. Starting Position. To start the program, the Marine begins with the Table 1 and executes each exercise for the required number of repetitions as indicated. The Marine controls the starting level and progression. When the Marine executes the maximum repetitions for Table I within a 15-minute exercise period, he progresses to Table II. To maintain this level of development, the Marine should also execute the maximum repetitions for Table II within a 15-minute exercise period. The Marine should keep substitution of exercises to a minimum. However, it he completes a full 15 minutes of strenuous exercise and exercises all muscles, then there should be no appreciable difference in the overall development.

e. Bench Conditioning, Table I

(1) Exercise 1: Side-Straddle Hop. This is a two-count warmup exercise done at moderate cadence. The starting position is the position of attention. On count ONE jump slightly into the air, swinging the arms out to the sides and up to a vertical position, hands touching.(See fig. 3-40, A.)At the same time, spread the feet wider than shoulder-width apart. On count TWO, using a slight flexing of the knees and ankles, jump slightly into the air and return to the starting position by swinging the arms back down to the sides. Twenty repetitions of this exercise is the standard dosage throughout the program.

(2) Exercise 2: Hand Walk. Remove the lower horizontal bar. Adjust the upper horizontal bar so that it is high enough to permit a "dead hanging" position, with the feet off the ground.(See fig. 3-40, B.j From the "dead hanging" position, release one hand and drop the arm to the side of the body. Then raise that arm and regrasp the horizontal bar. Release the bar with the other hand and drop that arm to the side. Repeat this as many times as possible.

(3) Exercise 3: Situps. Lie down with the fingers interlocked and placed behind the head. Hook the toes under the foot braces. Raise the trunk and upper body to an upright sitting position, twisting it to the left and then forward and downward until the right elbow touches the left knee.(See fig. 3-40, C.)Lower the body to the starting position .Sit up again but twist the body to the opposite direction as before, touching the left elbow to the right knee. Again lower the body to the starting position. The starting dosage is 20 situps. The Marine should continue the progression until he has attained 40 situps.

Figure 3-40 is available in the illustration section at the end of the book.

(4) Exercise 4: Double Step-upStarting at one end of the bench, step up onto the bench, and walk across it. Step down from the other end; turn around and repeat the process to return to the starting point.(See fig. 3-40, D.)Each return to the starting point constitutes a repetition. The starting dosage is 20 repetitions. Maximum dosage is 35 repetitions. This exercise should be done at a rapid cadence.

(5) **Exercise 5: Isometric Bar Lift.** Adjust the lower bar so that it is slightly higher than the beltline. Placing the feet on the footplates at the base of the frame, grasp the lower bar so that the hands are spread shoulder-width apart. Assume a crouched position and lift with maximum effort using the arms, back and legs.(See fig. 3-40, E.)Starting dosage is 4 repetitions of a stress time of 5 seconds followed by a 5-second rest prior to the next repetition. The Marine obtains progression by lengthening stress periods to 6 and later 7 seconds. Do not increase the number of repetitions.

(6) **Exercise 6: Knee Lift**. Adjust the upper bar to the same height used in Exercise 2.Adjust the lower bar so that is stops rearward movement of the hips when the "dead hanging" position is assumed.(See fig. 3-40, F.) Keeping the arms extended, flex the legs and raise the knees as high as possible. Hold this position for 5 seconds, then return to the starting position. After 2 seconds in the starting position, raise the knees again. Each return to the starting position constitutes one repetition. The dosage is five repetitions. Progression is the same as in Exercise 5.

(7) **Exercise 7: Isometric Pull.** Adjust the lower horizontal bar to a position where it is slightly higher than the beltline. Grasp the handles and pull outward.(See fig. 3-40, G.)Apply maximum effort and hold for approximately 5 seconds. Relax for 5 seconds between repetitions; perform four repetitions. Moving the body closer to or farther away from the bar will change the stress from the upper arms to the forearms. Progression is the same as in Exercise 5.

(8) **Exercise 8: Isometric Compression.** Maintain the position as in exercise 7.(See fig. 3-40, H.)Grasping the handles in the same manner, press in **with** maximum effort and hold for approximately 5 seconds. Relax for 5 seconds between repetitions. Progression is the same as in Exercise 5.

(9) **Exercise 9: Isometric Press.** Remove the lower horizontal bar. Adjust the upper horizontal bar until it is about 6 inches lower than the extended arms can reach. Stepping on the footplates at the bottom of the frame, grasp the bar with both hands and push up.(See fig. 3-40,1.)Keep both the legs and arms slightly flexed and the back straight. Apply maximum effort for 5 seconds then relax for 5 seconds. Complete four repetitions. Progression is the same as in Exercise 5.

(10) **Exercise 10: Pushups.** Grasping the foot braces with both hands, assume the front leaning rest position.(See fig. 3-40, J.)Keeping the back and legs straight, lower the body until the chest is lower than the **hands,** then return to the starting position. The Marine should complete the maximum possible number of repetitions.

f. Bench Conditioning, Table II. There is no limit on the maximum number of repetitions attainable in Exercises 3, 4, and 6 of Table II. The only limit imposed is that the entire program of 10 exercises outlinted in either table should not exceed 15 minutes.

(1) **Exercise 1: Side-Straddle Hop.** This is a two-count warm up exercise done at a moderate cadence. The starting position is the position of attention. On count ONE, jump slightly into the air, swinging the arms out to the sides and up to a vertical position, hands touching.(See fig. 3-41, A.)At the same time, spread the feet wider than shoulder-width apart. On count TWO, using a slight flexing of the knees and ankles, jump slightly into the air and return to the starting position by swinging the arms back down to the sides. Twenty repetitions of this exercise is the standard dosage throughout the program.

(2) Exercise 2: Pull-up. Adjust the horizontal bar so that is is high enough to permit a "dead hanging" position with the feet off the ground. Grasp the bar with both hands, palms facing forward. By flexing the arms, raise the body to a position where the chin is higher than the bar.(See fig. 3-41, B.)Then lower the body to the "dead hanging" position. Repeat as many times as possible.

(3) Exercise 3: Bench situps. Sit on the bench and hook the feet under the foot braces. With the fingers interlocked behind the head, lean back until the head touches the floor.(See fig. 3-41, C.)Return to the starting position. The starting dosage is 15 situps.

(4) Exercise 4: Step-Up. Face the bench and step up on it with one foot, bringing the trailing foot up next to the leading foot. Step back down again, leading with the same foot used first in stepping up.(See fig. 3-41, D.)Perlorm half of the total repetitions, then change the sequence of moving the feet to use the other leg in stepping up, and repeat the same amount of exercise. The starting level is a total of 40 step-up. **This exercise should be done at a rapid cadence.**

(5) Exercise 5: Isometric Bar Lift\djust the lower bar so that it is slightly higher than the beltline. Placing the feet on the footplates at the base of the frame, grasp the lower bar so that the hands are spread shoulder-width apart. Assume a crouched position and lift with maximum effort using the arms, back, and legs.(See fig. 3-41, E.)Starting dosage is four repetitions of a stress time of 8 seconds followed by a 5-second rest prior to the next repetition. Progression is obtained by lengthening stress periods to 10 seconds. Do not increase the number of repetitions.

Figure 3-41 is available in the illustration section at the end of the book.

(6)Exercise 6: Leg Lift. Adjust the bars and assume the starting position as shown in figure 3-41, F. Keeping arms and legs extended, raise the legs to a horizontal position and hold in that position for 2 seconds. Then lower the legs slowly to the starting position. Five repetitions is the starting level.

(7) Exercise 7: Isometric Pull. Adjust the lower horizontal bar so that it is slightly higher than the beltline. Grasp the handles and pull outward.(See fig. 3-41, G.)Apply maximum effort and hold for approximately 8 seconds. Relax for 5 seconds between repetitions; perform four repetitions. Moving the body closer to or farther away from the bar will change the stress from the upper arms to the forearms. The Marine obtains progression by lengthening the stress period to 10 seconds.

(8) Exercise 8: Isometric Compression. (See fig. 3-41, H.)Maintain the position as in exercise 7.Grasping the handles in the same manner, press in with maximum effort and hold for approximately 8 seconds. Relax for 5 seconds between repetitions; perform four repetitions. The Marine obtains progression by lengthening the stress period to 10 seconds.

(9) Exercise 9: Isometric Pres. Remove the lower horizontal bar. Adjust the upper horizontal bar until it is about 6 inches lower than the extended arms can reach. Stepping on the footplates at the bottom of the frame, grasp the bar with both hands and push up.(See fig. 3-41,1.)Keep both the legs and arms slightly flexed and the back straight. Apply maximum effort for 8 seconds, then relax for 5 seconds. Complete four repetitions. The Marine obtains progression by lengthening the stress period to 10 seconds.

(10) Exercise 10: Inclined Pushup. Assume the front leaning rest position with the feet on the bench.(See fig. 3-41, J.)Keeping the back and legs straight, lower the body until the nose touches the ground. By extending the arms, raise the body to the starting position. Repeat as many times as possible.

3903. THE 6-12 PLAN

a. Description and objective. The 6-12 plan of physical fitness has been developed to assist in regulating quantity and progression and to provide a convenient set of exercises. This is a basic program and will take 18 weeks to complete if you follow the moderate progression as to the time prescribed for each level of achievement. The time can be shortened as explained below. This plan consists of six basic exercises a day which can be completed in 12 minutes. There are six tables of six exercises each, thus allowing you to progress from table to table. The plan is progressive, fits any age group, contains balance and variety, and applies the principle of overload in a safe and gradual manner. Begin at Table I, Progression Guide, with the number of repetitions as indicated by age.

b. Formation. If just starting an exercise program, do not rush through the first table. Remember, individuals should remain at each level for about a week before moving upward. The time allotment stated for each exercise at the bottom of the tables is a guide; some people may take more and some less time on the individual exercises. At the end of a one-week period, when the individual can comfortable perform the six exercises in 12 minutes, move on to the next level. To a certain degree, the individual must be the judge of his ability to progress form level to level and table to table. If starting with a certain degree of fitness, some of the beginning tables may present little challenge.

c. Starting Level **and Progression.** There are three levels of achievement for each age group, indicated as A, B, and C. Start at the C level for the appropriate age group. At the end of a one-week period, or when the individual can do all exercises at that level within 12 minutes, progress to the B level. At the end of the second week, or when the individual can accomplish that level within 12 minutes, progress to the A level. At the conclusion of the third week or when the individual can achieve the A level within the time limitation, move on to table II.(See Tables I through VI, Progression Guide.)

d. Maintenance Level. Attempt to work through all six tables. If this proves to be too difficult, then maintain exercise at the—

(1) A-level on Table IV, Progression Guide it in the 45 to 49, 50 to 59, or over 60 age group.

(2) A-level on Table V, Progression Guide if in the 17 to 29, 30 to 39, or 40 to 44 age group.

e. Precautions. To achieve the maximum benefit, perform each exercise exactly as specified. Read the descriptions and study the illustrations. Do not slight the movements. Use a sensible approach and follow these four points as they apply before starting or during your exercise program.

(1) If you have the slightest doubt about your ability to participate in this exercise program, consult a physician.

(2) Stop immediately if you notice unusual breathlessness or chest pain while taking part in these exercises. If these conditions persist, consult a physician.

The following Tables of Progression Guides and Exercises are the progression of the 6-12 Plan program.

Table I. Progression Guide is available in the illustration section at the end of the book.

Table I. Exercises.is available in the illustration section at the end of the book.

Table II. Progression Guide.is available in the illustration section at the end of the book.

Table II. Exercises.is available in the illustration section at the end of the book.

Table III. Progression Guide.is available in the illustration section at the end of the book.

Table III. Exercises.is available in the illustration section at the end of the book.

Table IV. Progression Guide.is available in the illustration section at the end of the book.

Table IV. Exercises.is available in the illustration section at the end of the book.

Table V. Progression Guide.is available in the illustration section at the end of the book.

Table V. Exercises.is available in the illustration section at the end of the book.

Table VI. Progression Guide.is available in the illustration section at the end of the book.

Table VI. Exercises.is available in the illustration section at the end of the book.

(3) Unless you have exercised regularly and know yourself to be in good physical condition, start at table I with the C level appropriate to your age.

(4) If you are out of shape, admit that fact to yourself. Hide your pride; after all, you are in the privacy of your own quarters. Set your goal for the longer, steadier pull toward fitness. Resist the urge to pass over the lower numbered tables to find a table that will test your fitness. You are not trying to test, but rather to develop.

3904.WEIGHT TRAINING

Weight or barbell training should not be confused with the more common types of weight lifting used as a competitive sport. Weight lifting is designed to develop specific muscle groups so that the individual is capable of lifting a large amount during a single lift. In contrast, weight training is the systematic development of all the major muscle groups by the use of calisthenics reinforced with weight to provide resistance.

3905.WEIGHT LIFTING

a. Description and Objective. The weight lifting program is progressive and applies the principle of overload in a safe, gradual manner. The exercises of the table can be completed in 15 minutes. The objective of the exercises is to develop strength and muscular endurance, and muscle tone of the five major muscle groups:legs, arms, back, trunk, and shoulder girdle.(See fig. 3-42.)

b. Warmup. A warm up exercise is important to prepare the body for the more vigorous exercises that are to follow. Ten repetitions of the high jumper exercise are excellent for a warm up period. Also, an exercise which will require fast body movement is needed.To provide such exercise, 3 to 5 minutes of rope skipping is recommended to increase the individual's development potential.

c. Formation. Marines should take care in the completion of these weight lifting exercises. They should ensure that their back is straight during the lifting phase of all exercises. When exercises require assuming the standing position with the weight, they should always grasp the weight while in a squatting position and then rise to a standing position.

d. Progression. Each exercise has a starting number of repetitions and specified pounds of weight. After each 4th or 5th day of exercise, the Marine should increase the repetitions by one until he has reached the maximum of 10 repetitions. At this time, the Marine should increase the weight by 5 pounds. Then he should repeat the process again with the initial number or repetitions.

e. Weight Training

(1) Exercise 1:Squat(See fig. 3-42, A.)Starting level—six repetitions, 50 pounds (commonly called the flatfoot deep knee bend). Place the bar upon the shoulders. Stand with feet about 18 inches apart. Keeping the feet flat, lower the body into the low squat position. Stand erect and repeat. Exhale while lowering into the squat position and inhale while standing. This constitutes one repetition.

(2) Exercise 2:Waist Bender(.See fig. 3-42, B.)Starting level-6 repetitions, 40 pounds. Assume the standing position with the bar across the shoulders and the feet a shoulder-width apart .Bend forward at the waist until the upper body is parallel to the ground; return to the starting position. Each return to the upright position constitutes one repetition.

(3) Exercise 3:Curl (See fig. 3-42, C.)Starting level—6 repetitions, 40 pounds. Grasp the barbell with the palms facing to the rear and assume the standing position with the feet a should-width apart. With the barbell held in front of the hips, flex the elbows and lift the weight until the bar touches the upper chest. Lower the barbell back to the hip level position. Inhale deeply with the upward movement and exhale on the downward movement. Each time that the bar touches the chest will constitute one repetition.

(4) Exercise 4:Side Bender(.See fig. 3-42, D.)Starting level—6 repetitions per side, 40 pounds. Assume the standing position, with the bar across the shoulders, with feet a shoulder-width apart. Bend to the left as far as possible and return to the starting position. Repeat six times and then execute the same procedure to the right for six repetitions.

(5) Exercise 5:Standing Press(See fig. 3-42, E.)Starting level—6 repetitions, 45 pounds. Grasp the bar with the palms facing forward and assume the starting position. Curl the weight to the upper chest position. Inhale deeply and press the bar upward to an overhead position. Exhale while lowering the bar to the chest position. Each time that the bar is pressed upward constitutes one repetition.

(6) Exercise 6:Upward Row. .see fig. 3-42, F.)Starting level—6 repetitions, 40 pounds. Grasp the bar, hands close together, palms to the rear, and assume the standing position. Starting with the bar held in front of the hips, flexing the elbows and the shoulder girdle muscles, lift the bar straight up to an overhead position. Inhale deeply while lifting the bar. Exhale while lowering the bar to the hip position. Each time that the bar returns to the hips constitutes one repetition.

(7) Exercise 7:Shoulder Curl. See fig. 3-42, G.)Starting level—6 repetitions, 25 pounds. Grasp the bar, palms down, and assume the standing position. Keeping the elbows locked, curl the bar, pivoting the arms at the shoulders until the bar is in an overhead position and as far to the rear as possible. Return the bar in the same manner to the hip position. Each time that the bar returns to the hip position constitutes one repetition.

Figure 3-42 is available in the illustration section at the end of the book.

3906.ISOMETRIC EXERCISING

a. Description and Objective. Isometric exercising is the application of maximum effort during an exercise period. The isometric principle is to apply force gradually over a 5- to 10- second period until the maximum effort is applied. Relaxation follows for approximately 5 seconds and then force is applied again. The Marine continues this process at the prescribed level for each exercise for a period of 15 minutes or less. The objective of isometric exercises is to create muscle growth. Isometric exercises are the fastest means of creating muscle growth. They are founded on the fact that a muscle will grow only so fast regardless of the type or duration of the activity. The principle of overload—that the muscles develop commensurate with demand—reinforces that fact. However, isometrics will not develop cardiovascular or muscular endurance. Consequently, Marines who choose to use isometrics in their exercise routine must also include running or some other type of aerobic activity.

b. Area and Equipment isometric exercises may be designed to be performed with or without equipment.

c. Use With Other Programs. The isometric exercises presented in this section will provide a variety from which to choose. In addition to the exercises contained in this section, the application of isometric force is used in some of the exercises in section II.

d.Door Frame Exercises. The following exercises are designed for use with a standard door frame found in all offices or barracks.(See fig. 3-43.)

(1) Exercise l:Arm Press. Stand in the doorway with the legs straight, knees locked. Using the arm muscles, press hard upward against the top of the door frame. Repeat for three repetitions applying gradual effort to maximum contraction.

(2) Exercise 2:Leg Press. Stand in the doorway with the hands on the top of the door frame, elbows locked .With knees bent, press hard with the leg muscles .Repeat for three repetitions beginning with a gradual effort and increasing to maximum contraction. A low platform may be necessary to reach the top of the door frame and still maintain a bent knee position.

(3) Exercise 3:Side Press. Extend both arms to the side of the doorway. Palms are shoulder high, facing outward. With both arms, press hard against the sides of the door frame. Repeat for three repetitions. Begin gradually and increase to maximum contraction.

(4) Exercise 4:Lateral Raise. Extend both arms to the sides of the doorway, arms down, palms facing inward. With the back of the hands, press hard against the sides of the door frame .Repeat for three repetitions. Begin with a gradual effort and increase to maximum contraction.

Figure 3-43 is available in the illustration section at the end of the book.

(5) Exercise 5:Neck Press?lace the forehead against the door frame, hands clasped behind the back Using the neck muscles, press hard against the door frame. Repeat for three repetitions, then reverse position so that the back of the head is resting on the door frame. Again do three repetitions .Begin gradually with both exercises and increase to maximum contraction.

(6) Exercise 6:Door Pull Stand facing the edge of the open door and grasp the doorknobs. Pull outward with both arms (if doorknobs are not available grasp the edge of the door).Applying outward pressure, move the body toward and away from the door .Repeat the three repetitions. Begin with gradual effort and increase to maximum contraction.

Chapter 4
COMBAT WATER SURVIVAL
Section I. Marine Corps Water Survival Program

4101.MILITARY SWIMMING

Military swimming emphasizes strokes that result in staying power rather than speed. Marines need to swim easily, quietly, and with adequate vision. Strokes should be used that allow the Marine to carry basic combat equipment and to tow or push a wounded buddy while keeping the face out of the water to allow breathing. The best strokes to use are the side stroke and the breast stroke. Marines must be proficient in floating, drown-proofing, and in using their basic combat equipment to make flotation devices. They must be drilled on using the Kapok and Mae West buoyancy compensators and on emergency egress from helicopters, landing craft, and amphibious assault vehicles.

4102.PROGRAM DEVELOPMENT

Commanders of the Landing Force Training Commands (Atlantic/Pacific) are tasked with developing and maintaining the individual water survival and swimming training program. They are responsible for training all Marine Corps water safety and survival instructors and integrating this training in to their combat readiness training program. All Marines are deployable and as soldiers of the sea must constantly practice the skills necessary to achieve mission accomplishment. Commanders must be imaginative in program development, integrating basic skills development which enhances the total mission accomplishment. An example is teaching waterproofing of the Marine's backpack so that it can be used as a flotation device in streams or rivers. Once learned, this skill would be practiced first in a pool and then in a stream or a river. Routine inspections during deployments would ensure that backpacks could be used as flotation devices.

4103.QUAL STDS AND TRAINING GUIDELINES

The inherent nature of Marine Corps operations and training requires that Marines achieve an ability to survive in water. Water survival and swimming training is designed to reduce fear of water, instill self-confidence, and develop a Marine's ability to survive in water. As a minimum, every Marine should be qualified as swimmer, third-class (S3). This qualification should be met during recruit training or Officer Candidate School. Once qualified, a Marine need not be requalified unless requalification at a specific level (S3 or higher) is required by unit mission, military occupational specialty or duty assignment, or for other reasons determined by the commander. The following classifications and standards apply to Marine Corps water survival and swimming training. Abbreviations will be used for service, medical, and training record entries citing MCO 1510.29A, Individual Water Survival and Swimming Training, as authority.

a. Qualification standards and test procedures are described in paragraph 4104.
b. Swimmer classification and abbreviations are:

(1) Unqualified - UQ.

(2) Swimmer, third-class (S3).Minimum aquatic skill level for all Marines.

(3) Swimmer, second-class (S2)-

(a) Minimum standard for all naval aviators, naval flight officers, and aircrewmen with written waiver from the commanding officer.

(b) Requirements for military occupational specialty 1803 (assault amphibian vehicle officer), 1833 (assault amphibian vehicle crewman), 1302 (engineer officer), 1371 (combat engineer), 1381 (shore party specialist), and 1379 (engineer operations chief).

(4) Swimmer, first-class (S1)-

(a) Naval aviator, naval flight officer, and aircrewman qualification requirement.

(b) Requirements for military occupational specialty 0321 (reconnaissance man), 8652 (reconnaissance man, parachute jump qualified), 8653 (reconnaissance man, self-contained underwater breathing apparatus qualified), 8654 (reconnaissance man, parachute/self-contained underwater breathing apparatus qualified), 9952 (self-contained underwater breathing apparatus Marine (officer/enlisted)), and 9953 (parachutist/self-contained underwater breathing apparatus Marine (officer/enlisted)).

(5) Water Survival Qualified (WSQ).Ultimate water survival/swimming goal for Marines.

c.Instructor and assistant instructor classifications are-

(1) **American Red Cross Advanced Lifesaver.** Individuals who are currently certified as American Red Cross advanced lifesavers may assist qualified instructors during Marine Corps water survival/swimming training and testing. Certification is valid for 3 years.

(2) American Red Cross Water Safety Instruction. Individuals who are currently certified as American Red Cross water safety instructors may conduct swimming training and qualify individuals as 3d, 2d, and 1st class swimmers. Certification is valid for 2 years.

(3) United States Marine Corps Water Safety and Survival Instruction. Those personnel certified as water safety and survival instructors by one of the landing force training commands may conduct Marine Corps water survival and swimming training and American Red Cross swimming and water safety training. These instructors may also participate as assistant instructors during the conduct of landing force training command-sponsored water safety and survival instructor courses. Certification is valid for2 years.

(4) American Red Cross Water Safety Instructor Traine. Individuals qualified who have been certified as instructor trainers by the American Red Cross may train American Red Cross water safety and basic swimming instructors. Certification is valid for 2 years.

(5) United States Marine Corps Water Safety and Survival Instructor Traine. Individuals qualified who are certified as instructor trainers through special training courses conducted jointly by the American Red Cross and the Marine Corps, under the direction of the Commandant of the Marine Corps, may train water safety and survival instructors and American Red Cross water safety and basic swimming instructors. Certification is valid for 2 years.

d. Training Guidelines Instructors engaged in training Marines as swimmers will use the current edition of the American Red Cross swimming and water safety courses instructors' manuals as the authoritative textbooks for swimming procedures, strokes, breaks, and holds.

4104.WATER SURV/SWIM QUAL STD AND TEST PROCS

a. The following qualification standards and test procedures are applicable to Marine Corps water survival and swimming training. The utility uniform without cover, boots/oxfords or socks will be worn for all swimming tests through swimmer, first class. The utility uniform with boots/oxfords will be worn as prescribed for the water survival qualification.

(1) Swimmer, Third-Class (S3) To qualify as a swimmer, third-class, a Marine must enter the water feet first from a minimum height of 5 feet and remain afloat for 5 minutes. During this time the Marine must swim 50 yards using any stroke or combination of strokes. This test should be taken by all Marines as early as possible in their initial training. Those unable to pass this test will be classified as UQ and should be given instruction in fundamental swimming skills. Those who barely meet requirements are swimmers who need help and should receive additional instruction in fundamental swimming skills, whenever possible.

(2) Swimmer, Second-Class (S2.) To qualify, a Marine must enter the water feet first from a minimum height of 10 feet and remain afloat for 10 minutes. The Marine must swim 100 yards and use the three basic survival swimming strokes (side stroke, backstroke, and breast stroke) for a minimum distance of 25 yards each.

(3) Swimmer, First-Class (SI)A prerequisite to qualification as swimmer, first-class, is successful completion of the test for swimmer, second-class. To qualify as a swimmer, first-class, a Marine must be able to do each of the following:

(a) Approach a person of approximately the same size while in the water, demonstrate a release, get that person in a carry position, and tow the "distressed swimmer" 25 yards.

(b) Enter water feet first and immediately swim underwater for 25 yards. Swimmer is to break the surface twice for breathing during this distance at intervals of approximately 25 feet.

(c) Remove trousers in water, inflate for support, and remain motionless for a minimum of 1 minute.

(d) Swim 220 yards using any survival swimming stroke or any combination of survival strokes.

(4) Water Survival Qualification (WSQ)A prerequisite to water survival qualification is successful completion of the test for swimmer, first-class. To be water survival-qualified, a Mannemust:

(a) Enter water by jumping from a height of 10 feet.

(b) Stay afloat in full utility uniform (less cover) for 1 hour. Boots/oxfords will be removed after 5 minutes in the water but will be retained by the swimmer.

(c) Traverse 75 yards in deep water in full utility uniform (less cover) with boots/oxfords.

b. All water survival and swimming training will include instruction in the following:

(1) Employment of safety lines or other comparable expedients to assist in swift stream/river crossings.

(2) Adverse physiological effects caused by cold water (hypothermia) and the precautionary measures to be taken prior to exposure to such an environment.

(3) Employment of a standard and expedient flotation devices in a water survival situation.

(4) Removal of individual field equipment after unscheduled water entry. Entry will be made from a minimum height of 10 feet with a complete set of field equipment, properly packed as an expedient flotation device and waterproofed. The ALICE pack will then be removed, the rifle laid across the top of the pack, the helmet remaining or put back on the individual's head, and the individual will traverse 25 yards to the side of the pool or water's edge.

c. The additional water survival qualification (flight physiology training) and requalification required of personnel assigned to flight status will be conducted in accordance with OPNAVINST 3710.7_,Promulgation of NATOPS General Flight and Operating Instructions. Swimming qualification (first-class, SI) is a prerequisite to water survival (flight physiology) training and is a one-time requirement.

Section II. Water Sports and Competitive Activities

4201.GENERAL

The commander, when developing the combat readiness program, needs to look carefully at the mission. If the mission requires Marines who must perform mission-essential water-related activities of long duration, he may want to look to water sports/competitive activities as a method of developing endurance. Endurance and confidence can only be developed by extended time in the water. While defined training can accomplish this goal, the goal can be enhanced through training which is integrated with the physical conditioning program. This approach will provide variety and enthusiasm. Two activities which will accomplish this goal are water relays and water polo. The important consideration is that the unit achieve total participation by all members of the unit.

4202.WATER RELAYS

Development of water relays will generally conform to the guidelines in chapter 5, section IV. When contemplating relays, the commander should review chapter 5 and adapt the principles to the size of the unit and the facility to be used.

4203.WATER POLO

The official rules provided in this paragraph will need to be adapted to the unit's unique situation, its size, and the size of the facility. Commanders should make every effort to have all members of the unit participate.

a. NCAA Rules This style of play is based on two popular sports: swimming and basketball. Playing is done on the surface of the water by teams of seven players each. The size of the playing area is 25 yards by 14 yards, which is the size of the typical high school or college indoor pool. The deeper the water, the better. If the entire playing area is deep-6 feet or more—this is ideal, although most indoor pools have a shallow end. At each end of the pool is a goal. In deep water, the goal is 10 feet across and 3 feet high. In shallow water, where the goalie can stand on and jump from the bottom, the goal is 10 feet across and 5 feet high. One of the seven players on each team is the goalie. He is the only player on the team who can stand on or jump from the bottom (if it is shallow enough to permit this), catch and pass the ball with both hands at the same time, or hit the ball with a clenched fist. However, the goalie may not swim across or pass the ball across the middle of the pool.

b.Equipment. The goalie on the visiting team must wear a white cap with the number 1 clearly marked on it; his teammates, who are designated as guards and forwards, must wear white caps numbered 2 through 7, with the substitutes wearing white caps numbered 8 through 21.The goalie on the home team must wear a dark-colored cap with the number 22 clearly marked on it; his teammates designated as guards and forwards, must wear similar dark-colored caps numbered 29 through 42.The guards and forwards on each team may swim freely up and down the pool as they see fit, interchanging positions as often as they wish. They may not stand on or jump from the bottom, touch the ball with more than one hand at a time when catching or passing or shooting it, or enter inside the opposing team's 2-yard line unless preceded by the ball. The ball is similar to a soccer ball, except that it is yellow and covered with a waterproofing substance that makes it easier to handle with one hand.

c. **Playing Time.** In intercollegiate competition the game consists of four 7-minule quarters, the teams changing ends after every period of play. In high school competition the game consists of four-,five-, or six-minute quarters, depending on the ages of the participants. Here again the teams change ends after every period. There is a very brief interval between each quarter. At the start of each period the teams line up at their respective ends, and with ablast of the whistle, the head referee throws the ball into the middle of the pool. The fastest swimmers on each team then race to gain possession.

d. Officials. The head referee walks along one side of the pool watching for infractions. He is aided by the assistant referee, who patrols the opposite side of the pool. The referees are both equipped with whistles and two official's flags. These flags should be 12 inches square and should correspond in color to the caps worn by the respective teams. Whenever a referee sees an infraction, he blows his whistle and signals with his flag indicating that the player nearest to the ball with the appropriate colored flag will be given possession of the ball. Whenever the whistle is blown, all players should react by looking immediately to see which color flag is being waived which will indicate which team should be moving into the defense. Other officials are a timer and a scorer. As in all sports, it is important that the officials be skilled and competent.

e. **Technical Fouls.** The following are some of the common technical fouls: starting before the referee blows his whistle to open play; holding onto or pushing off from the side of the pool during play; taking hold of the ball underwater when tackled by an opposing player; swimming inside an opponent's 2-yard line unless preceded by the ball; touching the ball with both hands at the same time (goalie excepted); standing, walking on or jumping from the bottom when taking an active part in the game. When a technical foul occurs, thereferee blows the whistle and with his flag awards possession. The player on the team awarded possession nearest the point of infraction then has 5 seconds to put the ball into play by passing to a teammate or by dropping the ball into the water and swimming after it.

f. Personal Fouls. The following are some of the more common personal fouls: committing any of the technical fouls mentioned above for the purpose of scoring or preventing a goal; holding, ducking, pulling, pushing off from, swimming over, or impeding the arm or leg movement of any opponent who is not touching the ball; and splashing water in the face of an opponent. When a personal foul occurs, the referee will follow the same procedure as with a technical foul. While doing this, the referee will also call out clearly and loudly the number of the player who was guilty of the infraction and a personal foul will be marked against that player by the official scorer. When a player accumulates five personal fouls, he fouls out of the game and must be replaced by a substitute.

g. Penalty Shot. A penalty shot can be awarded by either of the officials or the scorer when:

o An offensive player inside the opponents' 4-yard line not touching the ball is held, ducked, pulled-back, kicked, or struck.
o A team has accumulated a total of ten personal fouls. When the former occurs, the head or assistant referee should immediately blow the whistle and by holding a two-flagged stick in a vertical position above his head, signal that a penalty shot has been awarded. When the latter occurs, the scorer should use a buzzer located at the scorer's tabic to signal that one team has accumulated ten personal fouls, thereby entitling the other team to a penalty shot. A penalty shot is taken from the 4-yard line in front of the goal. All players except the defending goalie must leave the 4-yard line until the shot is taken. No player can be within 1 yard of the shooter. After ascertaining that the shooting player is on the 4-yard line and the goalie is onThegoal line, the referee will ask the shooter to lift up the ball. When he docs, the referee will give a sharp blast from the whistle. At the whistle, the shooter must shoot without delay and without any faking at the goal. The goalie may try to block the shot. If he does, or the shot is otherwise missed, it is immediately in play and action continues.

h. When a Goal is Scored. When a goal is scored, either from a shot taken by a player out in the field or by penalty shot, the defending goalie must then pass the ball to the nearest referee. The referee will then pass the ball back to the goalie, who puts it into play by a pass to one of his teammates as soon as possible.

Section III. The Battle Swimming Test

4301.GENERAL

The battle fitness test is the evaluation contained in the Guidance for Basic Warrior Training (BWT) as part of the Basic Warrior Training Concept Plan.

4302.THE BATTLE SWIMMING TEST

The battle swim test consist of the following:

a. Be able to climb a three meter high board, while wearing a field uniform, carrying a M-16, web gear, and four canteens. Jump off blindfolded, recover and swim 100 meters with the M-16.

b. Repeat the same process, without the blindfold, and upon entering the water a second time, recover and retrieve all submerged gear.

c. Remain afloat and tread water for five minutes.

d. Demonstrate a knowledge of how to cross a stream using field expedient techniques and a single strand rope bridge.

e. Successfully complete helicopter egress training.

Chapter 5
COMPETITIVE CONDITIONING ACTIVITIES
Section I. Organization of Competitive Activities

5101.LEADERSHIP OF COMPETITIVE ACTIVITIES

a. Place in the Program. Competitive conditioning activities consist of dual combatives, relays, team contests, and team sports in which individuals or teams compete against an opponent to win. Competition is one of the best ways of maintaining interest in the physical training program during the sustaining stage. Organized competition provides enjoyable, vigorous physical activity that has proved to be one of the best supplements to conditioning drill activities. The benefits of competition are the development of aggressiveness, unit pride and identity, teamwork, and the will to win. These activities help develop cardiovascular and muscular endurance, strength, and coordination. Scheduling competitive activities in an orderly and progressive manner is desirable. The progression should be from relays to dual combatives, to team contests, and finally to team athletics. These activities should be conducted as part of the program after a basic level of conditioning has been developed, usually during the slow-improvement stage of conditioning. Muscles and joints should be strengthened by preconditioning to withstand the strain placed upon them by sudden stops and turns, body contact, bearing of weight, and falls. Competitive activities, however, should not be allowed to dominate the physical training program.

b. Area and Equipment. Some of the competitive activities included in this chapter require specific types of areas and equipment; others do not. The area requirement can usually be satisfied on available training fields. When items of equipment are required or specific courts or field layouts are to be marked off, such information is included.

c. Leadership. The principal factor for success as a leader of competitive activity is an energetic, dynamic, enthusiastic approach. The leader's attitude is reflected by the group, so the commander must carry on the activity in a snappy and vigorous manner. Confidence on the part of the commander will create an impression of decisiveness and certainty. Confidence grows out of experience and a thorough knowledge of the activity. Mastery of subject is the first step in developing confidence, assurance, and poise.

(1) The following suggestions are offered for leaders of competitive combatives, contests, and sports:

(a) Get the activity underway quickly by selecting and teaching only the essentials.

(b) Use rules to add to the enjoyment of the activity and do not allow them to interfere with the spirit of competition.

(c) Stop the activity before interest begins to lag.

(d) In team contests, clearly distinguish sides by attempting to maintain unit identity.

(e) Always insist on fair play, enforcing the rules impartially.

(2) The following procedure is recommended for presenting a competitive activity.

(a) Name the activity.

(b) Briefly explain the objective of the activity and give only the pertinent rules.
(c) Have a demonstration at slow speed and answer questions.
(d) Organize groups into teams and appoint captains.
(e) Arrange teams in the proper starting positions.
(f) Conduct the activity.

d. Competitive Units. Units for competition should be the same organizations in which Marines train: battalions, companies, platoons. In most situations, the unit is the squad.

e. Provisions for Instruction. One of the most effective methods of maintaining interest and participation in competitive activities is to provide instruction in activities with which most Marines are unacquainted. Such instruction can be conducted during regular physical training periods. Careful planning is required to keep all individuals continuously engaged in vigorous activity. The materials in the following sections should be used as a guide for instructional purposes.

f. Officiating and Control. Every effort should be made to provide good officials for all competitive activities. Poor officials quickly cause dissatisfaction among participants in team activities and create a situation which can turn an organized activity into a brawl. Each company should have or develop several competent and qualified officials available for games on company and platoon levels.

5102.MILITARY FIELD MEETS

When units reach the latter part of the slow-improvement and sustaining stage of training, interest in the program may lag. A change in course content can arouse the desire to participate. An event which does not require a high degree of skill yet demands strenuous activity is ideal for the military field meet. This is a series of team contests conducted on a station-to-station basis during a given period of time. Team contests carried on simultaneously provide essential training, vigorous exercise, and stimulating competition in an atmosphere that is enjoyable for all. Because of the healthy rivalry that a field meet arouses, it is an excellent form of inter-unit competition.

a. Objective and Advantages. The objective of the military field meet is to provide activity for everyone in the participating units. Activities are chosen that will develop aggressiveness, teamwork, a will to win, and competitive spirit, and that will stimulate interest and build esprit de corps. The military field meet can be included as part of the physical training program or as part of the off-duty recreational program. It is a form of contest that can be conducted in nearly all circumstances because it can be easily modified, requires a minimum amount of equipment, and can be readily organized.

b. Level of Competition. The military field meet is flexible. It can be adjusted to large or small groups. A company-size unit is the most desirable, but it may also be administered within a larger unit. If it is conducted within a company, the participating units will be squads. If it is held within a larger unit, platoons compose the teams. A larger-size unit requires more extensive organization and administration.

c. Selection of Events. In organizing a military field meet, select events that are simple and easy to administer. All rules and regulations should be clearly understood by everyone. No event should require previous practice. In selecting the events, consider the interest and capabilities of the Marines and available equipment and facilities. Select events which will require that all members of the units participate. Events should not be dominated by the athletes; rather, MOS-related events which have significance in producing skills and combat readiness should be used.

d.Equipment and Facilities. The site must be large enough to permit events to be grouped about a central control point. Use available facilities such as parade decks, Softball fields, or tracks. A public address system is desirable at the control point for the initial orientation of teams and for subsequent announcements of time lapses, cumulative scores, and final standings. A tally board is necessary so that all units can constantly monitor how the events are progressing.

e.Administrators. Efficient administration of the military field meet depends on the referees, judges, and scorers. For these positions, choose individuals within the units who have had athletic or officiating experience. Prior to the day of the event, all administrators should be briefed and assigned a specific task in order that they may become familiar with the rules and organization of the contests they will conduct. The following officials should be available:

(1) A primary instructor or supervisor who is in charge of the control center and who is responsible for the successful operation of the athletic carnival. The supervisor must have an assistant to act as a timer and scorer.

(2) One assistant instructor in charge of each event. This person should be-

(a) Familiar with the rules of the event.

(b) Effective as a leader to ensure proper supervision and control over the participating teams.

(c) Enthusiastic to provide proper motivation.

(d) Self-confident of the ability to judge infractions of the rules. The assistant instructor must be fair in judgment and penalize without hesitation when infractions occur.

(3) Runners between event stations and the control point. These individuals collect and deliver scores.

f. Team Organization for Competition. The size of the teams is determined by the level on which the field meet is organized. Maintaining the integrity of the unit promotes esprit de corps, but this does not preclude grouping two squads into one team.

g.Conduct of Events

(1) The assistant instructor at each station takes charge of the group and gives a brief explanation of the major rules of the event. The assistant instructor speaks clearly and distinctly from a position to be seen and heard by all. A short demonstration is desirable if it will help clarify the event.

(2) The assistant instructor should make certain that teams can be clearly distinguished by the use of, for example, T-shirts and fatigue jackets, or caps and no cap, or colored jerseys. The assistant instructor should teach a whistle response (teams stop play immediately upon hearing whistle), get the event started as quickly as possible, and make any necessary corrections as the contest progresses. The rale of good officiating is to use a minimum of calls, yet maintain control of the contest. Penalize when necessary, but refrain from disqualifying contestants or teams.

(3) Keep the activity moving as rapidly as possible. When the central control point sounds the whistle to stop the play, all competition ends immediately. The assistant instructor then assembles the group, forwards the team scores to the central point, and, upon a signal from the central control point, rotates the teams to their next station. It is essential that rotation and orientation of teams be quick and orderly to reduce time spent between contests.

(4) Upon completion of the final event, assistant instructors move their teams to the central control point for the announcement of winning teams, presentation of awards (if any), and final critique.

h. Scoring System. The system for determining the winner of the military field meet should be simple and efficient. At the completion of competition, the scorer totals the points that each team has scored in all contests. The scorer then subtracts the number of points scored against a team (penalty points for nonparticipation,etc.) from the total. The resulting scores are placed in a column with the highest score at the top and the lowest at the bottom. The team with the highest total is the winner. This type of scoring system encourages a team to ensure total unit participation.

Section II.Combatives

5201.DESCRIPTTON AND OBJECTIVE

Combatives are strenuous, short competitive contests in which two individuals attempt to overcome each other in a bout of skill and strength. These contests help to develop Marines' resourcefulness, confidence, strength, agility, coordination, and will to win. Any level ground area can be used. However, extremely hard ground should be avoided as some of the combatives require ground contact. A whistle is needed to control the bouts during the competition since voice commands may go unheeded. The extended rectangular formation is used for dual combatives. Combatives are conducted on an informal basis. Marines are allowed to remain at ease between activities and are allowed to brush themselves off after being on the ground. To get the most out of combatives, individuals must be urged to overcome their opponents as quickly as possible as would be required in combat.

a. Benefits for Marine Training. The possibility of close contact with an enemy in combat faces Marines at all times. They must be trained to react aggressively and violently. Combatives may be used as an introduction to such hand-to-hand contact and should be followed by hand-to-hand combat training. Marines enjoy competition, and this type of activity is a welcome change from formal conditioning activities.

b. Instructor Responsibility

(1) The instructor tells the individuals that all combatives begin and end on the whistle signal. The instructor demonstrates each activity before having it performed, explaining it in simple terms.

(2) After stopping one activity, the instructor gets everyone in place for the next.

(3) The instructor must closely supervise combatives to insure that contestants do not use unfair or unsportsmanlike tactics. To avoid unnecessary injury, instructors must see that the bouts are closely controlled and opponents equally paired. Adjustments should be made in apparent cases of mismatched abilities.

5202.COMBATIVES TABLES

There are three tables of combatives, each of which can be completed in 15 minutes. These tables become progressively more difficult from lower to higher numbers.

a.Combatives Table I

(1) Open-Hand Slap Boxing(See fig. 5-1, A. individuals assume a boxer's stance, palms open, fingers extended and joined. Each contestant tries to slap the opponent about the head and upper body with the open hand. This is a good warm- up activity.

(2) Wrist Tug-of-War(See fig. 5-1, B.)Two individuals sit on the ground with the soles of their feet in contact. Each grasps the opponent's wrists so that the hands are directly over the feet. At the whistle, each individual tries to pull the opponent from sitting to standing position.

(3) Arm-Lock Wrestling.(See fig. 5-1, C.)Two individuals sit back-to-back with legs spread and arms locked at the elbows. Each right arm is inside the opponent's left. Each tries to force the opponent's left arm or shoulder to the ground. The individual who first wins three bouts is the winner.

(4) Bulling(See fig. 5-1, D.)Two individuals assume the Westmoreland wrestling hold, each grasping the opponent's neck with the right hand and the opponent's right elbow with the left hand. Each tries to force the other to move one foot by pushing, pulling, or otherwise manipulating. The individual who first wins two boutsis the winner.

(5) Indian Wrestling(See fig. 5-1, E.)Two individuals lie on the ground, side by side, with their heads in opposite directions. They link right elbows. On the instructor's signal or by mutual agreement, each individual raises the right leg approximately straight and far enough to engage the opponent's heel. To start the contest, each person usually raises the leg three times rhythmically and on the third time engages the opponent's heel. Each tries to roll the other over backward. The right leg is used for three bouts, then the left leg for three bouts.

Figure 5-1 is available in the illustration section at the end of the book.

b. Combatives Table II

(1) Wrist Bending.(See fig. 5-2, A.) Opponents pair off, face each other, raise their arms forward, and, with palms forward, interlock their fingers. At the starting signal, each individual attempts to bend the opponent's wrist. The arms are kept up and forward and are not allowed to swing around and down to the sides. The individual who first wins two bouts is the winner.

(2) Back-to-Back Push.(See fig. 5-2, B.)Two individuals stand back-to-back with elbows locked. Each right arm is inside the opponent's left arm. At the starting signal, each pushes backward, trying to move the opponent. Opponents are not allowed to lift and carry each other; only pushing is permitted. The one who pushes the opponent the farthest wins the bout. The individual who first wins two bouts is the winner.

(3) Hop and Pull Hand.(See fig. 5-2, C.) The individuals are matched in pairs. Each grasps the opponent's right hand and, hopping on the right foot, attempts to pull the opponent off balance. Contestants automatically lose if they touch their free hand or their lifted foot to the ground. For successive bouts, they alternate hands and feet.

(4) Westmoreland Wrestling. (See fig. 5-2, D.)Each contestant grasps the back of the opponent's neck with the right hand and the opponent's right elbow with the left hand. In this position, each attempts to pull, push, or force the opponent to touch the ground with any part of the body other than the feet. The individual who first wins two bouts is the winner.

(5) Crab Fight.(See fig. 5-2, E.)Two individuals sit on the ground facing in opposite directions with their hands on the ground behind them. At the whistle, they raise their hips and push with their shoulders and bodies, each trying to make the other's hips touch the ground. The individual who first wins two bouts is the winner.

Figure 5-2 is available in the illustration section at the end of the book.

c. Combatives Table III

(1) Hand Wrestling.(See fig. 5-3, A.) Two individuals stand facing each other. Their right feet are forward and braced side by side. They grasp right hands for the first bout, left hands for the second bout. Each pulls, pushes, makes sideward movements, and otherwise maneuvers to force the opponent to move one or both feet from the original position. The contestant who first wins two bouts is the winner.

(2) Back-to-Back Tug.(See fig, 5-3, B.) Two individuals stand back-to-back with both arms linked at the elbows. Each has the right arm inside the opponent's left arm. At the starting signal, each attempts to pull the opponent. Lifting and carrying are permitted. The contestants must maintain their original direction and keep their arms linked. After a predetermined time, the player pulled or carried the farthest is the loser.

(3) Wrestling to Lift off Feet.(See fig. 5-3, C.) Contestants face each other. Each places the right arm under the left arm of the opponent and around the body. The left arm is over the opponent's right shoulder. Each tries to lift the other off the ground. The individual who first wins two bouts is the winner.

(4) Arm Pull Between Legs.(See fig. 5-3, D.) Two individuals are paired off, back-to-back. Each bends forward and, extending the right arm between the legs, grasps the opponent's right wrist. At the starting signal, each person attempts to pull the opponent After a predetermined time, the player who has pulled the opponent the farthest is the winner of the bout. The one who first wins two bouts is the winner. Repeat with the left hand and then both hands.

(5) Rooster Fight.(See fig.5-3, E.) Each contestant grasps the left foot with the right hand from behind, and the right arm with left hand. Each hops on the right foot, and, by shoulder-butting the opponent or by feinting and sudden evasions, forces the opponent to let go of the foot or arm. The contestant who first wins two bouts is winner.

Figure 5-3 is available in the illustration section at the end of the book.

Section III .Relays

5301. DESCRIPTION AND OBJECTIVE

Relays are races in which each member of a team runs one leg of the race. The team effort decides the winner. Relays provide stimulating competition and contribute to the conditioning of personnel. They also develop aggressiveness, team spirit, and the will to win. Relays should be dispersed throughout the program for short periods of time to provide a change of activity.

a. Team Organization. Relays are conducted most efficiently in platoon-size groups. Teams of equal size must be organized. Competitive spirit is encouraged and teams are organized better by basing teams on units such as squads, crews, or sections. Team captains should be designated. Extra Marines may be used as officials. The number of individuals on a relay team should be limited to squad size. If larger teams are used, the runners will spend too much time awaiting their turns and too little time actually participating. Two to six teams are ideal for relay competition. It is difficult to keep track of winners when more teams compete.

b. Administration of Relays

(1) The time spent on any one relay should be relatively short. If one team achieves a substantial lead in a long relay, the competitive spirit and enthusiastic participation of the other teams may decrease. Several short relays are generally better than one long relay.

(2) To maintain competitive spirit throughout a number of relays, determine the teams that win, place, and show in each relay and their total points for all relays. This can be done by awarding points to all teams on the basis of position at the finish of each relay. The team with the greatest number of points is the winner of the entire set of relays.

(3) Difficulties commonly encountered in conducting relays may be avoided by the following procedures:

(a) The last player in a relay race should be conspicuously identified. For example, the last runner can tie a handkerchief around the head or arm, take off the shirt, put on a hat or take it off, or use some other means.

(b) Another way to keep track of the progress of the race is to have each player sit or squat as soon as each is finished.

(c) Judges at the starting line can keep the runners from starting too soon.

(d) To prevent contestants from turning before they run the full distance, they should be required to run around a peg, pole, or assistant instructor.

(e) Batons, handkerchiefs, tent pegs, or other objects should be passed from one runner to the next when relays are run on a circular track.

(4) Before a unit's first participation in a relay, inform participants of the rules and scoring system. Violation of the rules should not result in disqualification. Instead, impose point penalties. A point penalty is imposed by subtracting a point from the team total at the conclusion of the relay.

(5) Careful administration will prevent most violations. For efficient conduct of relays, follow this procedure:

(a) Announce the name of the relay.

(b) Form the individuals in relay position.

(c) Briefly explain the relay and the rules for running it.
(d) Demonstrate.
(e) Have a definite finish line, and ensure that all know where it is.
(f) Answer questions.
(g) Conduct the relay.
(h) Determine winner and award points.

5302. RELAY TABLES

a. Events. The following relays are grouped into a table of activities. Each relay table can be completed in 15 minutes. Thus, relays can be used as a sole activity or as a part of a longer period. Each table is planned for a platoon-size group (30-60Marines). Adequate warmup for participants is provided by conducting several repetitions of exercises 1 and 2 of a conditioning drill. The recommended relay tables require an area 40 by 60 meters in size. Each table provides a variety of activity. The tables are progressive in the overload applied and should be scheduled in numerical order although not necessarily on successive days.

b. Relay Table I
(1) **60-Meter Lane Relay** (See fig. 5-4, A.) Each team is assembled in single file behind the starting line. On signal, the first individual of each team runs to the turn-around line 30 meters away, then runs back and touches the next teammate waiting at the starting line. The winning team is the first team to get its last member across the finish line. If an individual starts before being touched by the preceding runner, the team may be penalized.
(2) Wheelbarrow Race.(See fig. 5-4, B.) The players of each team pair off and line up in single file behind the starting line. The first individual walks on the hands while the partner grasps the ankles. They advance to the distance line (25 meters) behind which they exchange positions and return to the starting line. The rear individual must always hold the partner's ankles. After the first pair returns across the starting line, the next pair starts.

Figure 5-4 is available in the illustration section at the end of the book.

(3) Squad Front Relay .(See fig. 5-4, C.) The teams form in a line along the starting line with a 10-foot interval between teams. The members of each team lock elbows so that they are linked together. At the starting signal, the teams run to the distance line (20meters) where the left flank individual acts as a pivot. The team swings around on the pivot and returns to the base line. If a team breaks its links, it must reform before continuing. The first team to completely cross the base line intact is the winner.
(4) Crab-Walk Race.(See fig. 5-4, D.) The players of each team line up in single file. The first person of each team assumes the crab-walk position with feet forward on the starting line, hands on the ground behind it. At the starting signal, contestants move forward to the distance line (10 meters). They touch the line with their feet and then return to the starting line in the reverse position with the head and hands leading. The second person may not start until the first individual touches the finish line.

c.Relay Table II

(1) 100-Meter Lane Relay.(See fig. 5-5, A.)This relay is conducted exactly as the first relay of table 1 except that the start and turn-around line are 50 meters apart. This relay provides progression in sprinting.

(2) Frog-Jump Relay. (See fig. 5-5, B.) Each team lines up in single file. The first individual assumes a squatting position on the starting line. At the starting signal, the contestant progresses to the distance line (15 meters) and back by leaping forward, catching the weight on the hands, and bringing up the legs to the squat position for the next leap forward.

(3) Simple Relay. (See fig. 5-5, C.)Each team lines up in single file. Place a marker on the distance line (20 meters) in front of each team. Each team member, behind the first one, grasps the belt of the individual ahead. At the starting signal, each team runs as a unit to the marker, circles it, and returns to the starting line. The first team to completely cross the starting line intact is thewinner.

(4) Fireman's Carry Relay.(See fig. 5-5, D.)The players of each team line up in pairs, one behind the other. One individual in each pair carries the partner to the distance line (30 meters), using the fireman's carry. At the distance line, individuals exchange places and return to the starting line. As a variation, the person to be carried lies on the ground. The one carrying lifts the partner to the proper position. This relay may be performed with the other carries described in guerrilla exercises in chapter 3.

d.Relay Table III

(1) 200-Meter Circle Relay .(See fig. 5-6, A.)A course is laid out in a circular, rectangular,or oval pattern that is 200 meters around. Each team provides one runner on the starting line. On signal, the runner races around the 200-meter track and touches the next teammate, waiting at the starting line, who runs the same course. Each team member runs one lap of the course.

Figure 5-5 is available in the illustration section at the end of the book.

(2) Bear-and-Crab Race.(See fig. 5-6, B.) Each team lines up in single file. At the signal to start, the first contestant in each column assumes the bear-walk position (hands on ground in front), walks to the distance line (15 meters), and then runs back to the starting line. The returning contestant touches off the second individual and goes to the rear of the line. The second contestant assumes the crab-walk position (hands on ground behind) and crab-walks with feet leading to the distance line. This contestant also runs back to the starting line and touches off the next individual, who walks bear fashion. The rest of the members of each team alternate in this manner. The relay ends when the first person is back at the head of the line.

(3) Pilot Relay.(See fig. 5-6, C.) The players are grouped in threes, arms interlocked at the elbows, and outside players facing backwards. The middle person runs forward; the two outside individuals run backward. They run to the turning point (15meters), where they start back, this time with the middle person running backward and the two outside individuals running forward. The next set of three players starts when the first set crosses the starting line.

(4) Saddle-Back Relay.(See fig. 5-6, D.) Mark two parallel lines 15 meters apart. Each team selects a rider. The remaining members of each team count off. The even-numbered players from each team form in single files behind one line and the odd-numbered players from each team form in single files behind the other line directly across from their teammates. At the starting signal, the rider mounts the back of the first player of the team who carries the rider across the other line. There the rider changes mounts to the first player in the second line without touching the ground. This person carries the rider to the next player waiting in the first line. The relay continues until all of the mounts have carried the rider. A rider who falls off must mount again at the point of the fall. A rider who falls in changing mounts must get back on the original mount before making the change.

e. Relay Table IV

(1) 100-Meter Circle Relay (See fig. 5-7, A.)A course is laid out in either a circular, rectangular, or oval pattern that is 200 meters around. Each team is divided in half with each half positioned at starting lines on opposite sides of the track. Each runner races halfway around the track and touches a teammate who completes the lap. Each runner then waits in file at the halfway point until touched by the runner and then completes the second half of the lap. The first team to return all runners to their original starting line is the winner.

Figure 5-6 is available in the illustration section at the end of the book.

(2) In-and-Out Relay.(See fig. 5-7, B.) Each team lines up in a file with players 2 meters apart. At the starting signal, the first player runs back through the column in a zigzag fashion. He alternates, going to the right of one teammate and to the left of the next. Upon completing the run, the player lines up 2 meters behind the last one in the column. As soon as the first runner has passed the second person, the latter starts to run. This continues until all the players have realigned their original order. The team that finishes first is the winner. It may be desirable to have this relay continue until all individuals have run through their entire team two or three times in succession.

(3) Circle Race.(See fig. 5-7, **C.**) Each team forms a circle and holds hands with all individuals facing out except one who faces in and is the "driver."At the starting signal, the teams race to the distance line (20 meters) and back, keeping the circle intact. All the individuals in the circle must completely cross the distance line. The "driver" gives directions and orders. When the circle breaks, it must be reformed before it can continue. The first team completely over the starting line is the winner.

(4) Horse-and-Rider Relay.(See fig. 5-7, D.)Each team lines up in a single file. At the signal to start, the second player in each column leaps upon the back of the first who carries the rider across the distance line (30 meters). At the distance line, the rider dismounts and runs back to the starting line. There the rider picks up the third individual in the column, and carries this player to the distance line where the first player has remained. This continues until the last person is carried across the distance line.

Figure 5-7 is available in the illustration section at the end of the book.

Section IV. Team Contests and Athletics

5401. DESCRIPTION AND OBJECTIVE

Team contests are competitive activities in which Marines as a team compete with another team to win. They are guided by simple rules and organization. The function of team contests is to provide competition and an opportunity for body contact and to contribute to the development of physical readiness. In competing and working together as a team, individuals develop aggressiveness, the will to win, and teamwork. Team athletics deserve a prominent place in the physical training program because they contribute to increased combat efficiency. Because of the competitive nature of athletics and their natural appeal, individuals take part with enthusiasm. Athletic teams formed at intramural and higher levels are a strong unifying influence and provide one of the best means of developing esprit de corps.

a. Area and Equipment A level training field is sometimes the only area required. Many contests need no equipment. In contests requiring equipment, the need is for standard items such as logs, balls, nets, goals, and similar types of equipment. Specific requirements for area and equipment are listed with each contest description.

b. Progression. Team contests of a strenuous nature should be introduced after a basic period of conditioning has been completed and individuals are in the slow improvement stage of conditioning. Progression can take place from the less active to the more vigorous contests, and then from the non-contact to the contact or combative-type contests.

c. Necessity for Preconditioning Individuals must undergo conditioning prior to participation in athletics. Muscles, organs, joints, and ligaments not accustomed to stress and strain from sudden stops and starts, falls, body contact, rapid turns, prolonged running, and other rigors of athletic competition are subject to injury. Although athletics should not be introduced until players are physically prepared, there is still opportunity to engage in competition through lead-up contests. Individuals learn many of the skills required for athletics while participating in team contests.

d. Benefits. Athletics are beneficial primarily to sustain interest in the program and to maintain an achieved level of physical fitness. Athletics are a supplement, not a substitute for other types of conditioning activities which should continue. All of the desirable traits of physical fitness cannot be developed through athletics, yet their contribution is significant. For athletics to make a proper contribution to physical conditioning, the selected sports must bevigorous. The team contests recommended here are carefully selected for their simplicity, aggressiveness and applicability to squad- teaching teamwork. They are generally strenuous and involve the whole team throughout the entire duration of the contest.

5402. TEAM CONTESTS

a. Pushball

(1) **Players.** 10 to 50 players on a side. (See fig. 5-8.)

(2) Equipment A large pushball 5 to 6 feet in diameter.

(3) Area. A field 240 to 300 feet long, 120 to 150 feet wide. Mark a center line in the middle of the field parallel to the end lines. Mark a line 45 feet on either side of this center line and parallel to it, extending it across the width of the field. Mark another parallel line 15 feet from each end line, extending it across the width of the field.

(4) Game. Four 10-minute quarters are played. Give 2-minute rests between quarters and 5-minute rests between halves. The object of the game is to propel the ball over the opponent's goal line by pushing, rolling, passing, carrying, or any other way except kicking.

(a) The ball is placed on the center line with the opposing captains 3 feet from the ball. The rest of the players are 45 feet from the ball, on their half of the field. On the referee's starting whistle, the captains immediately play the ball with their respective teams coming to their assistance.

(b) At quarter time, the ball remains dead for 2 minutes at the spot where it was when the quarter ended. At half time, the teams exchange goals. The play is then started as it was in the beginning.

Figure 5-8 is available in the illustration section at the end of the book.

(c) Players may use any means of interfering with an opponent's progress except striking and clipping. Clipping is throwing the body across the back of an opponent's leg, while the opponent is running or standing. Legal use of force may be applied to all opponents whether or not they are playing the ball. For striking an opponent, the offender is removed from the game. The team penalty is half the distance to their goal. The penalty for clipping is the same.

(d) When any part of the ball goes out of bounds, it is a dead ball. The teams line up at right angles to the side lines and 3 feet apart at the point where the ball went out. The referee then tosses the ball between the teams.

(e) When, for any reason, the ball remains in one spot for more than 10 seconds, the referee declares the ball dead. The ball is then put into play as it is for an out-of-bounds situation.

(5) Scoring. A goal is scored when the ball, or any part of it, is propelled across the opponent's end line. A goal counts 5 points. The team scoring a goal has the privilege of trying for a point after the goal. To try for this extra point, the ball is placed on the opponent's 5-yard (or 15-foot) line. The teams line up on either side of this line separated by the width of the ball. Before the whistle blows, one player on each team may place hands on the ball. On the referee's signal, the ball is put into play for 1 minute. If any part of the ball is driven across the goal line in this 1-minute period, the offensive team scores 1 point. The defending team may not score during the opponent's try for the extra point.

b. Line Rush

(1) Players. Any number up to 50 on each side. (See fig. 5-9.)

(2) Equipment. None.

(3) Area. A field, 75 by 100 feet.

(4) Game. One team lines up behind one goal line and the other in midfield. On the starting signal, the team standing behind the goal line seeks to cross the goal line at the opposite end of the field within 30 seconds. The team in the center seeks to prevent it by catching and holding the runners. At the end of 30 seconds, the teams change.

(5) Scoring. Count the number of individuals who have crossed the far goal at the end of 30 seconds. After each team has had from three to five tries, the scores are added and the winner declared. A player scores 1 point when any part of the body is across the goal line.

Figure 5-9 is available in the illustration section at the end of the book.

c. Human Tug-of-War

(1) Players. 10 to 20 on a team.(See fig. 5-10.)

(2) Equipment. None.

(3) Area.40 to 60 feet.

(4) Game. Draw a line in the center of the area. Divide the players into two equal teams. Place them in single file on opposite sides of the center line facing each other. Each individual places the arms around the waist of the teammate in front. The two leaders of the opposing teams grasp each other around the waist. On signal, the teams try to pull each other over the center line within 30 seconds.

(5) Scoring. The team pulled across the center line loses. If neither team is pulled over the center line, but one team breaks its file, that team loses the match.

(6) Variation. Use a 3/4- or 1-inch rope. Space the leading players on each team 10 feet apart, each holding one end of the rope. The team pulled across the center line loses.

Figure 5-10 is available in the illustration section at the end of the book.

d. Master of the Ring

(1) Players. Any number. (See fig. 5-11.)

(2) Equipment. None.

(3) Area. A clearly marked circle large enough to contain all the players.

(4) Game. All the players stand inside the circle. At the signal, all players attempt to throw each other out of the circle. All tactics are fair except unnecessary roughness. When any part of the body touches across the line, the player is out and leaves the circle at once. Several officials are needed to spot the players who cross the line.

(5) Scoring The player who remains in the circle when all the others are out is the master of the ring.

(6) Variation 1. The players are divided into two equal teams. Each team is clearly marked. On signal, each team tries to throw the opponents out of the circle. The winning team is the team that eliminates all opponents from the circle.

(7) Variation 2.The players are divided into two equal teams. Each team sends only one individual into the circle. When one person has been forced out of the circle, the losing side sends in only one player. The team which eliminates all opponents is the winner.

(8) Variation 3.A pit, approximately 4-feet deep, is used rather than a circle on level ground. The contest may then be an individual or team activity.

Figure 5-11 is available in the illustration section at the end of the book.

e. Log Pivot Circle. Space teams far enough apart so each team can pivot in a circle without colliding with any other team. Each team holds a log in the bend of the arms in front of the chest. (See fig.5-12.) At the command **CIRCLE RIGHT, MOVE**,the left-flank individual holds the pivot while the log is carried around 360 degrees in a counter- clockwise motion, back to the original position. This movement may also be performed to the left by facing in the other direction and pivoting clockwise(**CIRCLE RIGHT, MOVE**).Other commands may be used such as: **CIRCLE HALF-RIGHT, CIRCLE HALF- LEFT** and so on. The first team to complete the prescribed movement is the winner.

Figure 5-12 is available in the illustration section at the end of the book.

f. Log-Rolling Race. Each team tries to roll its log a measured distance by pushing it with the hands and driving the body forward with the legs. (See fig. 5-13.)The first team to get the entire length of the log across the finish line wins.

Figure 5-13 is available in the illustration section at the end of the book.

g. Prone Pushing Contest.. Two teams lie prone, facing each other with a log between them. (See fig. 5-14.)Both teams place their hands against the log, keeping their arms straight. Then, by driving with the legs, each team attempts to push the other a measured distance to the rear.

Figure 5-14 is available in the illustration section at the end of the book.

h. Shuttle Relay Race. Each team in this race divides into two parts, A and B. Members of part A run 50 yards with a log held under their right arms. (See fig 5-15.)At the distance line, they give the log to their teammates in part B who bring it back to the starting line. The team pair finishing first is the winner.

Figure 5-15 is available in the illustration section at the end of the book.

5403.CROSS-COUNTRY AND DISTANCE RUNNING

a. Value of Running. Long-distance running gives some benefits that cannot be obtained in the same degree from any other sport. It builds powerful leg muscles, increases lung capacity, and develops endurance. For these reasons, cross-country and distance running should be included in the physical training program. These sports require only a few miles of open space which is generally available. Short cross-country runs and middle-distance runs can be used to supplement other activities, particularly team sports or the sports that develop precision or agility rather than endurance. Short cross-country runs can be scheduled once a week, gradually increasing the distance as the physical condition of the runners improves.

b. Cross-Country Runs. A cross-country run is a distance run held on a course laid out along roads, across fields, over hills, through woods, and on any irregular ground. A flat cinder or dirt track is not a suitable surface for cross-country running. Some runs are as long as 6 miles while others are as short as 3 or 4 miles. If cross-country running is being used to supplement other activities, the 3-mile course is long enough for most. Cross-country runs should be used only after individuals reach the sustaining stage of conditioning. This type of running should then be scheduled occasionally to provide variety. Cross-country running has the advantage of allowing mass participation. Interest can be stimulated by putting the runs on a competitive basis.

c. Practice Methods. Conditioning is essential to distance and cross-country running. Championship distance running depends on stamina, which can be developed only through constant training. An individual of only average ability can become an outstanding distance runner by steady and careful training. Hiking is an excellent method for getting into condition before the season opens. Long walks build up leg muscles. During the first month of the season, training should be gradual, starting with short distances, and increasing day by day. At first the legs will become stiff, but the stiffness gradually disappears if running is practiced for a while every day. To prevent strain, it is essential to limber up thoroughly each day before running.

(1) In training a large group, leaders should be stationed at the head and the rear of the column. They should make every effort to keep the runners together. After determining the abilities of the unit in cross-country running, it is advisable to divide them into three groups. The poorest-conditioned group is started first, the best-conditioned group, last. The starting time of the groups should be staggered so that all of them come in at about the same time.

(2) In preliminary training, running is similar to ordinary road work in that it begins with rather slow jogging, alternating with walking. The speed and distance of the run is gradually increased. As conditioning improves, occasional sprints may be introduced. Atfirst, the distance run is from one-half to 1 mile. It gradually increases to 2 or 3 miles. On completing runs, require runners to walk for 3 or 4 minutes before stopping to permit a gradual cooling off.

d. Facilities and Equipment. There should be at least one timer with a stopwatch (preferably three) for timing the runners. A course 3 or 5 miles long should be measured and marked by one of the three following methods:

(1) Fasten directional arrows to the top of tall posts and place them at every point where the course turns. Such signs should also be placed at every other point where there may be doubt as to the direction of travel.

(2) Place a lime line on the ground over the entire course.

(3) Place flags, clearly visible to the runners.

- **(a)** A red flag indicates a left turn.
- **(b)** A white flag indicates a right turn.
- **(c)** A blue flag indicates the course is straight ahead.

e. Rules/Scoring. A cross-country team consists of seven individuals, unless otherwise agreed. In dual meets, a maximum of 12 on one team may be entered, but 7 or fewer are scored. First place scores 1 point, second place 2, third place 3, and so on. All who finish the course are ranked and tallied in this manner. The team score is then determined by totaling the points scored by the first five of each team to finish. The team scoring the least number of points is the winner. If fewer than five (or the number determined prior to the race) finish, the places of all members of that team are disregarded. If two or more teams score the same number of points, the event is called a tie.

5404.TEAM ATHLETICS

Team athletics add variety to the combat readiness program. At no time should they be allowed to replace intramural activities or to provide practice time for intramural or base teams. If team athletics are integrated into the program, the goal should be total participation and competition. Therefore, the games selected can be played by the skilled or the unskilled. They require limited equipment, yet will teach competitiveness, aggressiveness, and teamwork while they improve conditioning.

a.Soccer. Soccer is one of the best athletic activities for developing endurance, agility, leg strength, and great skill fulness in using the legs. The most popular sport in the world, soccer is the national game of many European, Asian, and Central and South American countries. In recent years, it has become popular in the United States. A soccer ball is the only equipment needed for the game, and Marines can learn to play it easily. Players do not need much skill to participate, but the amount of skill they can develop is almost unlimited.

(1) Place in the Program. Soccer should be introduced during the latter part of the slow improvement stage of physical training. It can be used as a competitive activity in the sustaining stage. It is primarily a spring or fall sport. Any level field is suitable for competition. The boundaries for the soccer field are similar to the dimensions for a football field. (See fig. 5-16.) Goal posts are essential, but they are easily constructed and usually temporary so that they may be removed when not in use.

Figure 5-16 is available in the illustration section at the end of the book.

(2) Basic Skills

(a) Passing. Passing with the feet is the primary means of moving the ball. Short passes are easier to control. Emphasis should be continually placed on the skill of passing.

(b) Dribbling. The ball is dribbled by a series of kicks with the inside or outside of the foot. Do not kick with the toe. Keep the head over the ball when kicking and propel it only a short distance at a time. Keep it close to the feet. If a player lets a ball move far from the feet, an opposing player can easily take it away.

(c) Instep Kicking. The instep kick is the basic soccer kick. The toe does not come in contact with the ball. The toe is pointed downward, and the instep (the area under the shoe laces) is applied to the ball with a vigorous snap from the knee.

(d) Inside-of-the-Foot Kicking. The ball is kicked with the inside of the foot, and the leg is swung from the hip. The toe is turned outward, and the sole of the foot is parallel with the ground as the foot strikes the ball. This kick is used for short passes and for dribbling.

(e) Foot Trapping. The foot trap is the method of stopping the ball by trapping it between the ground and the foot. Place the sole of the foot on top of the ball at the instant it touches the ground, keeping the foot relaxed. This is an effective way to stop a fast-moving ball.

(f) Shin Trapping. .The shin trap is a method of stopping the ball with the shins. Stand just forward of the spot where the ball should strike the ground and allow it to strike the shins in flight or on the bounce. Use either one or both legs from the knee down so as to absorb the bounce.

(g) Body Trapping. .Intercept the ball with any part of the upper body except the arms and hands. Keep the body relaxed and inclined toward the ball. To keep the ball from bouncing, move backwards from it as it strikes the body. This drops the ball at the feet in position for dribbling or passing.

(h) Heading. Heading is a technique for changing the direction of the flight of a ball by butting it with the head. Tense the neck muscles and jump up to meet the ball. Butt the ball with the forehead at about the hairline to reverse its direction. Use the side of the head to deflect it to the side.

(3) Offensive and Defensive Positions. Forwards usually play on the offensive half of the field. Fullbacks usually play on the defensive half of the field. Halfbacks are the backbone of the team. They move forward on the offense and back on defense. The goalkeeper almost always remains within a few feet of the goal.

(4) Abridged Rules

(a) A soccer team is composed of 11 players. (See fig. 5-16.)

(b) The player propels the ball by kicking it with the feet or any part of the legs, by butting it with the head, and by hitting it with any portion of the body except the arms or hands.

(c) The goalkeeper is the only individual allowed to place hands on the ball, but only in the goalkeeper's area. The term "hands" includes the whole arm from the point of the shoulder down.

(d) A goal is made by causing the ball to cross completely the section of the goal line lying between the uprights and under the cross bar.

(e) Each goal scores 1 point for the team scoring the goal.

(f) The penalty for a foul committed anywhere on the playing field (except by the defensive team in its penalty area) is a free kick awarded to the opposing team.

(g) All opponents must be at least 10 yards from the ball when a free kick is taken.

(h) The penalty for a foul committed by the defensive team in its penalty area is a penalty kick.

(i) A penalty kick is a free kick at the goal from the spot 12 yards directly in front of the goal. The only players allowed within the penalty area at the time of the kick are the kicker and the defending goalkeeper.

(j) An official game consists of two 30-minute halves,

(k) Teams change goals at the end of every quarter.

(l) In the event of a tie, an extra quarter may played. This may be followed by a sudden death period. If the tie still exists, the game can be decided by penalty kick.

(m) After a team has propelled the ball across a side line and out of play, the ball is put back into play by a throw-in from the side line by a member of the opposing team. The ball is thrown in from the point where it crossed the side line as it went out of bounds. The ball must be thrown in with both hands using an over-the-head motion, keeping both feet on the ground.

(n) When the offensive team propels the ball across the defensive team's goal line but not in the goal, the defensive team is awarded a goal kick-a free kick taken from within its own goal area. The ball must come out of the penalty area to be in play.

(o) When the defensive team causes the ball to go behind its own goal line, excluding the portion between the goal posts, the offensive team is awarded a corner kick—a free kick taken by a member of the offensive team at the quarter circle, at the cornerflag post nearest to where the ball went behind the goal line. The flag post must not be removed.

(p) The game is started and, after a goal has been scored, is resumed by placing the ball in the center of the midfield line. Players must be on their side of the line until the ball is kicked. The ball must be kicked forward and must move at least 2 feet to be legal. The first kicker may not touch the ball twice in succession at the kick off. The opposing team must be 10 yards from the ball until it moves.

b.Speedball. Speedball offers vigorous and varied action with plenty of scoring opportunities. It is easy to learn and provides spontaneous fun. Little equipment is needed-a ball is all that is absolutely necessary. Speedball combines the kicking, trapping, and intercepting elements of soccer; the passing game of basketball; and the punting, drop kicking, and scoring pass of football.

(1) Place in the Program.Speedball, like soccer, should be introduced into the physical training program during the latter part of the toughening stage and used as a competitive activity in the sustaining stage. Speedball may be played any time weather permits but is primarily a spring or fall activity.

(2) Method of Play.The game is played as follows:

(a) Two teams of 11 each play the game under official rules, but any number of players may successfully constitute a team.An inflated leather ball, usually a soccer ball, is used. The playing field is a football field with a football goal post at each end. (See fig. 5-17.)

(b) The game starts with a soccer-type kickoff. The kicking team tries to retain possession of the ball and advance it toward the opposite goal by passing or kicking it. Running with the ball is not allowed, with the result that there is no tackling or interference. After the ball touches the ground, it cannot be picked up with the hands or caught on the bounce but must be played as in soccer until it is raised into the air directly from a kick. Then the hands may be used.

(c) When the ball goes out of bounds over the sidelines, it is given to a player of the opposing team and is put into play with a basketball throw- in. When it goes over the end line without a score, it is given to a player of the opposing team who may either pass or kick it onto the field.

(d) When two opposing players are contesting possession of a held ball, the official tosses the ball up between them as in basketball.

(e) Points are scored by kicking the ball under the crossbar of the goal post, drop kicking the ball over the crossbar, or completing a forward pass into the end zone for a touchdown.

Figure 5-17 is available in the illustration section at the end of the book.

(3) Basic Skills

(a) Shared Skills. Skills include kicking, passing, heading, and trapping (from soccer); punting, dropkicking and forward passing (from football); and passing, receiving, and pivoting (from basketball).

(b) Kickups. The kickup is a play in which a player lifts the ball into the air with the feet so as to legally play the ball with the hands. The kickup generally makes the transition from ground play to aerial play. The technique of making the play depends upon whether the ball is rolling or stationary. If the ball is rolling or bouncing toward the player, the foot is held on the ground with the toe down until the ball rolls onto the foot. Then the foot is raised, projecting the ball upward. If the ball is stationary, the player rolls it backward with one foot. Then the player places a foot where the ball will roll onto it. The player then raises the foot, projecting the ball as before. If a ball is rolling away, the player should stop it with a foot and play it as a stationary ball. The player can also raise the ball by standing over it with a foot on either side. The player presses the feet against the ball and jumps into the air, propelling the ball into the hands.

(4) Offensive and Defensive Play. The positions of players in speedball are much the same as in soccer. However, some of the positions have different names. Each team has 11 players. The forward line is composed of five players, the right end, right forward, center, left forward, and left end. The second line consists of right halfback, fullback, and left halfback. In the next line is the right guard and left guard. The player who defends the goal is the goal guard. The strategy employed in speedball during offensive play is very similar to that of soccer. There are two types of defensive formations in speedball: man-for-man and position defense. Man-for-man defense is recommended for beginning players.

(5) Abridged Rules

(a) Field. A speedball field is 100 yards long with additional 10-yard end zones on each end and 55 yards wide, the same as a regulation football field.

(b) Players. Eleven players are on a team. The goal guard has no special privileges.

(c) Time. Ten-minute quarters. Two minutes between quarters. Ten minutes between halves. Five minutes for extra overtime periods.(Begin the first overtime by a jump ball at center, same goals; change goals in the event of a second overtime period.)

(d) Winner of Toss. The winner of the toss has the choice of kicking, receiving, or defending a specific goal.

(e) Kickoff. A kick off starts the first and second half of a game and starts play after a score. The kick off is made from any place on the 50-yard line. The kicking team must be behind the ball when it is kicked. The receiving team must stay back of its restraining line (10 yards distance) until the ball is kicked. The ball must go toward the opposing team's goal before it can be played. Kick off out of bounds goes to the opponent at that spot. A kick off touched by the receiving team before going out of bounds, with no impetus added, still belongs to the receiving team. A kick off, in possession and control of the receiving team and then fumbled out of bounds, belongs to the kicking team at that spot.

(f) Starting Second and Fourth Quarters. The ball is given to the team that had possession at the end of the previous quarter, from out of bounds, as in basketball.

(g) Second Half. The first-half receiving team kicks off at the beginning of the second half.

(h) Fly and Ground Ball. The most characteristic playing rule of speedball is the difference between a fly ball (or aerial ball) and a ground ball. A player is not permitted to touch a ground ball with the hands and must play it as in soccer. A fly ball is one that has risen into the air directly from the foot of a player (for example, a punt, drop kick, place kick, or kickup).Such a ball may be caught with the hands provided the catch is made before the ball strikes the ground again.

(i) Kickup and Overhead Dribble. A kickup is a ball that is so kicked by a player that he can catch it himself. A bounce from the ground may not be touched with the hand because it has touched the ground since being kicked. This rule prohibits the ordinary basketball dribble, but one overhead dribble (throwing the ball into the air and advancing to catch it before it hits the ground) is permitted.

(j) Out of Bounds. If a team causes the ball to go out of bounds over the sidelines, a free throw-in (any style) is given to the opposing team. When the ball goes over the end line without scoring, it is given to the opponents who may pass or kick from out of bounds at that point.

(k) Tie Ball. In case two players are contesting the possession of a held ball, even in the end zone, a tie ball is declared. The ball is tossed up between them.

(6) Scoring

(a) Field Goal(3 points).A soccer-type kick, in which a ground ball is kicked under the crossbar and between the goal post from the field of play or end zone, is a field goal. (A punt going straight through is not a field goal for it is not a ground ball. The ball must hit the ground first.) A drop kick from the field of play that goes under a crossbar does not count as a field goal. A drop kick from the end zone that goes under the crossbar counts as a field goal;if it goes over the crossbar, it is ruled as a touch back, and the other team takes possession of the ball.

(b) Dropkick(2 points). A scoring drop kick must be made from the field of play and go over the crossbar and between the uprights. The ball must hit the ground before it is kicked (usually with the instep).

(c) End goal(1 point). This is a ground ball which receives its impetus (kicked or legally propelled by the body) from any player, offensive or defensive, in the end zone and which passes over the end line but not between the goal posts.

(d) Penalty Kick(1 point).A ball kicked from the penalty mark that goes between the goal posts and under the crossbar is a penalty kick. The penalty mark is placed directly in front of the goal at the center of the goal line.

(e) Touchdown(1 point). A touchdown is a forward pass from the field of play completed in the end zone. The player must be entirely in the end zone. If the player is on the goal line or has one foot in the field of play and the other in the end zone, the ball is declared out of bounds. If a forward pass is missed, the ball continues in play, but it must be returned to the field of play before another forward pass or drop kick may be made.

(7) Substitutions. Substitutions may be made any time when the ball is not in play. A player who is withdrawn may not return during that same period.

(8) Timeout. Three legal timeouts of 2 minutes each are permitted each team during the game.

(9) Fouls

(a) Personal Foul. Kicking, tripping, charging, pushing, holding, blocking, and unnecessary roughness of any kind, such as running into an opponent from behind, are personal fouls. Kicking at a fly ball and thereby kicking an opponent is a personal foul. Four personal fouls disqualify a player from the game.

(b) Technical Foul. Illegal substitution, more than three timeouts in a game, unsportsmanlike conduct, and unnecessarily delaying the game are technical fouls.

(c) Violation. Traveling with the ball, touching a ground ball with the hands or arms, double overhead dribble, violating tie ball, and kicking or kneeing a fly ball before catching it are violations.

(d) Penalties. The offended player shall attempt the kick.

(e) Officiation of Fouls

1 Fouls in the field of play allow no follow-up while fouls in the end zone always allow follow-up.

2 On penalty kicks, with no follow-up, only the kicker and goalie are involved.

3 On penalty kicks, with a follow-up, the kicking side is behind the ball and the defending side behind the end line or in the field of play. No one is allowed in the end zone or between the goal posts except the goal guard. The kicker must make an actual attempt at goal and cannot play the ball again until after another player plays it.

c. Volleyball. Volleyball is a popular sport. The game entails much physical activity, yet it is not strenuous. It is, therefore, a game for young and old alike, for beginners and for skilled players. It may be played indoors or outdoors on any type of terrain. As an informal activity, volleyball can be played by any number; as an organized activity, it provides, as few other sports do, a game for 12 to play in a limited area. While volleyball requires no great skill to play, it permits a high degree of proficiency. Players naturally get more enjoyment when they know the game and play it well. For this reason, instruction in the basic skills should be provided.

(1) Place in the Program Volleyball may be used occasionally as a competitive activity during the sustaining stage. Though it is a year- round sport, it should be included in the physical training program only when a more strenuous activity is impractical. Volleyball is an excellent off-duty activity.

(2) Instruction. Usually during the first three or four classes, a 10- to 15- minute period of instruction, followed by scrimmage, is enough to teach the basic skills, rules, and techniques of volleyball. More time, if available, can be given to teaching basic skills, but emphasize competitive play rather than formal instruction. Divide the unit into 6-person teams, organized on the basis of ability .Teams should be as nearly equal as possible.

(3) Basic Skills

(a) Passing the High Ball. The chest pass is the most effective method of playing the ball. To receive the ball, the knees are fixed, and the body is tilted forward. The elbows are raised sideward to a point in line with the shoulders. The wrists are extended in line with the forearm. The arms, wrists, and hands are rotated inward. To pass the ball, the hands are chest-high, thumbs pointing inward. The fingers are flexed, forming a cup, allowing them to contact the ball. On contact with the ball, the wrists are snapped while the fingers and elbows are pushed upward, sending the ball upward. A high ball is much easier to handle than a low one.

(b) Passing the Low Ball. A ball that is lower than the waist is one of the easiest to hit, but it is also a frequent cause of fouls (holding or carrying the ball). The best position for handling a low ball is to have the knees flexed and arms flexed at the elbows and rotated so the thumbs are pointing outward, the palms up. When the fingers contact the ball, the entire body reacts in a lifting motion. The arms and hands swing upward in a scooping action. It is important that the fingers, not the palms, contact the ball, and that the ball is batted, not thrown.

(c) Underhand Serve. Take a position behind the back line facing the net, left foot forward, holding the ball in the palm of the left hand. The left knee is flexed; the right knee is straight. Swing the right arm back and at the same time move the left hand (holding the ball) across the body in line with the right hip. Then swing the right arm forward hitting the ball off of the left hand with the palm of the right hand, raising the hips and arching the back in the same motion. Be certain to swing the right arm in a straight line, or the ball will be difficult to control. When the opposition is in formation, the server should try to place the ball in the right or left back area, and not near the net.

(d) Setup. A setup is a ball hit into the air near the net by one player, so that a teammate may hit or "spike" it sharply downward into the opponent's court. The chest pass is the best pass to use. The ball is sent approximately 10 feet into the air toward the spiker so it will descend from 4 to 20 inches from the net.

(e) Spiking. The spike is a leap into the air and a sharp downward hitting of the ball into the opponent's court. A spiker must be able to spring easily from the floor, judge the movement of the ball, and strike it with a downward movement of the arm. To jump from the floor, step off with one foot and jump with the other. Stand with the right or left side to the net, facing the setup player. Much depends upon the setup player to place the ball in the proper position. The spiker jumps into the air and strikes the ball above its center so as to drive it downward. A snapping movement of the arm and wrist will drive the ball forward and downward with power and control. Aim for a weak spot in the opponent's defense.

(f) Blocking. The block is a defense technique used to prevent a spiker from driving the ball across the net. It is an attempt by one or more defensive players at the net to block a hard-hit shot by using the force of the ball to send it immediately back into the opponent's court. In an effective block, forwards on the defensive team spring into the air at the time of the spike, placing both hands and arms in the expected path of the ball. An effective block tends to upset the offense and presents another element for the spiker to worry about. To be effective, the blocker must anticipate the path of the ball and time the block with the spike.

(4) Abridged Rules

(a) The volleyball court is 30 feet wide by 60 feet long. (See fig. 5-18.)

(b) The top of the net is 8 feet high.

(c) A volleyball team consists of 6 players.

(d) A match consists of the best two out of three games.

(e) The first team scoring 15 points wins the game, provided that they have 2 points more than their opponents.

(f) A deuce game is a game in which both teams score 14 points. The game is continued until one team obtains a 2-point advantage over the other.

(g) Only the serving team can score. If the serving team commits a fault, it loses the serve to the opposing team.

(h) The team members who receive the served ball rotate one position in a clockwise direction.

(i) The ball is put into play by serving from behind the back line.

(j) A served ball touching the net results in the loss of the serve. At any other time during play, a ball touching the net is still in play.

Figure 5-18 is available in the illustration section at the end of the book.

(k) The ball is out of play when it touches the ground or goes outside one of the boundary lines.

(l) All line balls are good.

(m) The players must hit or bat the ball; they may not throw, lift, or grasp it.

(n) A player may not touch the ball with any part of the body below the knees.

(o) A player may not play the ball twice in succession. In receiving a hard-driven spike, a defensive player may make several contacts with the ball even if they are not simultaneous. All such contacts, however, must constitute one continuous play, and all must be above the knees.

(p) The ball may be touched no more than three times on one side of the net before being returned across the net to the opposing team.

(q) A player must not touch or reach across the net.

(r) A player may touch the line under the net, but may not cross it.

Chapter 6
EVALUATION OF PERFORMANCE DURING TRAINING

6001.OBJECTIVES OF COMBAT READINESS

The objectives of the Marine Corps physical fitness program are-

a. To contribute to the health and well-being of every Marine through regular exercise and health education.

b. To develop Marines who are physically capable of performing their duties in garrison and in combat.

c. To develop Marines with a reserve level of physical fitness that will help them win in combat.

d. To provide a medium for developing the self-confidence of the individual Marine and thereby enhancing overall discipline, morale, esprit de corps, unit efficiency, and the desire to excel within the Marine Corps.

6002. THE PHYSICAL FITNESS TEST

The physical fitness test (PFT) has distracted attention from the proper goal of the physical fitness program. Simply stated, the goal of the physical fitness program is the success of Marines in combat. Unit commanders must not allow the PFT to become an end in itself. The PFT should be used within the commander's program of combat readiness training as a standard, an index of the physical fitness of individual Marines at a given time and place. The PFT is not difficult, and it must be kept in perspective. All athletes should be first class, and first class is attainable for all Marines. The conduct of the PFT is described in Marine Corps Order 6100.3_.

6003. OTHER METHODS OF EVALUATION

The commander should use other ways of evaluation to provide variety and imagination. Intangible objectives, such as confidence and aggressiveness, are hard to measure. The use of either inspection or observation is necessary to evaluate objectives of this nature. The commander has several other methods available when evaluating the physical condition of the command. The additional methods of inspection, observation, medical examination, and testing are summarized as follows:

a. Formal Inspections. Formal inspections, using inspection officers and standardized rating criteria, may help evaluate unit physical fitness.

b. Daily Observation. Routine observation of physical performance and appearance can serve as an indicator of an individual's or unit's physical readiness. However, mere observation is not a totally reliable or accurate means of evaluation.

c. **Physical Examination.** Medical examination may detect any individual disability or detrimental physical condition. It may guide in application of remedial, therapeutic, or limited exercise programs.

d. Other Physical Fitness Tests. Some other tests are described in detail in this manual for information purposes. Commanders are cautioned, however, not to place too great an emphasis on testing.

6004. RESPONSIBILITY

The commander is responsible for the physical fitness of the command and for the measurement and evaluation of its physical readiness. The goal of combat readiness training remains to prepare Marines and units for the physical demands of combat. The bulk of the commander's combat readiness training program should be such activities as obstacle and confidence courses, endurance courses, and progressive load-bearing marches which will prepare Marines for combat. Group activities which are competitive and combative in nature are advantageous and add imagination and enthusiasm to the program. Activities such as orienteering, which is not only physically demanding but also competitive and which teaches essential subject skills, are excellent and motivate all concerned. A commander's goal should be to provide an integrated training program that is so aggressive and demanding of Marines and leaders alike that a PFT administered at any time will reflect that each and every Marine is both physically and mentally prepared for the demands of combat.

6005. COMMANDER'S PHYSICAL FITNESS PROGRAM

The Commandant of the Marine Corps has directed that the physical fitness test shall be the universal measure of adequate individual physical fitness. The focus of a commander's physical fitness training program, however, should not be to prepare Marines to pass the physical fitness test, but to prepare them for the physical rigors of combat.

a. Administration. The physical fitness program requires each Marine to spend a minimum of 3 hours per week in physical fitness training, to be tested semiannually, and to obtain a minimum standard of third class. The attainment of a higher score is a laudable individual goal which should be encouraged but should not become a unit objective. Overemphasis could be detrimental to training required to develop the complete Marine. All commanders should exercise sound judgment as well as positive, aggressive leadership in striving toward the training objective.

b. Variety of Activities. The program must include the following elements:

(1) Physical Readiness Training. physical readiness training is a complete physical training program which develops and maintains the strength, endurance, and physical skills needed to sustain the individual during combat.

(2) Remedial Physical Conditioning. Remedial physical conditioning is a process by which physically substandard Marines are conditioned to meet prescribed standards.

(3) Competitive Conditioning Activities. Competitive conditioning activities consist of teams or individuals competing against an opponent to win. This involves a combination of sports and military skills designed to foster competitive attitudes and develop unit pride and esprit de corps.

(4) Occupational Conditioning. Marching long distances, dry net training, etc., should not be ignored as a means to fulfill the time requirement.

c.Time. In order to develop the desired level of physical fitness, unit commanders are advised that a minimum program of 3 hours per week is mandatory. The program must encompass at least three exercise periods weekly. Each period should include calisthenics, running, or other forms of vigorous activity .The minimum desirable length for each exercise period is 60 minutes. Shorter periods may be authorized by the unit commander if dictated by the local training situation. Unit commanders are encouraged to use the normal working day to satisfy this training requirement. However, commanders are authorized to conduct required physical fitness training during off-duty hours when the mission, workload, personnel status or other significant factors preclude fulfilling it during duty hours.

d.Action. Marine Corps Order 6100.3_ requires that commanders-

o Establish and maintain an effective physical fitness program which ensures that all Marines maintain an acceptable level of physical fitness.

o Ensure that all individuals are medically qualified prior to participating in the physical fitness program.

o Establish a minimum physical fitness program of at least 3 hours per week. Physical training may be authorized on an individual basis at the discretion of the unit commander.

o Ensure that all Marines participate in physical fitness conditioning activities commensurate with their medical qualifications and limitations.

o Conduct physical fitness testing for all Marines.

o Place all Marines who fail to pass the PFT on a daily command-supervised remedial physical conditioning program until they pass the test.

o Ensure that results of physical fitness testing are entered on section A of the fitness report. Comments are placed in section C in connection with attainment of superior physical performance or a medical excuse from the PFT.

6006. OTHER FORMAL TESTS

Commanders must realize that the best index of combat readiness is personal observation of performance on conditioning marches of 15 miles while under load. This manual presents three additional indicators of unit fitness. These indices are the physical readiness test, which was formerly the Marine Corps standard and is now routinely administered at Officer Candidate School; the airborne trainee physical fitness test which many Marines who attend U.S. Army formal schools are required to take; and the battle fitness test, the ultimate test of the combat readiness of Marines. Commanders may wish to include these tests or portions thereof within their combat readiness programs.

6007. PHYSICAL READINESS TEST

a. Purpose. The purpose of this test is to measure the Marine's ability to meet the minimum standard of physical readiness for duty in the field. This test is currently administered to all male officer candidates at Officer Candidate School, MCCDC, Quantico, VA. Female candidates do not take this test.

b. Test Area. Administration of this test requires an area suitable for crawling, jumping, and running 160 yards to include a trench that is 8 feet wide.

c. Equipment. The equipment required for this test includes—
o A platform, rail on a ledge, or a box 18 inches high,
o A 20-foot climbing rope,
o A stopwatch.

d. Conduct of Test. Adequate timers and other supervisory personnel must be provided by the company staff. All events should be conducted in a single session of one morning or afternoon. The events may be run in any sequence.

e. Uniform. The uniform for the physical readiness test is as follows: full utilities, cartridge belt, belt suspender straps, two canteens (full of water), rifle, and helmet. Depending on the heat condition, the PT shirt may be worn in place of the utility jacket. All equipment will remain secure throughout the test. Except for emergencies, canteens are not to be emptied.

f. Events
(1) Event 1: Climbing Uphill (Step-ups)This event simulates marching uphill at a rapid and steady pace. The Marine stands in front of the step. On the command GO, the Marine places either foot on the step and steps up. Hand pressure on the knee may be used if desired. The Marine then stands erect and steps backward and down, on one foot at a time. The same leg may be used for each repetition or the legs may be alternated. For 100 points, Marines will complete 60 repetitions in the maximum allotted time of 1 minute and 20 seconds (80 seconds).
*(2) Event 2: Rope Climb. T*his event resembles entering and leaving a hovering helicopter, using ropes in house-to-house fighting, and ascending and descending landing nets. The Marine assumes a sitting position on the ground at the bottom of the rope with hands at the highest point that can be reached. On the command GO, the Marine jumps up and climbs to the top. The Marine must touch the 20-foot mark with one hand, then descend without sliding or dropping. Marines must reach the 20-foot mark in the maximum allotted time of 30 seconds or less.
(3) Event 3: Evacuation. This event simulates reaching and evacuating a wounded Marine under fire. The Marine assumes the prone position at a distance of 50 yards from a casualty who is of approximately equal weight. On the command GO, the Marine springs up and covers the distance in a sprint. The Marine lifts the casualty to a fireman's carry and returns to the starting point. Assistance by the casualty is permitted. The tested individual must carry all equipment; the casualty has none. Marines must negotiate the course in a maximum time of 47 seconds or less.
(4) Event 4.'Advance by Fire and Maneuven. This event simulates advancing and assaulting during an attack. The Marine assumes the prone position at the starting point. On the command GO, the Marine then alternately crawls, runs in a zigzag manner, rolls, jumps, and assumes different firing positions for the next 130yards. Before reaching the objective, the Marine must "hit the deck," roll over, and assume the firing position three times. While assuming the firing position, the Marine must pull back the charging handle, take aim, and squeeze the trigger before rising from the ground. After rising from the ground for the third time,the Marine continues running and jumps to clear the 8-foot trench in one leap. The Marine then finishes the course in

the maximum allotted time of 2 minutes. The Marine is allowed multiple attempts only within the specified time limit.

(5) Event 5: Forced March. This event is a combat run of 3 miles without halts and against time. The Marine must double-time. Marines must complete the 3-mile course in a maximum allotted time of 30 minutes or less.

g. Event Standards

(1) To successfully pass the physical readiness test, the Marine must pass each event with a minimum of 80 points for each. Points are then assessed for each event in accordance with the scoring tables. (See fig. 6-1.)

(2) If a Marine fails one event, the highest possible score that the Marine may receive for the entire test is 79 percent. (See example 1 in fig. 6-2.) If two events are failed, the highest possible score for the entire test is 69 percent, regardless of the total points accumulated. (See example 2 in fig. 6-2.) If three events are failed, the highest possible score is 59 percent. Failure in four or more events will result in a score of zero for the entire test.

(3) To convert the total points for all five satisfactorily passed events to a final percentage score, total the accumulated points for all five events and divide by five. (See example 3 in fig. 6-2.)

Figure 6-1 is available in the illustration section at the end of the book.

Figure 6-2 is available in the illustration section at the end of the book.

6008. AIRBORNE TRAINEE PHYSICAL FITNESS TEST

a. Use and Composition of the Test.

(1) Use. The airborne trainee physical fitness test is a means of determining the physical ability of the applicant for acceptance to and retention in the airborne course of instruction.

(2) Test Events The test battery consists of five events as follows: chinups, knee bender, pushups, situps, and an endurance run.

b. Method of Scoring and Standards

(1) Scoring. The Marine will be scored by a trained scorer who is thoroughly familiar with the minimum standards for the test events. The Marine will be scored on a pass or fail basis. The performance on each event may be recorded on the reverse side of DA Form 705 (Physical Fitness Testing Record).The examinee records the personal information on the face of the card as specified on the first, second, and third lines. The face of the card is identified by writing diagonally across the lower half of the card "Airborne Trainee PPT—See reverse side."Using line 7 on the reverse side of the scorecard, the scorer enters the test title in the block entitled "(Other (Specify))" and completes the required information on the remainder of that line. The "Remarks" section can be used to record the test event titles and the applicant's performance on each test event.

(2) Standards. To successfully pass the test, the examinee must reach the standard in each test event. The standards follow:

(a) Chinups

(b) Knee bender 80(2-minute period)

(c) Pushups 45(2-minute period)

(d) Situps45(2-minute period)

(e) Endurance runs:2 miles in 15:54 minutes or less in athletic gear and 4 miles in 32 minutes in utilities and running shoes.

c. Uniform for Testing

(1) Examinees. The prescribed uniform for test participation is boots and the work uniform of the season. No headdress is worn. When climatic conditions permit, jackets or outer shirts may be removed.

(2) Officials. Scorers and other test officials should be uniformly and distinctively dressed for contrast with Marines being tested.

d. Test Administration Procedure

(1) Preparation for Test. The administration of the test to a large group makes it mandatory that the test effort be organized and efficiently operated. All testing is not completed with large test groups; on certain occasions, individuals and small groups are tested. Care must be exercised to administer the test uniformly and to standardize the conduct of all elements of the test. Regardless of the size of the test group, the following elements of sound test administration should be included:

(a) An orientation to include the purpose, method of administration, scoring of the test, preparation of the score card, and required standards.

(b) A correct demonstration of each event to ensure that there is no misunderstanding of the proper form and required standards.

(c) Completion of all five test events in one test period with all Marines taking the events in the same sequence.

(d) Adequate rest periods between the test events to allow for recovery before the next event.

(2) Method of Administration with a Large Group. With a minimum of 12 lanes per test event, 14 officials can administer the test battery to 150 or 200 Marines in 2 hours. The officials are designated as follows: one officer in charge, one demonstrator,and 12 scorers. If more or fewer Marines are to be tested, a greater or lesser number of officials will be required. The following procedure is recommended:

(a) Conduct an orientation and ensure the examinees have properly completed their score cards.

(b) Assign Marines to lanes and caution them to remain in the same lane order throughout the test.

(c) Explain and demonstrate the chinup event, administer it, and score it. Then proceed to the knee bender and pushup events, and administer them in the same manner.

(d) Grant a 5- to 10-minute rest period after the pushup event. Advise against excessive consumption of water during the break period.

(e) Explain and demonstrate the situp event, administer it, and score it as prescribed. Then move to the run area, explain the running event, and complete it.

(f) Retain the scorecards at the completion of the running event.

(3) Method of Administration with a Small Group. A similar procedure is followed for the testing of individuals and small groups. The informality usually associated with small groups must not conflict with sound test administration. With fewer examinees, fewer officials are required.

e. Description and Explanation of Test Events

(1) Test Event 1:Chinups

(a) Purpose. This event is devised to test arm and shoulder flexor strength.

(b)Equipment There is one horizontal bar per lane, made of plumber's pipe or a gymnasium horizontal bar 1 1/2 inches in outside diameter. The bar must be rigidly supported at a height of 8 feet above the ground. The upright supports must be 5 feet apart. There must be a movable stand at each bar for short Marines to stand on to reach the bar.

(c) Officials There is one scorer per lane.

(d) Organization. The examinees, holding their scorecards, stand in order behind the restraining line in their respective lanes. The scorers take each scorecard when the Marine is called forward for the test.

(e) Starting Position The bar is grasped with the palms turned toward the face, the thumbs underneath the bar. The body is fully extended in a "dead" hanging position with the arms straight and the feet above the ground.

(f) Movement. Pull the body directly upward until the chin is placed over the bar. Lower the body until the elbows are completely straight and the body is again in the "dead" hanging position. Repeat as many times as required.

(g) Instructions. Explain and demonstrate the fully extended "dead" hanging position with the proper grasp. Show that the chin is placed over the bar at the top of the movement and that the arms are fully extended, the elbows completely straight, at the bottom of the movement (the hanging position). Explain that the body must be kept from swinging and that it is permissible to raise the legs and flex the hips when pulling up, but that any kicking, bicycling, or jerking motion with the trunk or legs is not acceptable. Inform the examinees that no penalty is exacted for hanging on the bar to rest in the bottom position but that this is not to their advantage. Tell them that half-completed chinups are not counted, and that the scorer will repeat the number of the last correct chinup when incorrect execution is detected.

(h) Administration and Scoring. Caution the examinees to assume the "dead" hanging position and wait for the scorer's command to begin. The scorer is at the examinee's left with a clear view of the bar. If the examinee begins to swing widely, the scorer should stop the swinging by extending the left arm across the front of the examinee's body, being sure not to hinder the execution of the chinup. The scorer counts aloud the number of chinups correctly executed. When a chinup is not correctly executed, the scorer repeats the number of the last correct one. The scorer records the number of correct chinups on the scorecard and returns the card to the examinee.

(2) Test Event 2:Knee Bender

(a) Purpose. This event measures the strength and endurance of the leg muscles.

(b) Equipment. None.

(c) Officials There is one scorer per lane.

(d) Organization. Marines stand in numerical order behind the restraining line in their respective lanes. The scorer takes each scorecard when the Marine is called forward for the test.

(e) Starting Position. The feet are spread less than shoulder width apart, hands on hips, thumbs in the small of the back, elbows back.

(f) Movement. Do a knee bend and at the same time bend slightly forward at the waist and thrust the arms between the legs until the extended fingers touch the ground. The hands are about 6 inches apart. The bend is approximately a three-quarters bend. From this knee bend position, recover to the starting position by moving the body upward, straightening the knees, and returning the hands to the hips. Repeat as many times as required.

(g) Instructions. Explain and demonstrate the correct starting position. Be certain examinees understand the correct knee bend and that only the tips of the fingers touch the ground. Tell them the scorer will repeat the number of the last correct knee bender when incorrect execution is detected. Some of the common errors are failure to correctly bend the knees, failure to touch the ground, and failure to assume the completely erect position after the bend has been executed.

(h) Administration and Scoring. The scorer stands to one side so as to see that the knees are properly bent and the fingers touch the ground as prescribed. From this position, the scorer can view the examinee to see that a properly erect position is assumed after each knee bend. The scorer counts aloud the number of correctly executed knee bends. When a knee bend is done incorrectly, the scorer repeats the number of the last correct one. The scorer records the number of correct knee bends on the scorecard and returns it to the examinee.

(3) Test Event 3:Pushups

(a) Purpose. Pushups measure arm and shoulder extensor strength.

(b) Equipment. None.

(c) Official. There is one scorer per lane.

(d) Organization. Marines stand behind the restraining line in their respective lanes until the scorer calls on them to perform. The scorer takes each scorecard when the Marine comes forward.

(e) Starting Position. The front-leaning rest position is the starting position. The body is straight from head to heels, palms are flat on the ground directly underneath the shoulders, and elbows are straight and locked. The body weight is supported on the hands and toes throughout the event.

(f) Movement. Bending only the elbows, lower the body in one straight plane until the chest touches the scorer's hand. Straightening and locking the elbows, raise the body in one straight plane, returning to the original front-leaning rest position. Repeat as many times as required, keeping the body in a straight line from head to heels.

(g) Instructions. Explain and demonstrate that the arms are straight at the beginning and completion of the movement and that the chest must touch the scorer's hand, but the stomach and thighs must not touch the ground. Also explain that the whole body must be maintained in a straight line as it is lowered and raised; that is, there is to be no breaking at the hips or shoulders so that any body part is lowered or raised in advance of the other or as a separate segment .Dipping or rolling through the shoulders is illegal, as is lowering or raising the body with one arm or shoulder at any time. Resting is not permitted during repetitions. Instruct

Marines that the scorer will repeat the number of the last correct pushup when incorrect execution is detected.

(h) Administration and Scoring. It is recommended that Marines assume a prone position while placing their feet and hands in the proper positions. This permits them to rest while the scorer gets into position and, at the same time, provides a feel of the body in a straight plane from head to heels. The scorer lies on the right hip and side to the right of the examinee. This gives the scorer a clear view of the examinee's body to see any errors. The palm of the scorer's right hand rests flat on the ground underneath the lowest part of the examinee's chest. By keeping the right forearm flat on the ground at an angle in front of the examinee's right arm, the scorer's position will not prevent the examinee from lowering the body completely. The scorer's left hand is free to test the straightening of the elbow at the completion of the movements and to point out body segments being lowered or raised separately. When in position and ready, the scorer has the examinee assume the starting position and begin doing pushups. The scorer counts aloud the repetitions done correctly and repeats the number of the last correct pushup if one is done incorrectly. There is no penalty if the contour of the examinee's body causes the hips to protrude slightly out of line, provided that the whole body is raised and lowered simultaneously. The scorer enters the number of repetitions on the scorecard and returns it to the examinee.

(4) Test Event 4: Situps

(a) Purpose. This event primarily measures abdominal strength.

(b) Equipment. None.

(c) Official. There is one scorer per lane.

(d) Organization Marines stand behind the restraining line in their respective lanes until the scorer calls on them to perform. The scorer takes each scorecard when the Marine comes forward.

(e) Starting Position. The examinee lies flat on the back with knees flexed, both feet flat on the ground. The correct angle of the thighs to the ground is 45 degrees. If the heels are too near the buttocks, the applicant will not be able to sit up. Legs are spread shoulder width apart. The examinee interlaces fingers and places them behind the head in contact with the ground. The feet are not held by another person.

(f) Movement. Bend forward at the waist and raise the upper body until the head is directly over the knees. Heels are not to leave the ground. Elbows remain in the same plane to the head and body throughout the event. The upper body is slowly lowered to the starting position until the head touches the ground. Repetitions are done at a slow cadence with no rest periods.

(g) Instructions. Explain and demonstrate the correct starting position and the proper execution of the situps to be sure that Marines understand the movement. Warn them that their knees must remain flexed during each situp, the heels cannot leave the ground at any time, and they may not roll up on one side and push up with one elbow. Tell them they must do the repetitions at a slow cadence with no rest periods. Instruct Marines that the scorer will repeat the number of the last correct situp when incorrect execution is detected.

(h) Administration and Scoring. When the performer is in position and ready, the scorer has the performer assume the starting position and begin doing situps. The scorer counts aloud the correct executions. When a situp is improperly done, the scorer repeats the number of the last correct one. No situp is credited if the hands are unclasped from behind the head, if the back is used to bounce up from the ground (which means the shoulders would not touch the ground), or if one shoulder or elbow is used to push up. The examinee is not penalized if the heels slide forward slightly so long as the knees remain flexed and the heels maintain contact with the ground. The scorer enters the number of repetitions on the scorecard and returns it to the examinee.

(5) Test Event 5:Endurance Run

(a) Purpose. This event measures cardiovascular endurance.

(b) Equipment One stopwatch or watch with a sweep second hand.

(c) Area. A large training field on which a quarter-mile track has been staked out or a level road over flat terrain may be used as a running surface. A 1-mile route is designated with wooden stakes marking the start point, finish point, and one-quarter mile intervals.

(d) Officials For large groups, a scorer times the event and controls the conduct of the run, and a guide runs with the group and sets the pace.

(e) Organization The run is conducted with groups of Marines in a column formation. Company-sized units may run at the same time with platoons serving as running groups. The scorer issues the command to assume double-time.

(f) Starting Position. Marines are assembled in proper column formation (column of twos, threes, or fours, as appropriate to the size of the group) with short men to the front. When all is ready, the column is moved forward a short distance before the running period is started.

(g) Movement. At the command DOUBLE-TIME, MARCH, the examinees retain their places in the column formation and execute the command. Length of steps is about 40 inches. The scorer has the group execute the run. The formation is maintained during therun.

(h) Instructions. Marines are instructed to maintain formation while running and are informed that the guide will set the proper pace. They will be instructed in the command used to control the column in the execution of the test. The scorer should announce the 4-minute, 2-minute, 1-minute, and 30-second remaining time intervals.

(i) Administration and Scoring. The event may be administered as previously prescribed to a large group, to several Marines, or to an individual. An individual examinee usually does not require a guide or pacer. If the event is administered on a training field, the scorer may stand in the center as the runner(s) circles about the field. This method of administration relieves the scorer of running with each group to be tested. Scoring is based on successful completion of the run as prescribed. The scorer should announce the remaining times as prescribed in (h).

6009. BATTLE FITNESS TEST

The battle fitness test is a physical fitness evaluation which is currently under development as part of the basic warrior training concept plan.

Chapter 7
THE HUMAN BODY

7001.GENERAL

To implement an effective physical conditioning program, a leader must understand how the human body functions and how exercise affects the body .This chapter provides only an introductory discussion of anatomy and functioning.

a. Body Functioning During the Stages of Conditioning. For more on the stages of conditioning, see chapter 1.

(1) During the toughening stage, the waste products of muscle activity (lactic acids) collect more rapidly than the blood can remove them. This acid waste builds up in the muscle tissue and irritates the nerves in the muscle fiber, causing pain and stiffness. As the exercise program continues, more blood is carried through the muscle, removing the waste materials more rapidly and eventually causing the soreness to disappear.

(2) During the slow improvement stage, the blood circulation in the muscles increases, and the body as a whole becomes a more efficient machine. The improvement is rapid in the first few weeks, but as a higher level of skill and conditioning is reached, the improvement becomes less noticeable. The body reaches its maximum level of performance after 6 to 10 weeks and should then be maintained at this peak.

(3) During the sustaining stage, it is possible to maintain this state of conditioning through 15 to 20 minutes of exercise a day, but the exercise must be quite strenuous.

b. Diet. Regular exercise increases the appetite. If the desire for greater amounts of food is satisfied by a balanced diet, the body benefits.

(1) There are two main types of foods: body-building and energy- producing.

(a) Body-building foods consist of proteins, which build up tissue and repair wear and tear.

(b) Energy-producing foods include carbohydrates and fats. Carbohydrates provide a quick source of energy, while fats act as a reserve store of energy.

(2) In addition, food contains vitamins, mineral salts, and water. During hot weather and strenuous training periods, the body requires greater fluid intake.

(3) Diet should be supplemented with proper rest to allow the digestive system to digest the food.

(4) Occasionally, especially during early stages of conditioning, strenuous exercise may cause vomiting. Although vomiting is not a frequent occurrence, it is not usually a cause for concern.

7002. SYSTEMS OF THE BODY

The systems of the body include the skeletal, muscular, circulatory, respiratory, endocrine, digestive, genitourinary, and nervous systems. Each has a different function, but all must work in cooperation with one another to insure a sound body. Of these systems, the first four are the most affected by exercise and are discussed in succeeding paragraphs.

7003. THE SKELETON

a. Bones. The skeleton (see fig. 7-1) is composed of about 206 bones, which-
o Provide structure for the body and support for the attachment of muscles,
o Protect vital organs, such as the brain, lungs, and heart,
o Manufacture red blood cells, which carry oxygen through the body.

Figure 7-1 is available in the illustration section at the end of the book.

b. Joints. A joint is a place of union between two or more bones. Joints can be-
o Immovable, such as in the face and head.
o Slightly movable, such as in the vertebrae or sternum.
o Freely movable, such as in the shoulder, hip, knee, ankle, and elbow. The bones in a freely movable joint are held in place by ligaments. Freely movable joints are of the greatest importance in physical training because exercise improves their mobility and stability.

c. Cartilage and Ligaments. The joints in the body, where bones connect, are supported by cartilage and ligaments.
o Cartilage is a tough, elastic, translucent tissue that acts as a shock absorber or buffer between bones. Examples are the discs between the vertebrae, the tissue attaching the ribs to the sternum, and buffers in the knee joints. Damaged cartilage does not heal.
o Ligaments are connective tissue that bind bones together. They contract and extend but are not elastic. Consequently, ligaments can be torn or strained. Damaged ligaments will heal to some extent.

7004. THE MUSCLES

a. General. Muscles are tissue, or an organ composed of tissue, which contract or extend to produce movement. Muscles are of three classifications: involuntary, voluntary, and cardiac. The voluntary and cardiac muscles are the most affected by physical training; physical training has little effect on involuntary muscles.(Seefigs. 7-2, 7-3, 7-4.)
o People have no control over the involuntary muscles, an example of which is the diaphragm,
o Voluntary muscles are the larger skeletal muscles which cause movement and which the individual can control.
o Cardiac muscle is found only in the heart and for all practical purposes is an involuntary muscle. Physical training has a significant effect on the fitness of cardiac muscles.

b. Muscle Structure. Muscle is composed of cells which are grouped into bundles called fibers. Groups of fibers form muscles. These bundles of fibers are held in place by sarcolemma, a thin, sheath-like material that surrounds the muscle bundles and secretes a fluid that lubricates the muscle tissue. The fused ends of the sarcolemma form the tendons which attach the muscles to the bones.

c. Attachment of Muscles. The arrangement of muscles on the skeleton provides the proper angle of pull to make movement possible. Voluntary muscles are usually attached to the skeleton in two places.

d.Action of Muscles. To produce motion and to do work a muscle usually shortens its fibers. The movement may be flexion such as bending the arm at the elbow, or extension, such as the straightening the arm. Some muscles raise the arms or legs, others depress the raised limb. Some muscles have the primary function of rotating the trunk from side to side. In addition, muscles act as stabilizers as well as prime movers.

Figure 7-2 is available in the illustration section at the end of the book.

Figure 7-3 is available in the illustration section at the end of the book.

Figure 7-4 is available in the illustration section at the end of the book.

7005. FUNCTIONING OF THE SKELETON AND MUSCLES

a. General. The skeleton and muscles function in close coordination. Muscles move the body, and bones support the muscles.

b. Effect of Exercise on Bones. Continuous exercise, particularly among younger people, usually strengthens the bones, allowing them to withstand greater stress and strain. Bones which are not used lose much of their strength, a factor which should be considered when individuals return to physical training after a prolonged period of inactivity. People in this category should refrain from activities which might result in bone breakage before the bone is exercised back to normal condition. The condition known as "march fracture" is usually a result of this condition.

c. Muscular Strength. When exercised vigorously to improve strength, muscle grows in size. In general, the larger the muscle, the stronger the muscle. Furthermore, trained muscle functions more efficiently than untrained muscle. Trained muscles contract more vigorously and with less effort. Regular and strenuous exercise also toughens muscle, making it firmer and able to stand more strain.

d. **Muscular Endurance.** Muscles naturally become fatigued during continuous, repeated work. Through specific exercises, such as chinups or situps, local muscle groups become exhausted long before an individual fatigues. Training to develop muscular endurance enables people to continue a relatively heavy load of exercise over a long period of time. Lengthened exercise periods improve muscular endurance.

e. Muscular Coordination. Improved speed and strength result in part from improved muscular coordination. An unskilled individual may use irrelevant muscles to perform a particular activity, increasing the amount of work without increasing the mechanical output. This increase in skill is highly desirable, but it should be offset by greater effort in duration or intensity to compensate for the loss in overload due to increased skill.

f. Muscular Fatigue. When the rate of work is increased, the energy required is proportionately much greater than the increase in rate. For example, if an individual doubles running speed, the amount of power demanded to do this is increased eight times.

g. Circulation in Muscles Regular, strenuous exercise causes the creation of new capillaries and the opening of inactive, latent capillaries, increasing blood circulation within the muscle by as much as 400 percent. This increases the supply of food materials and oxygen to the muscle, improving its endurance. This process takes about 8 to 12 weeks of regular conditioning in young adults and longer as age increases.

7006. THE CIRCULATORY AND RESPIRATORY SYSTEMS

a. The Circulatory System. The circulatory system transports blood to all parts of the body, removes waste products for disposal, and delivers protecting and repairing substances where needed. The circulatory system includes the heart and blood vessels.

(1) The Heart. The heart is a pump which forces blood through the blood vessels to the parts of the body. The heart is a little larger than a fist and is located in the left center of the chest between the lungs.

(2) Blood Vessels. The vessels carrying blood away from the heart are the arteries. They eventually divide into capillaries, the very small vessels through which diffusion and osmosis take place. The capillaries gradually increase in size, forming the veins which carry blood back to the heart.

b. The Respiratory System. The respiratory system performs the function of breathing, whereby oxygen is inhaled and carbon dioxide is exhaled. The respiratory system consists of the mouth, nose, trachea, lungs, and diaphragm.

o Trachea. The trachea, or windpipe, is a hollow, tube-like structure that carries air from the mouth to the lungs.

o Lungs. The lungs are elastic bags in the chest where the exchange of oxygen and carbon dioxide takes place.

o Diaphragm. The diaphragm is a thin, sheet-like muscle just below the lungs. During inhalation, the diaphragm flattens out and lowers, allowing the lungs to expand and fill with air. During exhalation, the diaphragm raises into a dome shape, helping to expel air from the lungs.

7007. CARDIOVASCULAR FUNCTIONING

a. General. Cardiovascular functioning is the combined functioning of the circulatory and respiratory systems. The chief organs of these systems, the heart and lungs, function together to provide oxygen to the body.(See fig. 7-5.)

Figure 7-5 is available in the illustration section at the end of the book.

b. Heart Action. The heart, the organ which propels blood through the blood vessels, is the chief organ of cardiovascular endurance. A person tires quickly if the heart lacks the capacity to circulate the blood through the body. The heart is a muscular organ and is developed by exercise just like any other muscle. The heart cannot be exercised alone; any exercise that puts a sufficient load on the heart, such as running or cycling, will also exercise other body parts. A conditioned heart tends to beat more slowly but pump more blood when at rest. This is known as an increase in "stroke volume,"a desirable condition because it enables the heart to pump more blood with a slower contraction rate.

c. Functions of the Lungs. Strenuous and regular exercise can improve the efficiency with which the lungs transmit oxygen to the blood by as much as 25 percent In a poorly conditioned person, some of the alveoli (air sacs) within the lungs are closed or collapsed. The forced breathing created by exercise over a period of weeks causes the air sacs to slowly expand, increasing the ability to absorb oxygen.

d. Relationship of Heart and Lungs. The heart pumps carbon dioxide-laden blood through the pulmonary artery to the lungs. In the lungs the carbon dioxide is exchanged for oxygen and the purified blood is returned to the heart by way of the pulmonary vein. The heart then pumps the blood through the aorta for circulation throughout the body. Blood moving into the muscles exchanges oxygen for carbon dioxide, after which time it is ready for the return trip to the heart.

e. Cardiovascular Functioning in High Altitudes. Marines to be employed in high altitudes should be acclimated in a similar area for 10 to 14 days prior to employment. Persons not accustomed to the rarefied air of higher altitudes tire more quickly and may collapse after rapid physical exertion. Air is much less dense at high altitudes than at sea level. Persons accustomed to sea level inhale only about 80 percent of the oxygen they are used to no matter how hard they breathe. Further, people accustomed to sea level or moderate altitude simply do not have enough red corpuscles in the blood to fulfill their needs at high altitudes. The red blood cells will increase over time.

f. Symptoms of Cardiovascular Malfunctioning. The brain is the first organ to react to a lack of oxygen. Unconsciousness results when the brain is denied sufficient oxygen. "Blacking out" is actually a defense mechanism to keep the body alive.

APPENDIX A
INSTRUCTOR HINTS AND INSTRUCTOR TRAINING

1. General. This appendix informs instructors on administration of exercise periods, commands, the extended rectangular and circle formations, methods of instruction and conduct of exercise activities, explanation of basic positions associated with the various activities, and instructor training.

2. Aim of the Physical Fitness Program. The aim of the physical fitness program is to prepare Marines physically for war. This can be conveniently split into two phases.

a. Preparation Phase

(1) Development of all-around physical fitness.

(2) Development of purposeful physical skills.

(3) Development of mental alertness.

(4) Development of character and leadership.

b. Maintenance Phase. Once the preparation is completed, the program must progress to a maintenance program. This phase revolves around the need to maintain the trained Marine's physical fitness for a particular role. This will include—

(1) Maintenance of a high standard of all-around physical fitness.

(2) Application of purposeful physical skills to combat skills.

(3) Development of leadership and self-confidence.

(4) Development of mental toughness and endurance.

3. Physical and Mental Development

a. Marines need to be developed physically and mentally. The physical training (PT) instructor must ensure that the program accomplishes this.

b. The system of the body functions best in response to progressive exercise. This means that Marines should master the easier exercises before they attempt the more difficult ones.

c. The body and mind are linked inseparably and must be considered as one in physical fitness training. Physical exercises assist in developing the mind.

d. The confidence in oneself and in one's comrades attained by physical achievement helps to promote the indomitable will to win. The Marine must be taught to realize this. It will help during arduous training, which at all times must be imaginative, realistic, and filled with enthusiasm. All physical fitness training must be associated with developing the qualities required of a Marine.

4. Planning and Preparation of ProgramsThe following factors must be considered when compiling a program (see also ch. 2)~

a. Aim or desired skill level.

b. Duration of the complete training program and number of physical training lessons.

c. Purpose of included exercises,drills, or practical applications.

d. Beginning/inventory fitness levels.

e. Medical facilities, medical personnel, and Marine Corps or local command regulations governing their proximity to the training.
f. Number of Marines to be trained.
g. Availability of instructors, assistants, and facilities.
h. Apparatus and equipment available.
i. Alternative training areas for varying weather conditions,
j. Stages of progression and tests of achievement.

5. Five Basic Principles of Program Design
a. Regularity. Regularity of exercise is far more important than the amount of exercise performed on a given day. A Marine should exercise a minimum of three times per week. Human muscle begins to atrophy after 3 days without exercise.
b. Balance. The program must be balanced in terms of total physical fitness. Imbalance is created when, for example, an individual merely runs. While the cardiovascular system is improved, general upper body strength may be lacking. Conversely, a person who just does weight training may develop strength, but not cardiovascular endurance.
c. Overload. If total body fitness is to be developed, then the muscular and cardiovascular systems must be made to perform more work than previously. This is done by increasing the frequency, intensity, and duration of the exercise. The body adapts to the amount of stress placed upon it and becomes stronger and more efficient.
d. Progression. To be successful, any type of training program must be progressive. This is an area where untrained persons conducting PT can cause injury to Marines resulting in failure of the PT program. Close supervision must be made of untrained instructors. The result of progressing too fast is that the class begins to reject PT because it is no longer enjoyable or motivating. The program should start at a slow pace, and the pace should increase as proficiency improves. Average persons will encounter three stages of progressiom the toughening stage, the slow improvement stage, and the sustaining stage. (See ch. 1.)
e. Variety. Some programs fail because the training becomes boring. Variety is essential and perhaps presents the greatest challenge to the commander and staff.

6. Grouping. Providing for different levels of physical fitness is particularly recommended in the early stages of conditioning. Older Marines and those in poorer physical condition should be expected to attain a group level of fitness, but they should be given more time to do it.
a. Homogeneous Grouping One simple method of providing for the difference in levels is to group them according to their condition. A two-group classification would divide Marines into highly conditioned and average groups. A finer classification could be obtained by dividing them into three groups—a highly conditioned, a moderately conditioned, and an unconditioned group.
b. Performance or Age Grouping. The segregation of Marines into different exercise groups should be based on physical fitness test scores or on the level of fitness they demonstrate. They may also be grouped at first according to age. A common classification by ages is under 30, 30 to 34, and 35 and above.

7. Preparatory Commands and Commands of Execution. The prepreparatory command describes and specifies what is required. The command of execution calls into action what has been prescribed. All preparatory commands are given with a rising inflection. The interval between commands is long enough to permit the average Marine to understand the first one before the second one is given.

8. Extended Rectangular Formation. The traditional formation for carrying on many physical training activities is the extended rectangular formation (fig. A-1). This formation is more compact than any other. It is the best type to use for large numbers of Marines because it is simple and easy to assume.

a. For the formation of oneplatoon, the base man paces off five paces from the stand, faces left and moves five paces, halts, and again faces left. With the base man positioned facing the stand, the platoon leader then commands. **FALL OUT AND FALL IN ON THE BASE MAN**. At this command, all Marines run to the designated area and reform. This procedure is preferred to marching the unit into position. If more control is desired, the unit may march at double time to the vicinity of the base man and then be directed to fall out and fall in on him. Much valuable time is wasted in the field by needless maneuvering of Marines at quick time in an effort to position the platoon or unit on the exact spot for the exercises.

b. A company-size unit assumes the extended rectangular formation from a column of threes or fours at normal intervals between squads. This extension can also be executed from a company mass without interval between platoons. In extending either a platoon or company-size unit, take your place at the head of the column and command:
(1) TAKE INTERVAL TO THE LEFT, MARCH. At this command, the Marines in the right flank file stand fast with arms extended sideward. All others turn to the left and run forward at double time. After taking a sufficient number of steps, all Marines face the front with both arms extended sideward. The distance between finger-tips is about 12 inches, and dress is right.

Figure A-1 is available in the illustration section at the end of the book.

(2) ARMS DOWNWARD, MOVE. At this command, the arms are lowered smartly to the sides.
(3) LEFT, FACE
(4) EXTEND TO THE LEFT, MARCH. At this command, those in the right flank file stand fast with arms extended sideward. All others turn to the left and run forward at double time. Spacing is the same as in (1) above, and dress is right.
(5) ARMS DOWNWARD, MOVE. Same as in (2) above.
(6) RIGHT, FACE.
(7) FROM FRONT TO REAR, COUNT OFF. At this command, the leading Marine in each column turns the head to the right rear, calls of **ONE** and faces the front. Successive Marines in each column call off in turn **TWO, THREE, FOUR, FIVE** in the same manner.
(8) EVEN NUMBERS TO THE LEFT, UNCOVER. At this command, each even-numbered Marine stride jumps sideward to the left, squarely in the center of the interval. In doing this, each one swings the left leg sideward, jumps from the right foot to the left foot, and smartly brings the right into position against the left.

c. To assemble the unit, you commandASSEMBLE TO THE RIGHT, MARCHAt this command, all return to their original position in the column at double time and reform on the base man.

d.It is recommended that the area for grounding equipment and arms be at the edge of, or well away from, the area to be used for exercising. To conserve time and ensure proper position of the unit, the base man or, if the unit is composed of several platoon-sizegroups, the various base men may precede the unit and establish their positions in relation to the instructor's stand.

9. Circle Formation.The circle formation is effective for conducting various exercise activities (fig. A-2). This formation has an advantage in that supervision of all Marines is facilitated and a moving formation is available which provides control. Guerrilla exercises, grass drills, and some forms of running are examples of activities which are more easily conducted in the circle formation than in the extended rectangular formation.

a. It is not advisable to have more than 60 Marines in a circle. When more must be accommodated, separate circles should be used.Concentric circles may be employed to reduce the size of the circle or to accommodate more persons. If concentric circles are formed, the different circles are made by designating squads for each circle.Each additional circle requires more Marines than the one inside it.For example, one squad of a platoon may form the inner circle and theremaimng three squads the outer circle.

b. When a platoon is to form a circle, the commands ai€IRCLEFORMATION, MARCH, FOLLOW MEUpon this command, the right flank squad of the column moves forward at double time with the leader of the platoon group gradually forming a circle in a counterclockwise direction. Each succeeding file falls in behind that on the right. After the rough outline of the circle is formed, the leader commandsPICK UP A **5**-YARD INTERVAL.

Figure A-2 is available in the illustration section at the end of the book.

c.The group may be halted and faced toward the center, or, if instruction is not necessary, the exercise activity may be executed without stopping the platoon.

lO.Leadership Techniques

a. Unless you experience all the exercises, you cannot appreciate how arduous they are, what movements are most strenuous and difficult, where the errors in performance are likely to occur, and what the proper cadence should be.

b. You must give everyone careful supervision and participate in the exercises to show that you can do them. When you participate, your assistant instructors should supervise because it is difficult for you to supervise and exercise at the same time.

c. Marines should never be kept too long in one position, especially a constrained one. They should never have to perform an exercise more times than they can do it without losing the proper form. You should insist upon proper form in the execution of all exercise activities.Even slight deviations from the proper form reduce the value of the exercise.

d. Avoid long explanations.As a rule, it should be necessary to give a full explanation of new exercises only. Explain the most essential features of an exercise first; add details later. Too many details at one time are more likely to confuse than to assist them.Minor corrections should be made to the entire class while the exercise is in progress (for exampHEADS **UP,** KNEES STRAIGHT). If necessary, follow this direction by the name of the person who is particularly at fault. If a Marine requires special attention, givethat person separate instruction by an assistant instructor to avoidwasting the time of the group.

e. The heavy demand on your voice can be lightened by training assistant instructors to assume some of the instruction. Using mass cadence is also an effective method of lessening the demand on your voice.

“Illustrations”

DAY	TIME	ACTIVITY
MONDAY	60 MIN	RUNNING ACTIVITY
TUESDAY	60 MIN	COMBATIVES
WEDNESDAY	60 MIN	CONDITIONING & LOG DRILLS
THURSDAY	60 MIN	CONDITIONING & RELAYS
FRIDAY	60 MIN	RUNNING & CONDITIONING
SATURDAY	60 MIN	RUNNING ACTIVITY

Figure 2-1. Sample A- -Single Activity.

DAY	TIME	ACTIVITY	REMARKS
MONDAY	20 MIN 20 MIN 20 MIN	RIFLE DRILL COMBATIVES RUNNING	PROGRESSIVE ACTIVITY SYSTEM
TUESDAY	30 MIN	GUERRILLA EXERCISES TABLE I	SINGLE ACTIVITY
WEDNESDAY	30 MIN	RIFLE DRILL	SINGLE ACTIVITY
THURSDAY	30 MIN	CIRCUIT INTERVAL TRAINING	SINGLE ACTIVITY
FRIDAY	20 MIN 20 MIN 20 MIN	STRENGTH CIRCUIT CONDITIONING DRILL CROSS-COUNTRY RUN	ROTATION ACTIVITY
SATURDAY	60 MIN	WARM UP RUN AND GRASS DRILL	SINGLE ACTIVITY

Figure 2-2. Sample B- -Various Time Periods, Activities, and Systems.

DAY	TIME	ACTIVITY
MONDAY	45 MIN	RUNNING – CROSS-COUNTRY
TUESDAY	75 MIN	TEAM ATHLETICS – TOUCH FOOTBALL
WEDNESDAY	45 MIN	RUNNING – CROSS-COUNTRY
THURSDAY	20 MIN 40 MIN	LOG DRILL DUAL COMBATIVES
FRIDAY	90 MIN	SPEED MARCH
SATURDAY	90 MIN	TEAM CONTESTS

Figure 2-3. Sample C--Sustaining Activities.

Route	Kilometers from SP	Recommended rate of march (kmph)	Remarks
SP: RJ 8th Div Rd—Superhighway	......	4	Bridge; hard surface; two guides.
RJ Jamestown—Hersey Rd	2.25	4	Traffic heavy; two guides; Jamestown Rd bears to right.
RJ Jamestown—Yankee Rd	7.10	4	Two guides
RJ Jamestown—Lightning Rd	10.50	4	Two guides
RJ Jamestown—Sunshine Rd	12.40	4	Light traffic; one guide Sunshine Rd; gravel, poor traction when wet.
RP; RJ Sunshine—Sedan Rd	14.80	4	Two guides

Figure 3-1.Example of Route Reconnaissance Report.

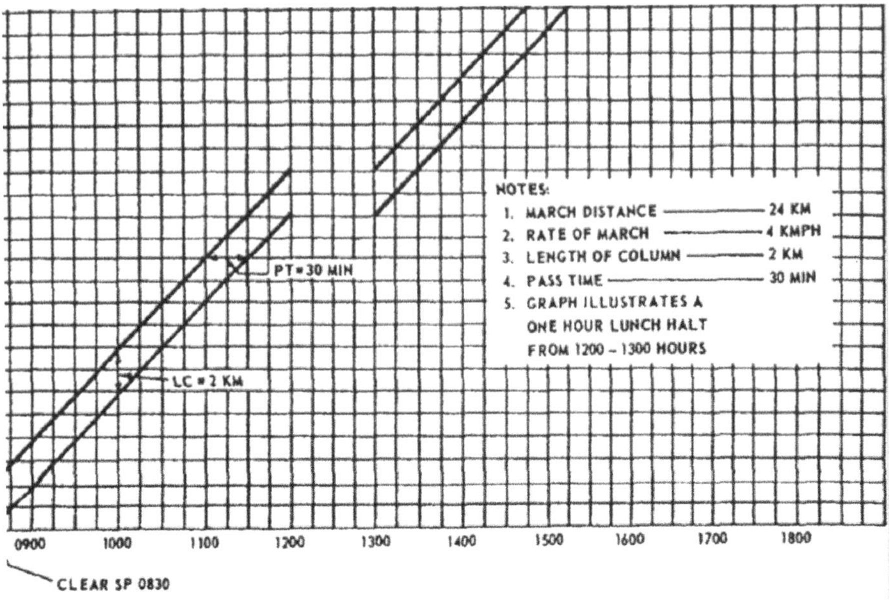

Figure 3-2.Road Movement Graph.

Figure 3-3.The Pace-Setter.

	SINGLE FILE	COLUMN OF TWO
2 m/MAN APART	2 4	1 2
5 m/MAN APART	5 4	2 7

GAP DISTANCES			
SERIAL:	(3-1) × 100 m	=	200 m
MARCH UNIT:	(12-1-2-) × 50 m	=	450 m
	TOTAL COLUMN GAP		650 m

Figure 3-4.Conditioning Drill One.

Figure 3-5.Conditioning Drill Two.

Figure 3-6.Conditioning Drill Three.

Figure 3-7. Rifle Drill.

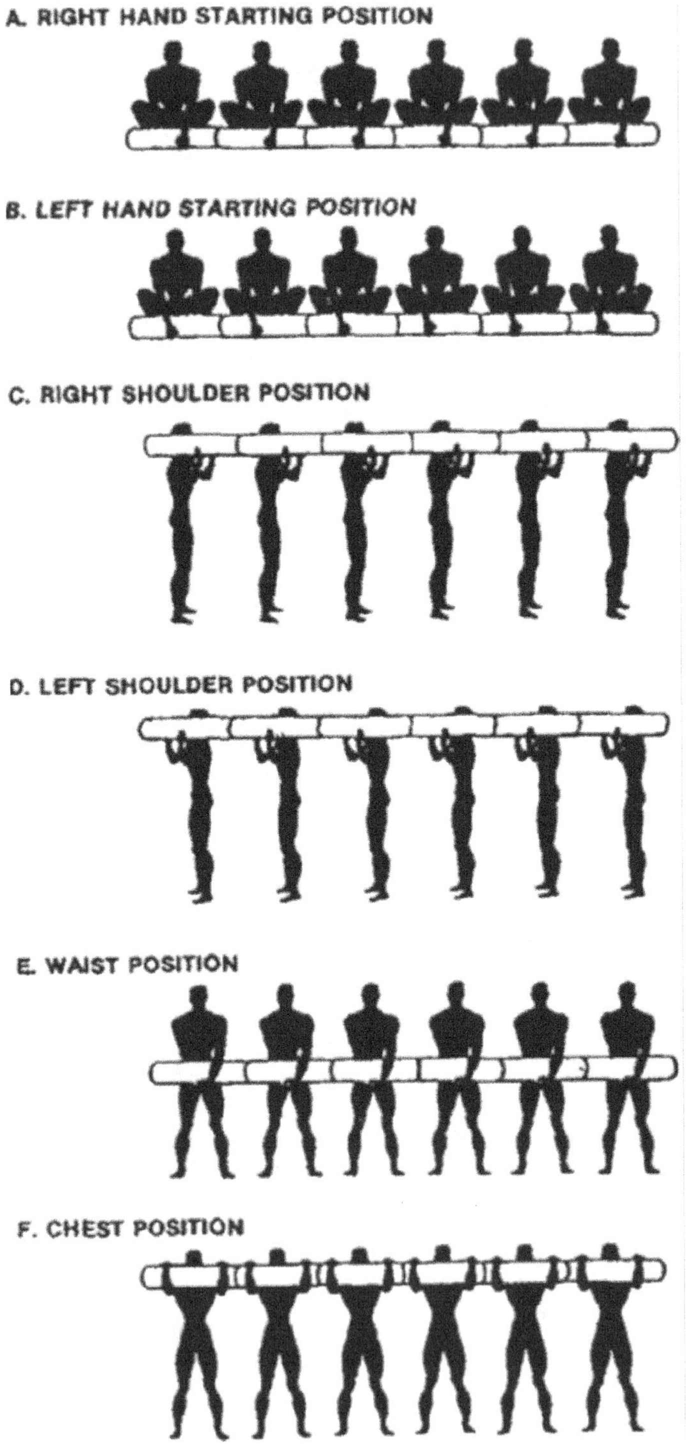

Figure 3-8. Starting Positions.

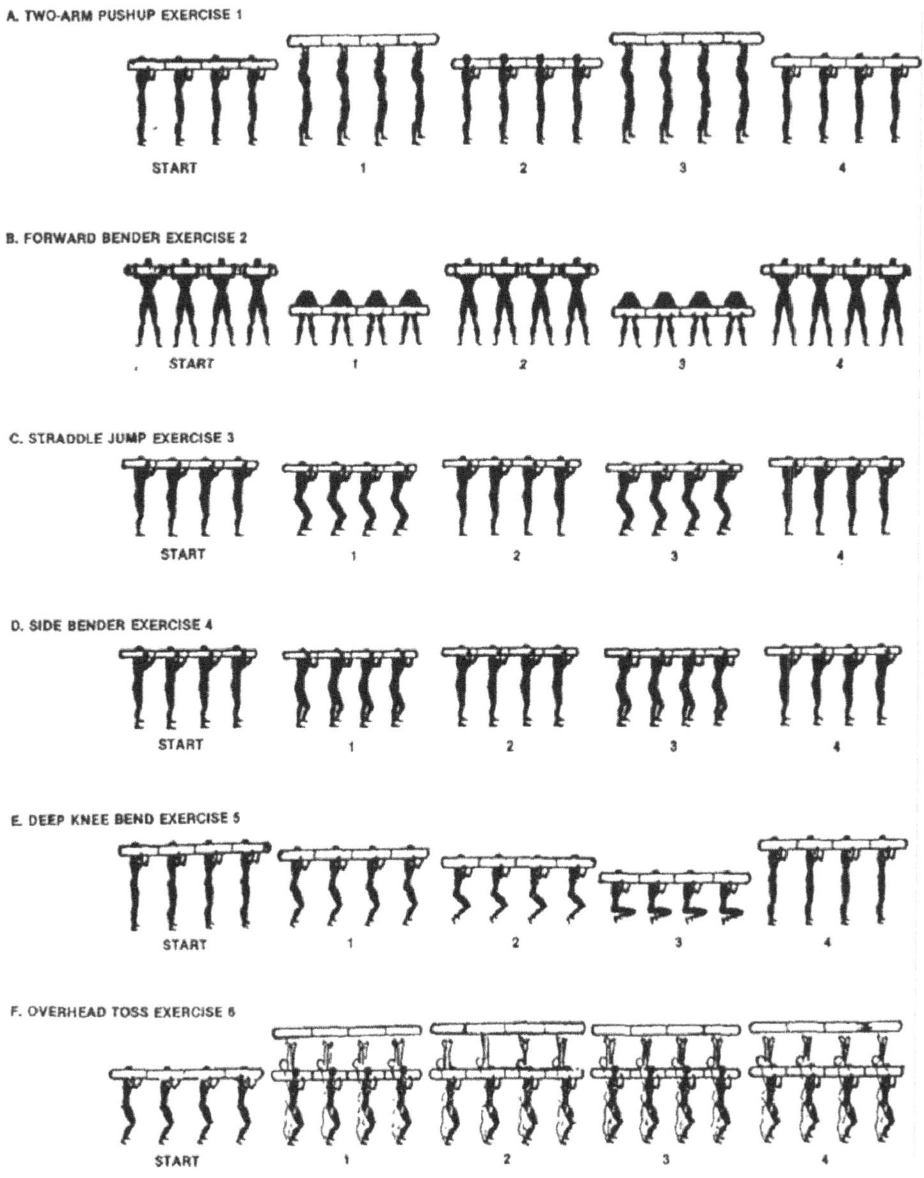

Figure 3-9. Log Drill

Figure 3-10.Basic Positions for Grass Drill.

Figure 3-11.Grass Drill One.

Figure 3-12.Grass Drill Two.

Figure 3-13.Guerrilla Table I.

Figure 3-14.Guerrilla Table II.

Figure 3-15.Proper Running Form.

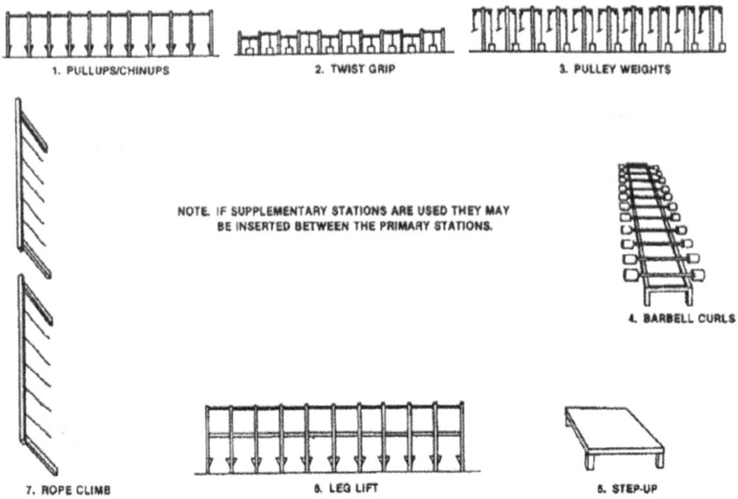

Figure 3-16. Fixed Strength Circuit.

STATION	ITEM	NO.	SPECIFICATION
1	BARBELL		1¼-INCH PIPE 5 FEET LONG WITH CONCRETE FILLED NO. 10 CANS
2	JUMP ROPE		¼- OR ⅜-INCH ROPE, 10 FEET LONG
3	TWIST GRIP		HANDLE 12 INCHES LONG, ROPE 4 FEET LONG, NO. 10 CAN CONCRETE FILLED.
4	INCLINE PLANE		¾-INCH PLYWOOD PLATFORM 2 FEET WIDE AND 6 FEET, 6 INCHES LONG, ELEVATED 10 INCHES AT ONE END. STRAP TO HOLD FEET DOWN.
5	WAR CLUB		HEAD IS 6 BY 12 INCHES, HANDLE IS 14 INCHES LONG BY 1¼ INCHES IN DIAMETER, ABOUT 20 POUNDS.
6	BICYCLE RIDE		PLYWOOD BOARD OR PLATFORM 2 BY 3 FEET WITH 2 BY 2 RUNNERS.
7	STEP – UP		A BOX OR STURDY PLATFORM 18 INCHES HIGH, 18 INCHES WIDE, 24 INCHES LONG.
8	ISOMETRIC PULL		TWO HANDLES 12 INCHES LONG WITH 4 FEET (BETWEEN HANDLES) OR LIGHT WIRE CABLE OR ¼-INCH ROPE

Figure 3-17.Movable Strength Circuit.

Figure. 3-18.Stations of Movable Strength Circuit.

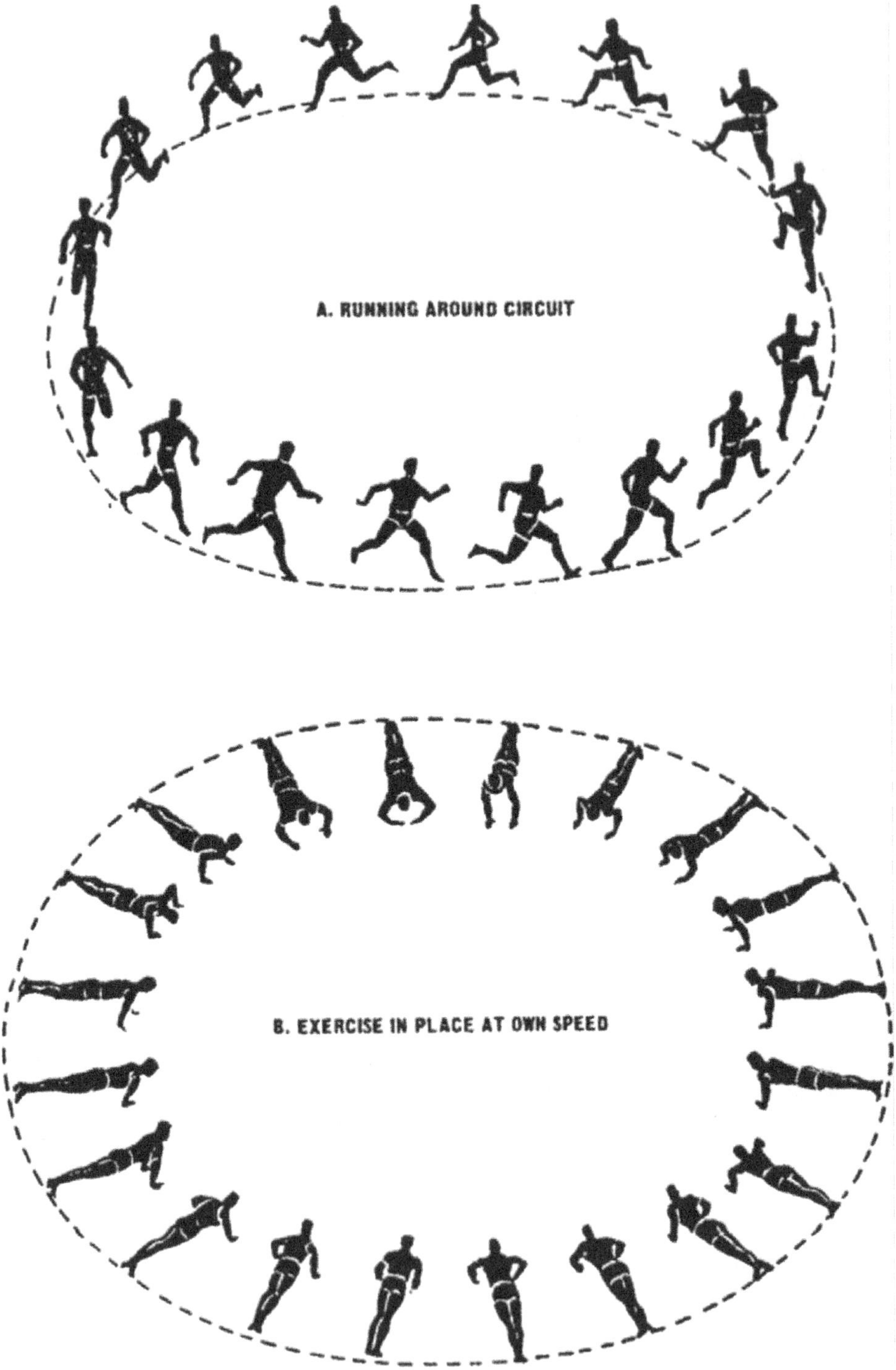

Figure 3-19. Circuit-IntervalTable.

Figure 3-20. Jumping.

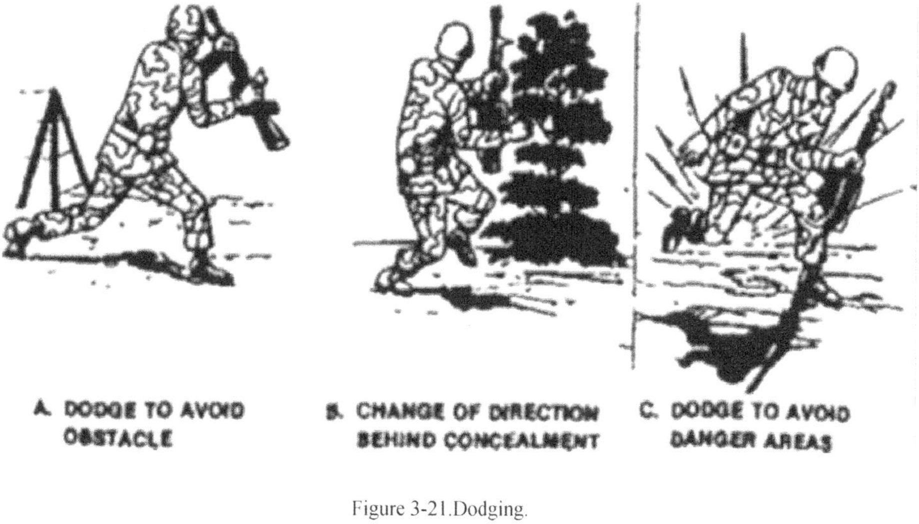

Figure 3-21.Dodging.

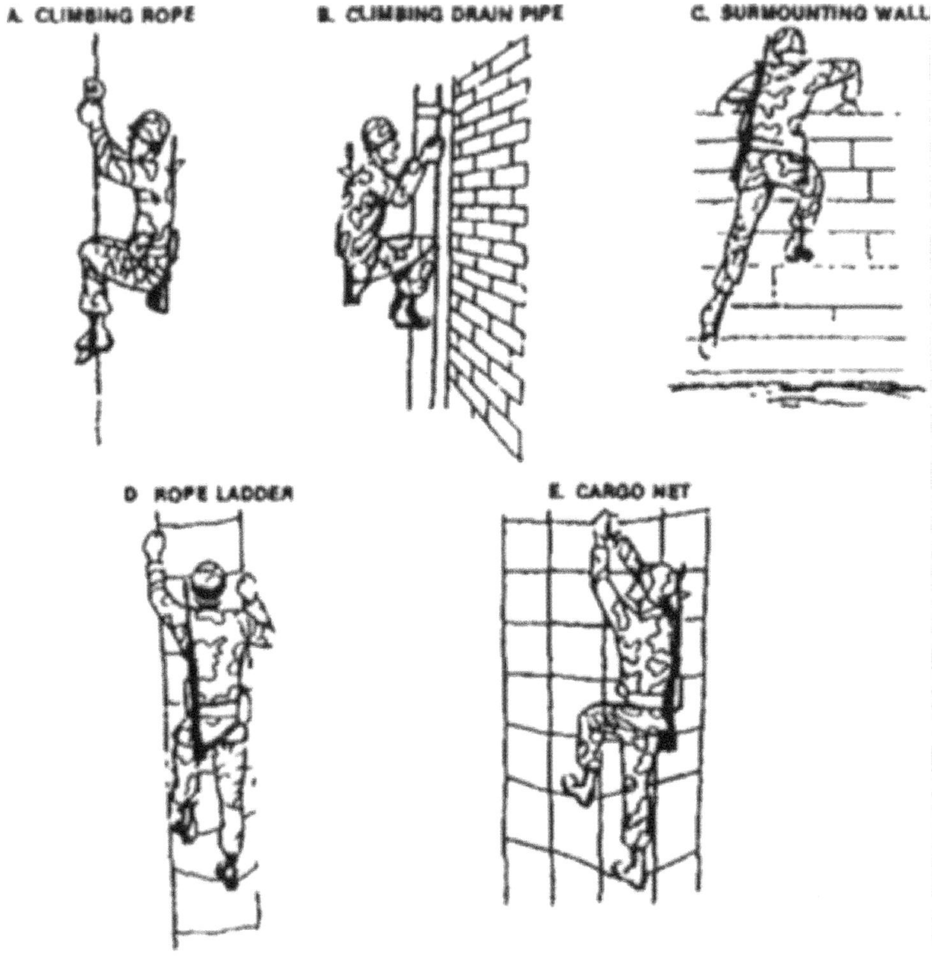

Figure 3-22.Climbing and Surmounting Vertical Objects.

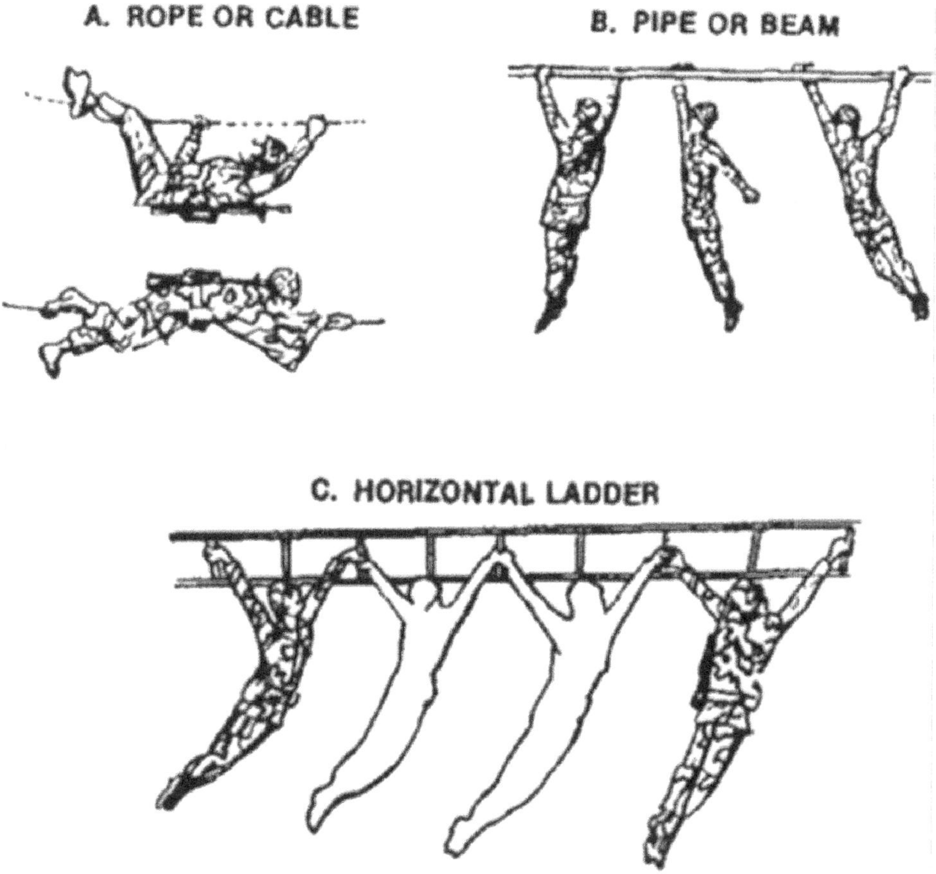

Figure 3-23. Traversing Horizontal Objects.

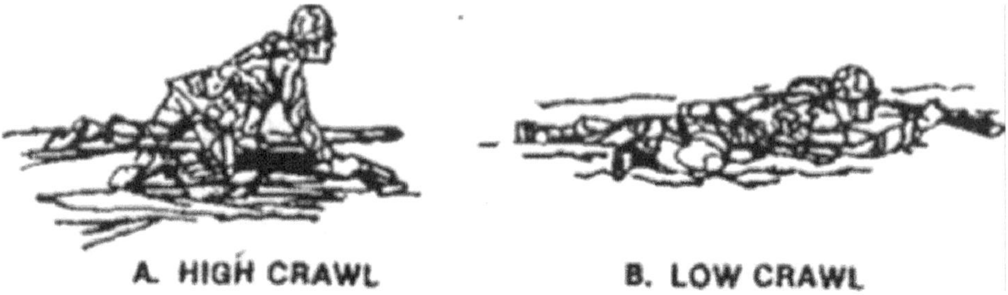

Figure 3-24.Crawling.

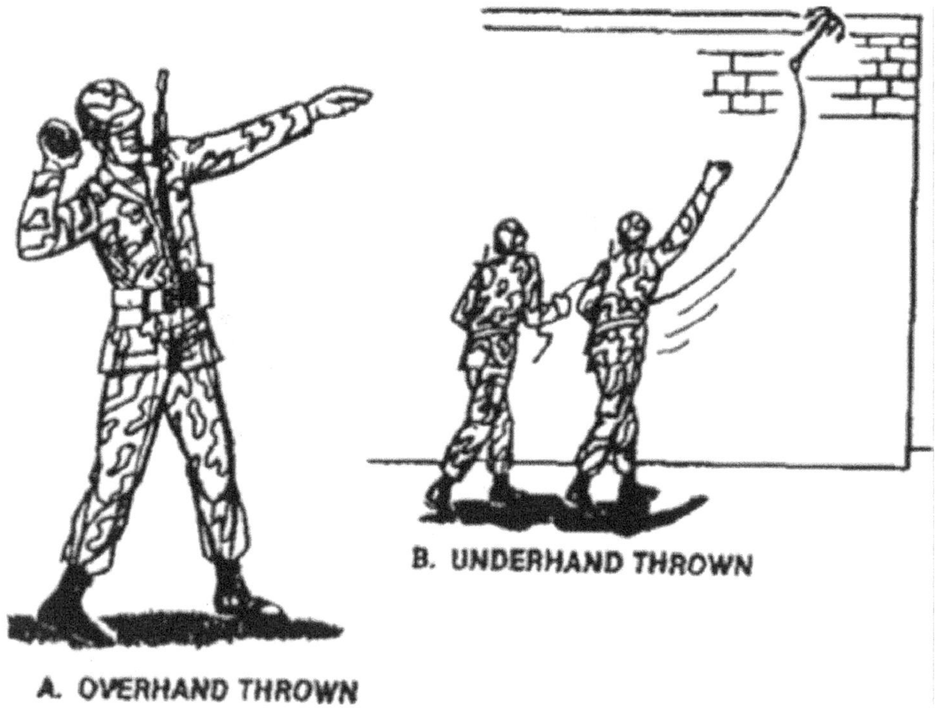

Figure 3-25.Throwing.

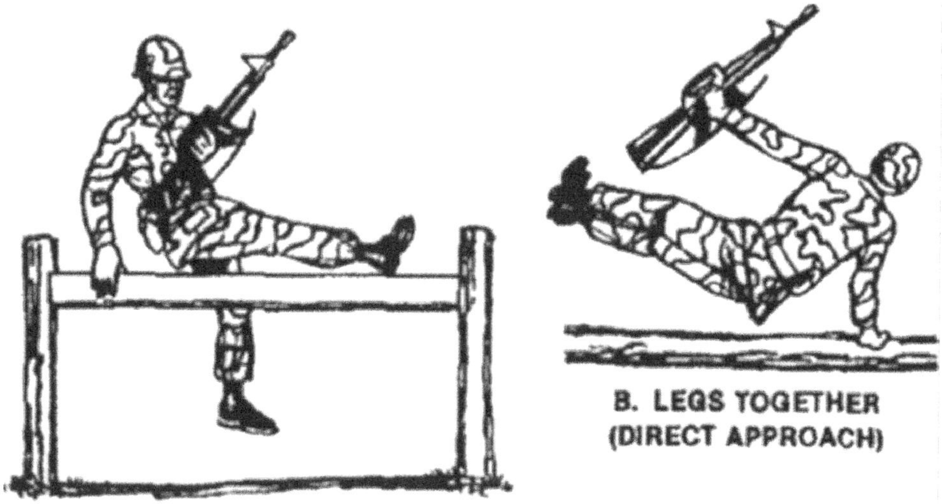

Figure 3-26. Vaulting

Figure 3-27.Balancing.

Figure 3-28.Falling.

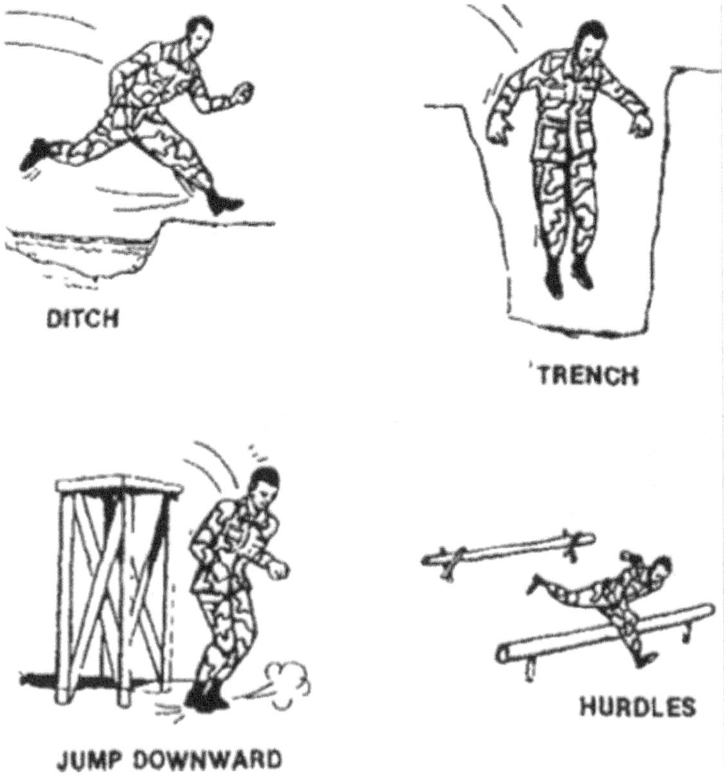

Figure 3-29.Jumping-Type Obstacles.

LANES TO GUIDE CHANGE OF DIRECTION

MAZES TO CAUSE CHANGE OF DIRECTION

Figure 3-30.Dodging-Type Obstacles.

Figure 3-31. Vertical Climbing Obstacles.

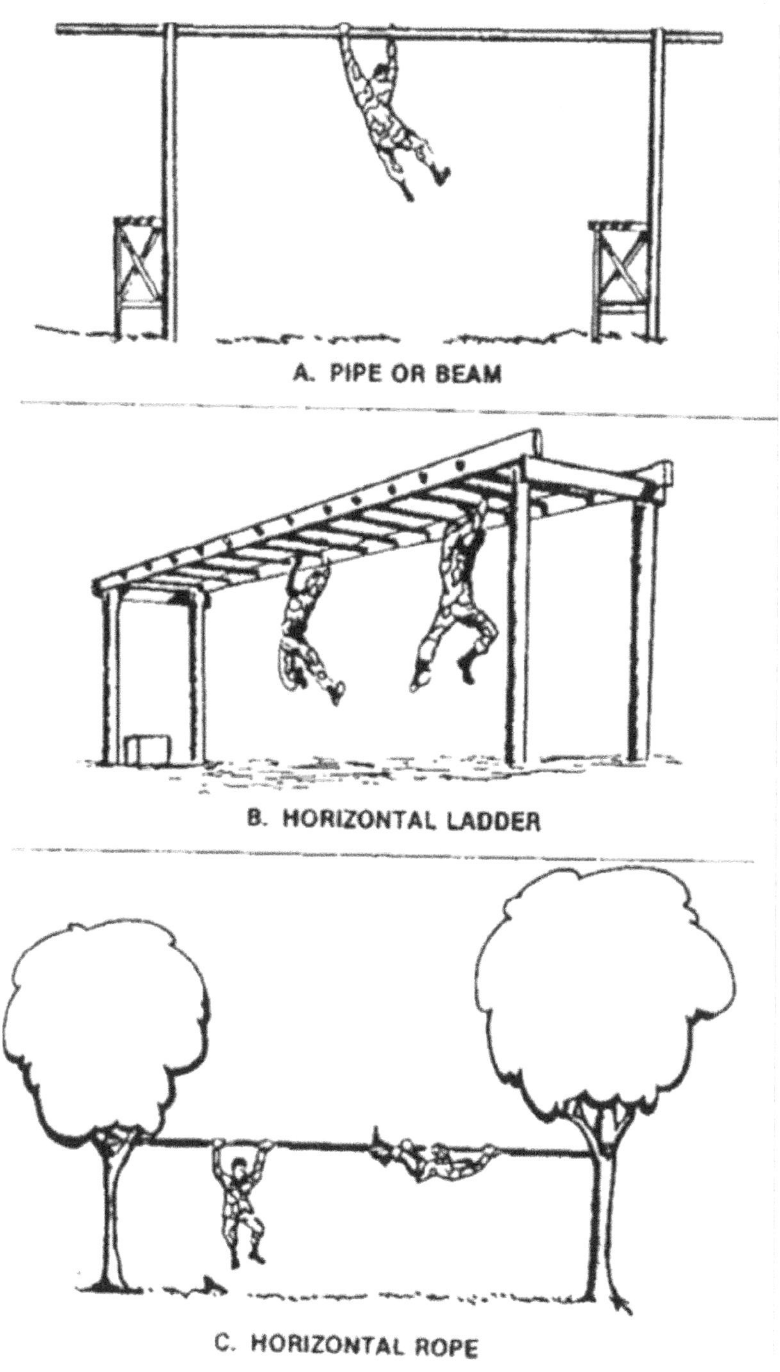

Figure 3-32.Horizontal Type Obstacles.

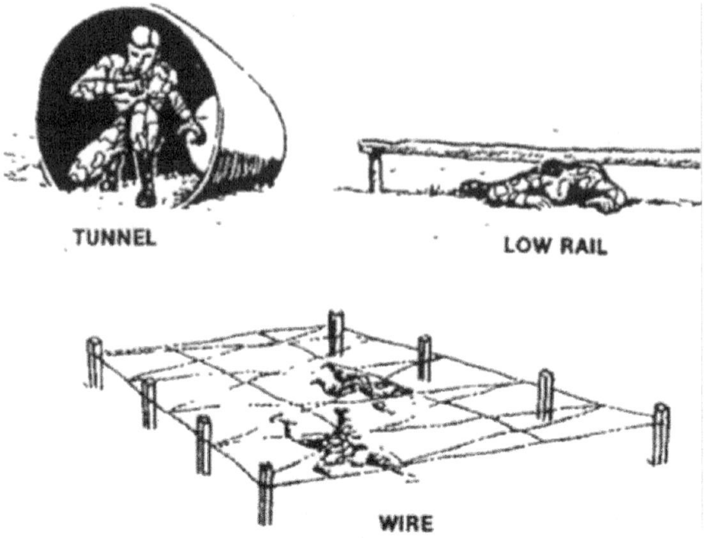

Figure 3-33.Crawling-Type Obstacles.

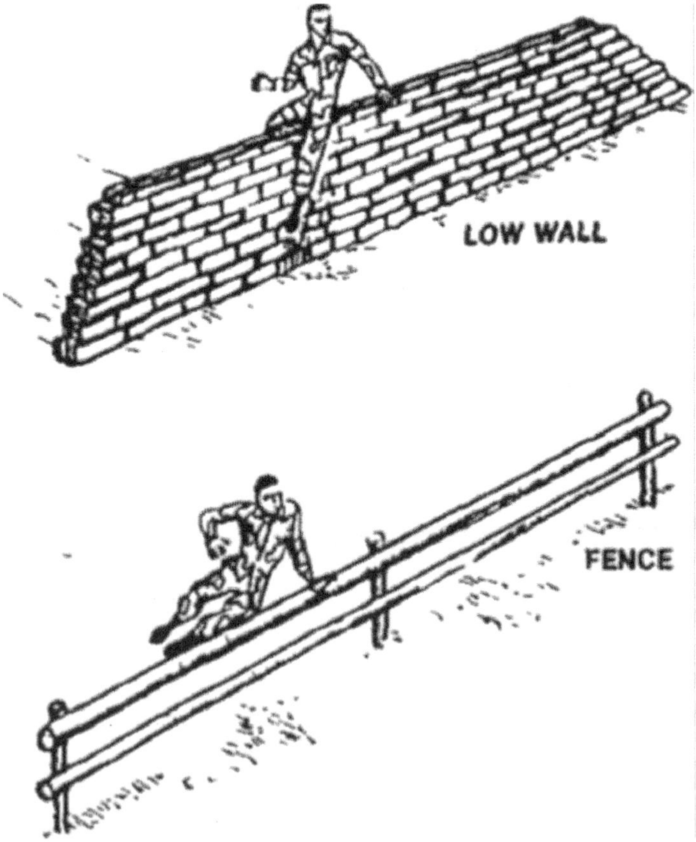

Figure 3-34.Vaulting Obstacles.

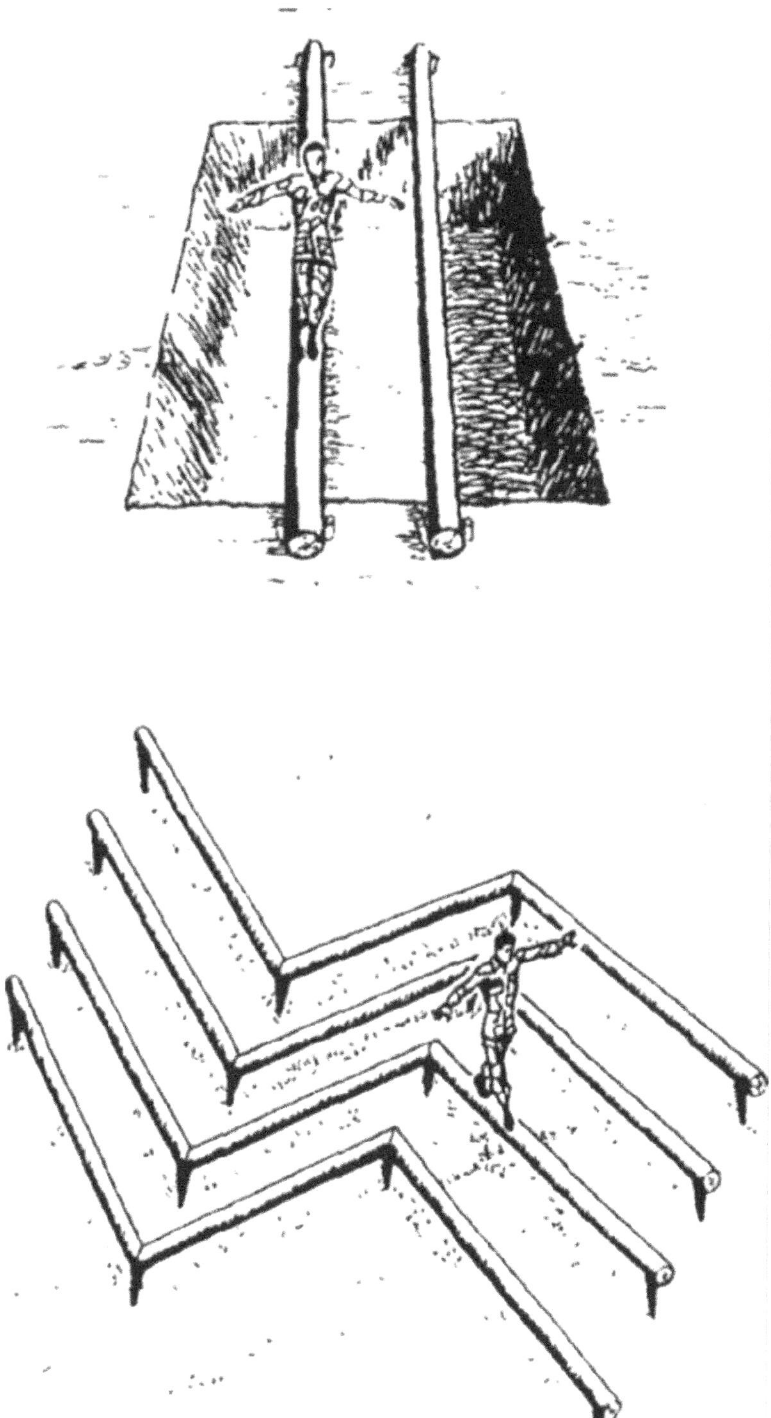

Figure 3-35.Balance-Type Obstacles.

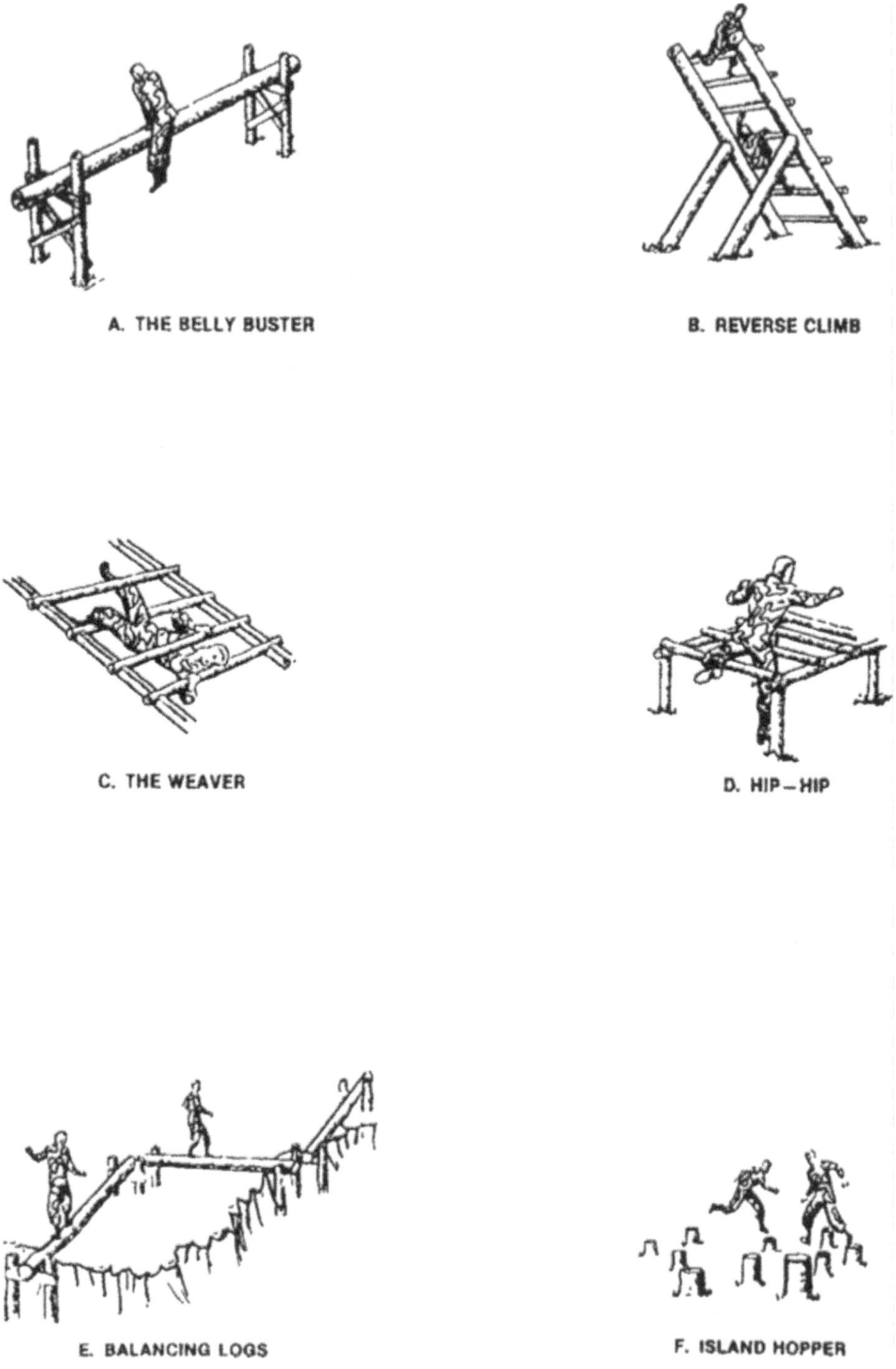

Figure 3-36.Red Group.

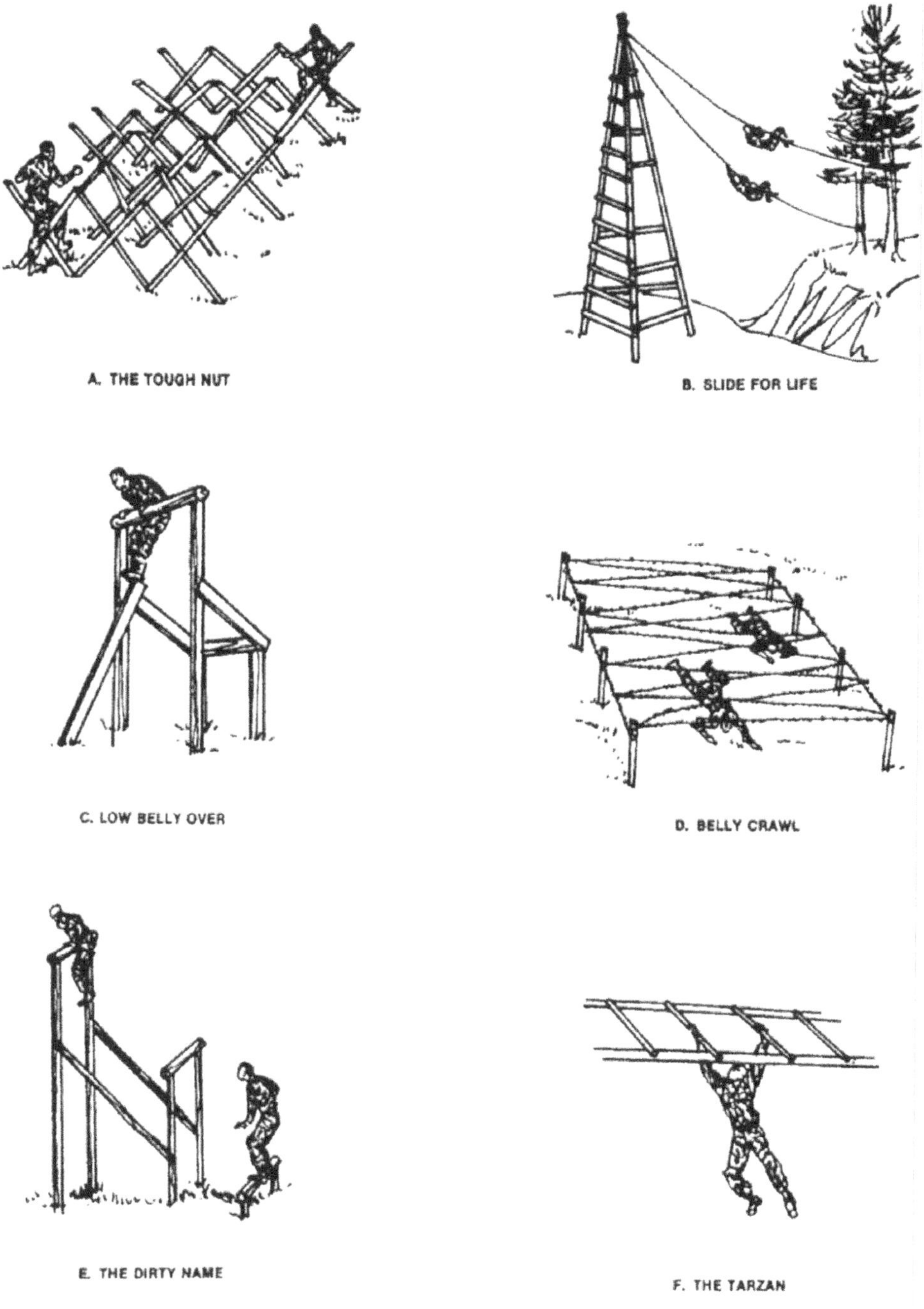

Figure 3-37.White Group.

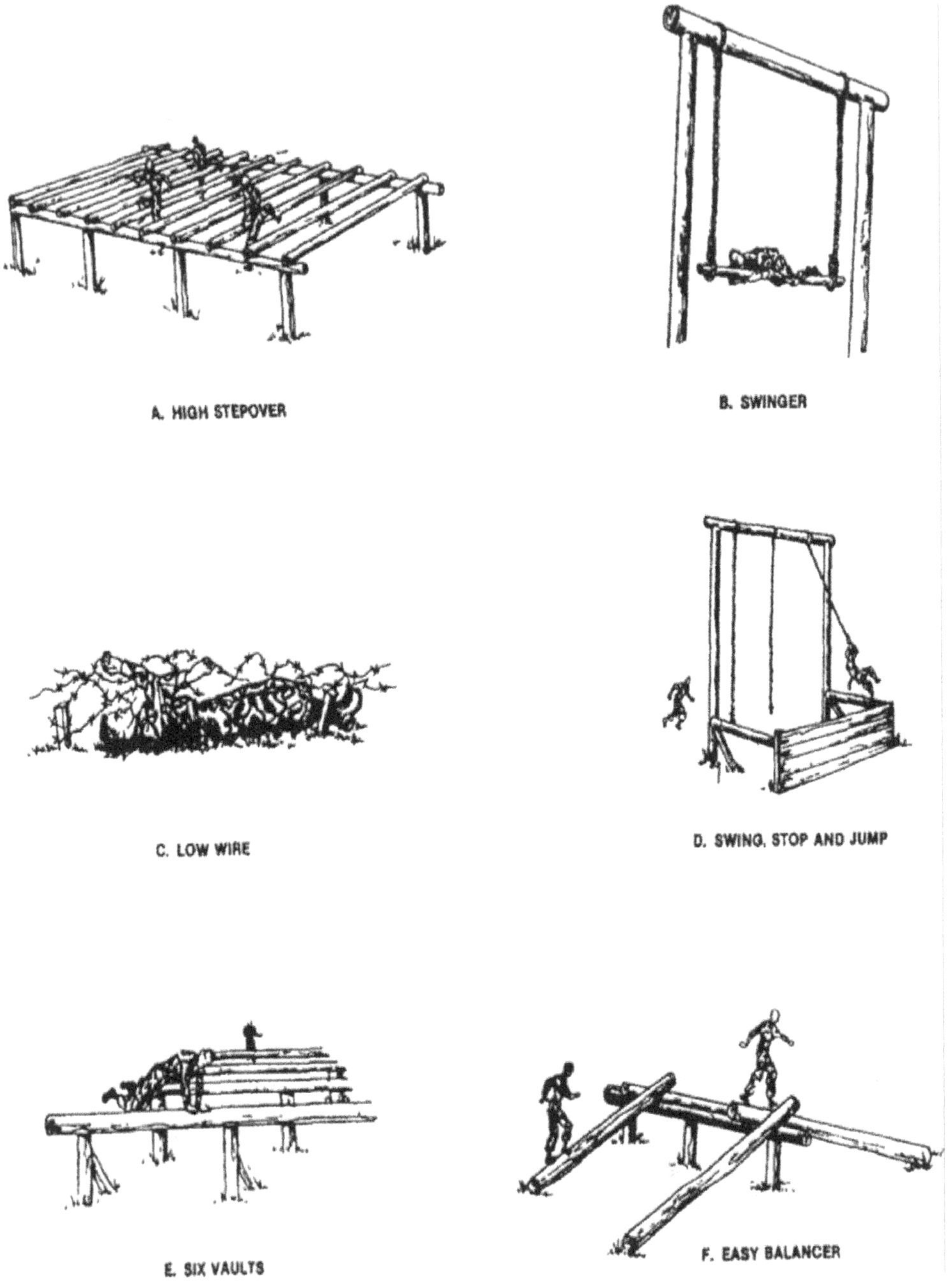

Figure 3-38.Blue Group.

A. INCLINE WALL

B. SKYSCRAPER

C. JUMP AND LAND

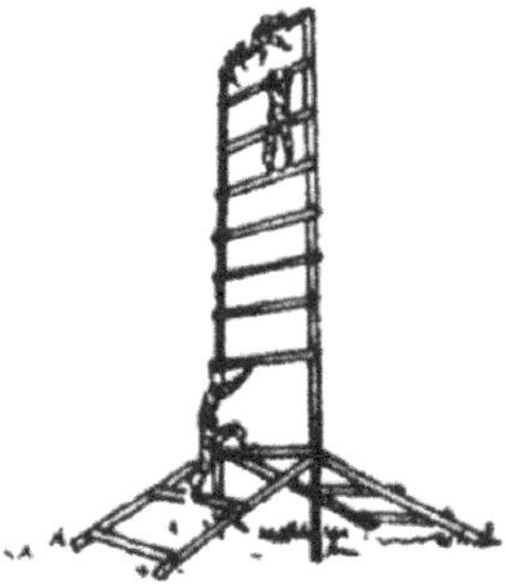

D. CONFIDENCE CLIMB

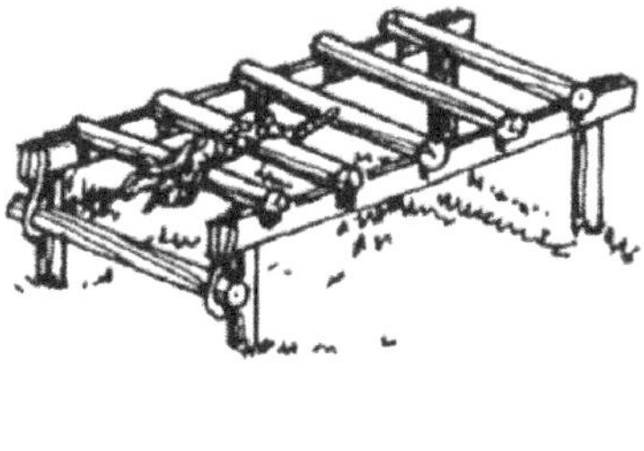

E. BELLY ROBBER

F. THE TOUGH ONE

Figure 3-39.Black Group.

Figure 3-40. Bench Conditioning, Table I.

Figure 3-41. Bench Conditioning, Table II.

PROGRESSION GUIDE

AGE GROUP	LEVEL	EXERCISES 1	2	3	4	5	6
17	A	15	18	14	15	15	250
to	B	13	16	13	13	13	235
29	C	11	14	12	11	11	[illegible]
30	A	13	14	12	13	13	200
to	B	11	13	11	11	11	185
39	C	9	12	10	9	9	165
40	A	11	11	10	11	11	150
to	B	9	10	9	9	9	135
44	C	7	9	8	7	7	120
45	A	9	8	8	9	9	100
to	B	7	7	7	7	7	90
49	C	5	6	6	5	5	80
50	A	7	6	6	7	7	75
to	B	5	5	5	5	5	70
59	C	3	4	4	3	3	60
60	A	4	5	4	4	4	50
and	B	3	4	3	3	3	40
over	C	2	3	2	2	2	30
Minutes for each exercise		2	1	1	1	2	5

Table I. Progression Guide.

Table I. Exercises.

PROGRESSION GUIDE

AGE GROUP	LEVEL	EXERCISES 1	2	3	4	5	6
17	A	17	17	17	9	19	300
to	B	15	15	15	8	17	270
29	C	13	1[illegible]	13	7	15	245
30	A	15	15	15	8	17	235
to	B	13	13	13	7	15	210
39	C	11	11	11	6	13	190
40	A	13	13	13	7	15	175
to	B	11	11	11	6	13	155
44	C	9	10	9	5	11	135
45	A	11	11	11	6	13	125
to	B	9	9	9	5	11	110
49	C	7	7	7	4	9	100
50	A	9	9	9	5	11	95
to	B	7	7	7	4	9	85
59	C	5	5	5	3	7	75
60	A	6	7	7	4	9	70
and	B	5	5	5	3	7	60
over	C	4	4	4	2	5	50
Minutes for each exercise		1	1	1	1 1/2	1 1/2	6

Table II. Progression Guide.

EXERCISE 1

EXERCISE 2

EXERCISE 3

EXERCISE 4

EXERCISE 5

EXERCISE 6

Table II. Exercises.

PROGRESSION GUIDE

AGE GROUP	LEVEL	EXERCISES 1	2	3	4	5	6
17	A	10	19	19	16	10	330
to	B	9	17	17	15	9	315
29	C	8	15	15	14	8	280
30	A	9	17	17	14	9	270
to	B	8	15	15	13	8	240
39	C	7	13	13	12	7	210
40	A	8	15	15	12	8	200
to	B	7	13	13	11	7	180
44	C	6	11	11	10	6	160
45	A	7	13	13	10	7	150
to	B	6	11	11	9	6	135
49	C	5	9	9	8	5	120
50	A	6	11	11	8	6	115
to	B	5	9	9	7	5	105
59	C	4	7	7	6	4	95
60	A	5	9	9	7	5	90
and	B	4	7	7	6	4	80
over	C	3	5	5	4	3	70
Minutes for each exercise		1 1/2	1	1	1 1/2	1	6

Table III. Progression Guide.

Table III. Exercises.

PROGRESSION GUIDE

AGE GROUP	LEVEL	EXERCISES 1	2	3	4	5	6
17	A	12	9	12	24	25	400
to	B	11	8	11	22	23	380
29	C	10	7	10	21	21	360
30	A	11	8	11	23	23	305
to	B	10	7	10	21	21	290
39	C	9	6	9	20	20	275
40	A	10	7	10	20	21	225
to	B	9	6	9	18	18	215
44	C	8	5	8	16	16	205
45	A	8	6	8	16	16	175
to	B	7	5	7	14	14	165
49	C	6	4	6	12	12	155
50	A	6	5	6	13	13	135
to	B	5	4	5	11	11	130
59	C	4	3	4	10	10	120
60	A	5	4	5	10	10	100
and	B	4	3	4	9	9	95
over	C	3	2	3	8	8	90
Minutes for each exercise		1	2	1	1	1	6

Table IV. Progression Guide.

Table IV. Exercises.

PROGRESSION GUIDE

AGE GROUP	LEVEL	EXERCISES 1	2	3	4	5	6
17	A	14	13	28	14	30	450
to	B	13	12	27	13	28	430
29	C	12	11	26	12	26	410
30	A	12	12	25	12	26	350
to	B	11	11	24	11	24	330
39	C	10	10	23	10	22	310
40	A	11	11	23	11	23	250
to	B	10	10	21	10	21	240
44	C	9	9	19	9	19	230
45	A	9	9	20	9	20	200
to	B	8	8	18	8	18	190
49	C	7	7	16	7	16	180
50	A	7	7	16	7	16	170
to	B	6	6	14	6	14	155
59	C	5	5	12	5	12	140
60	A	6	6	12	6	12	115
and	B	5	5	11	5	10	110
over	C	4	4	9	4	9	105
Minutes for each exercise		2	1	1	2	1	5

Table V. Progression Guide.

Table V. Exercises.

PROGRESSION GUIDE

AGE GROUP	LEVEL	EXERCISES 1	2	3	4	5	6
17	A	17	15	32	32	35	500
to	B	16	14	30	30	33	480
29	C	15	13	28	28	31	460
30	A	15	13	30	30	31	400
to	B	14	12	28	28	29	380
39	C	13	11	26	26	27	360
40	A	13	10	27	27	27	310
to	B	12	9	25	25	25	285
44	C	11	8	23	23	23	265
45	A	11	9	23	23	23	250
to	B	10	8	21	21	21	230
49	C	9	7	19	19	19	210
50	A	9	8	19	19	19	200
to	B	8	7	17	17	17	190
59	C	7	6	15	15	15	175
60	A	8	7	15	15	17	140
and	B	7	6	13	13	15	130
over	C	5	5	10	10	12	120
Minutes for each exercise		2	1	1	1	1	6

Table VI. Progression Guide.

Table VI. Exercises.

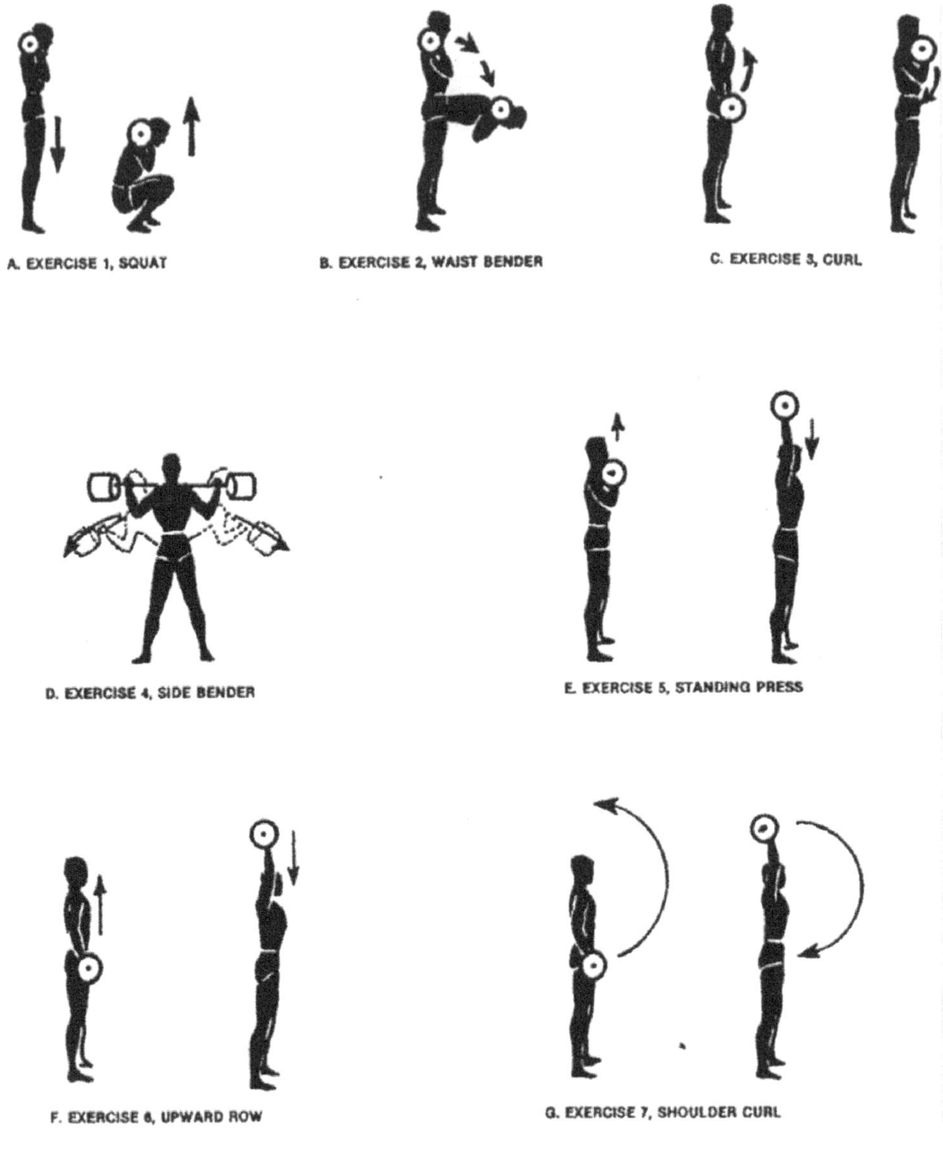

Figure 3-42. Weight Training.

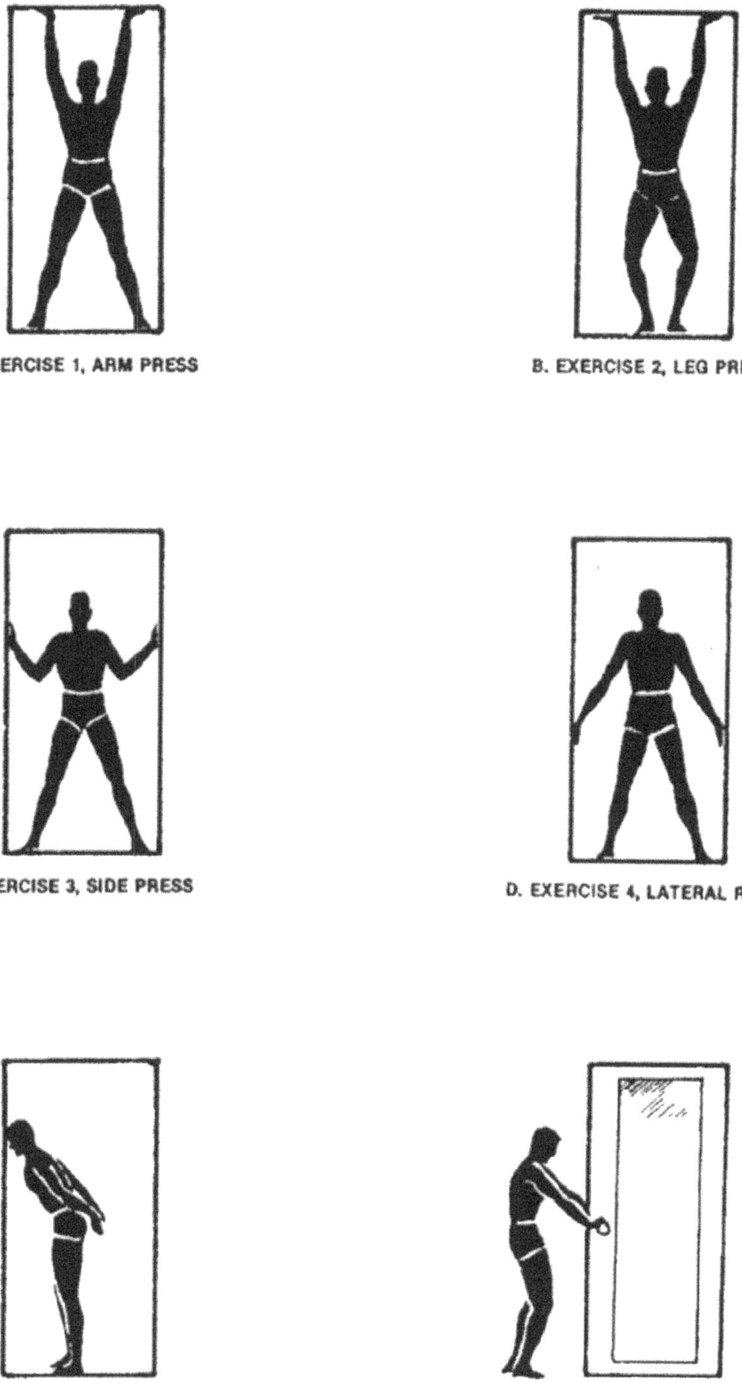

A. EXERCISE 1, ARM PRESS

B. EXERCISE 2, LEG PRESS

C. EXERCISE 3, SIDE PRESS

D. EXERCISE 4, LATERAL RAISE

E. EXERCISE 5, NECK PRESS

F. EXERCISE 6, DOOR PULL

Figure 3-43. Isometric Exercises.

Figure 5-1.Combatives Table I.

Figure 5-2.Combatives Table II.

Figure. 5-3.Combatives Table III.

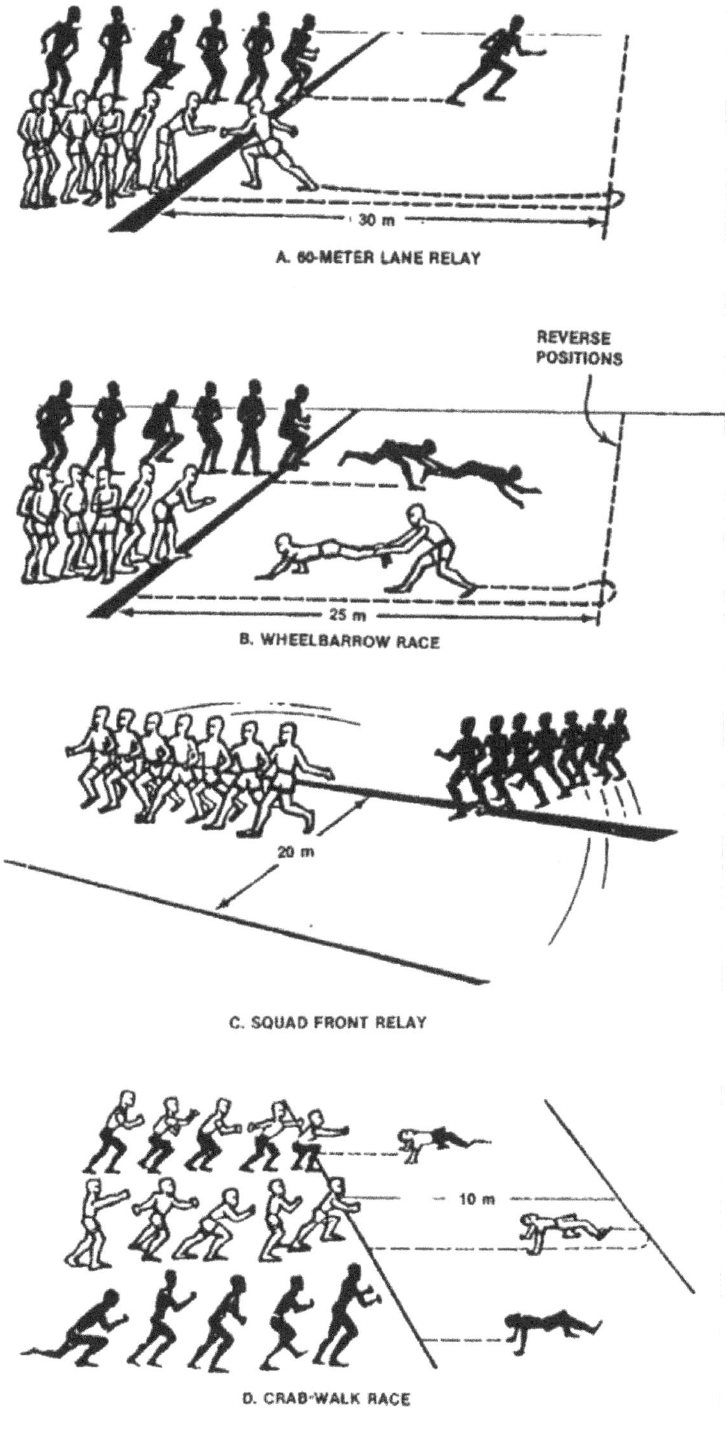

Figure 5-4.Relay Table I.

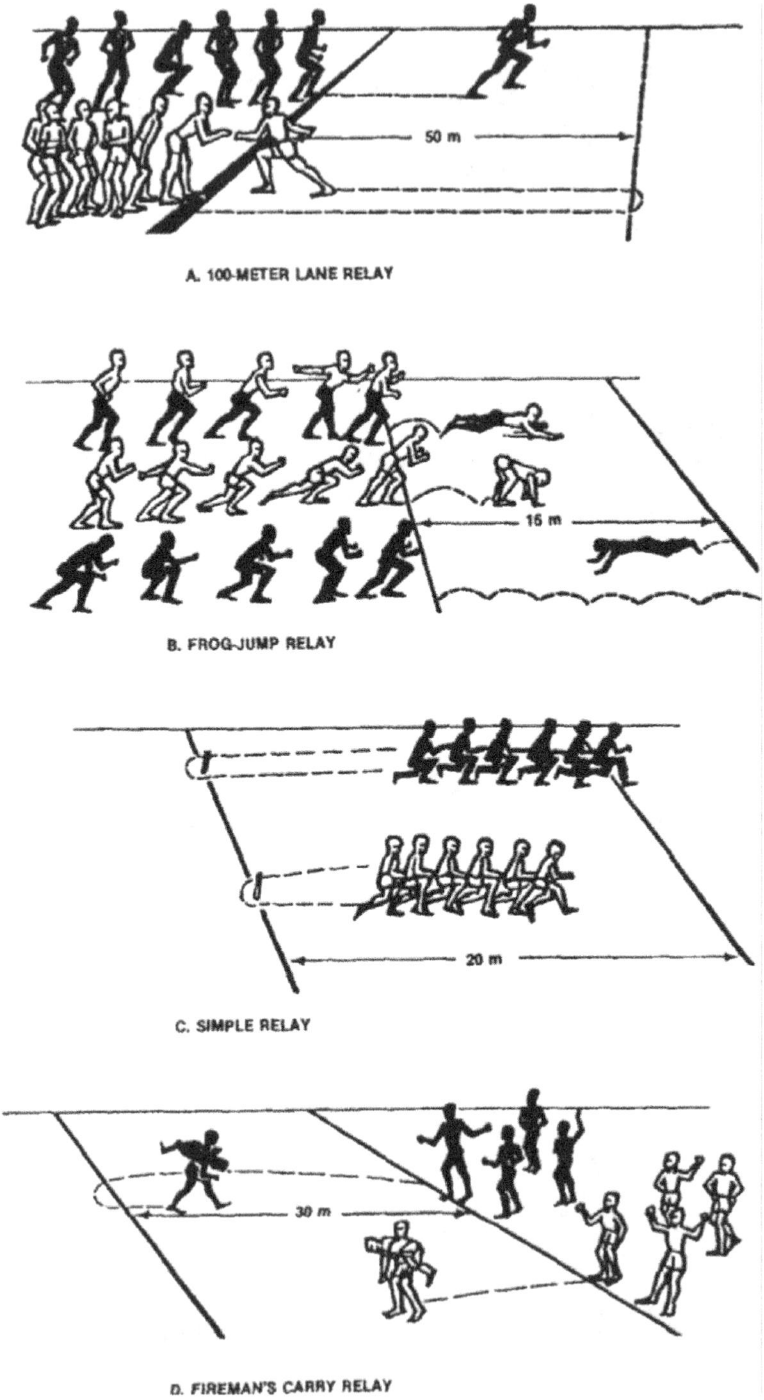

Figure 5-5.Relay Table II.

A. 200-METER CIRCLE RELAY

B. BEAR-AND-CRAB RACE

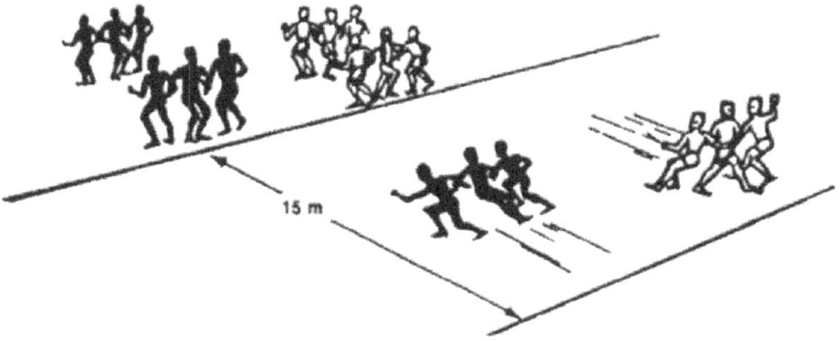

C. PILOT RELAY

D. SADDLE-BACK RELAY

Figure 5-6.Relay Table III.

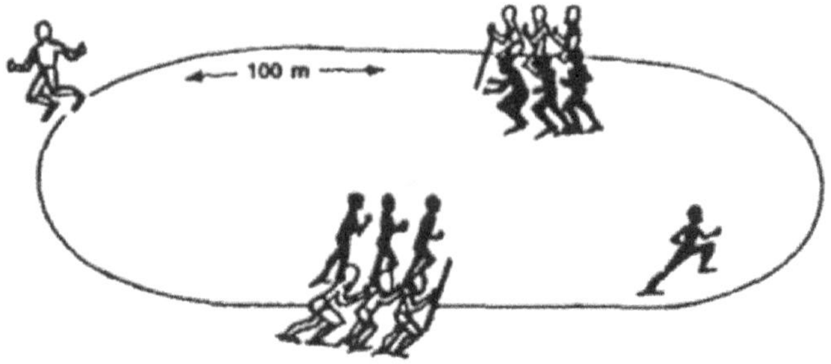

A. 100-METER CIRCLE RELAY

B. IN-AND-OUT RELAY

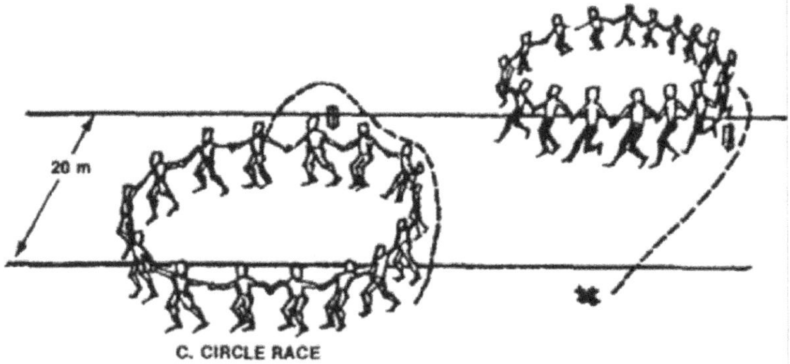

C. CIRCLE RACE

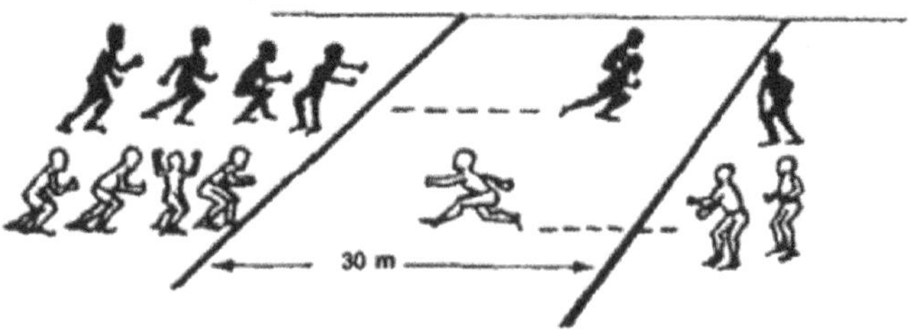

D. HORSE-AND-RIDER RELAY

Figure 5-7.Relay Table IV.

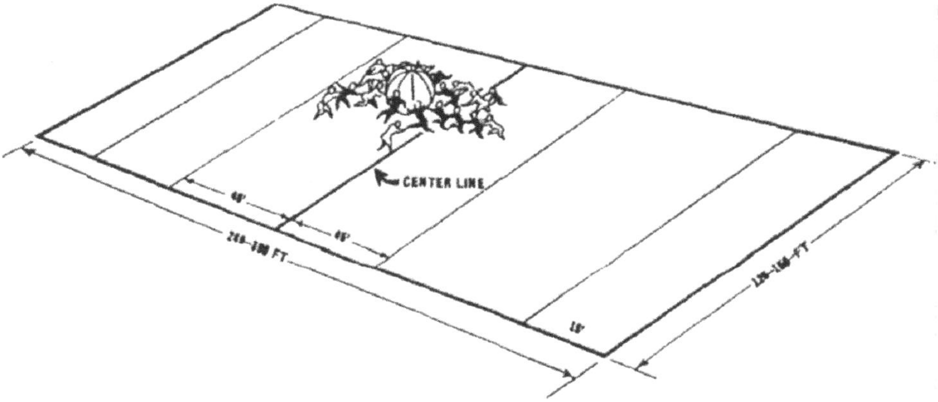

Figure 5-8.Pushball.

Figure 5-9.Line Rush.

Figure. 5-10. Human Tug-of-War.

Figure 5-11. Master of the Ring.

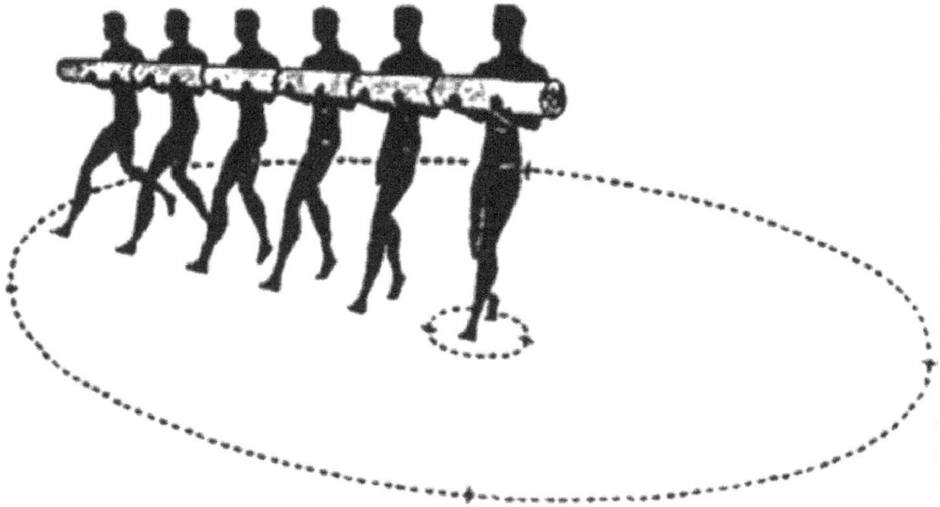

Figure 5-12.Log Pivot Circle.

Figure 5-13.Log-Rolling Race.

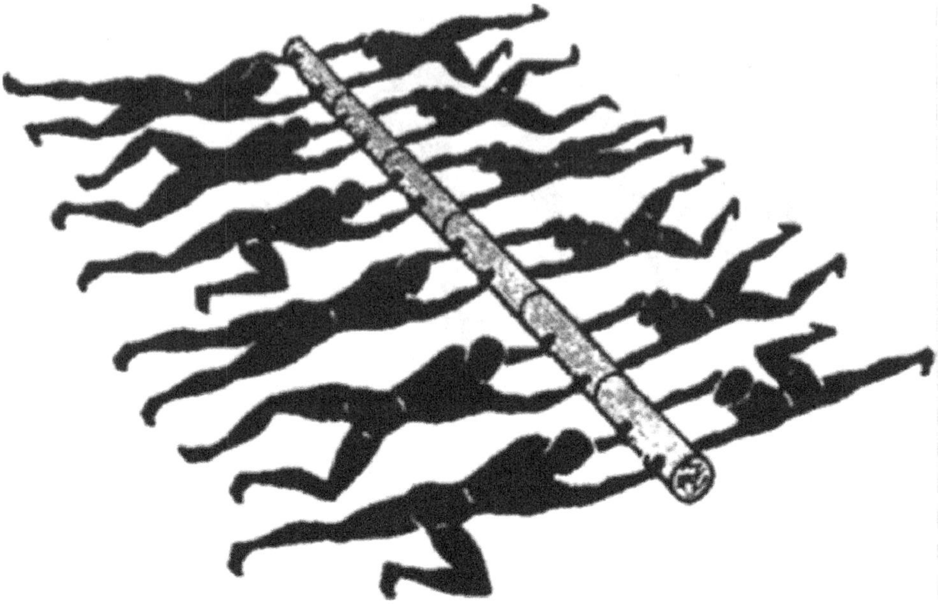

Figure 5-14. Prone Pushing Contest.

Figure 5-15. Shuttle Relay Race.

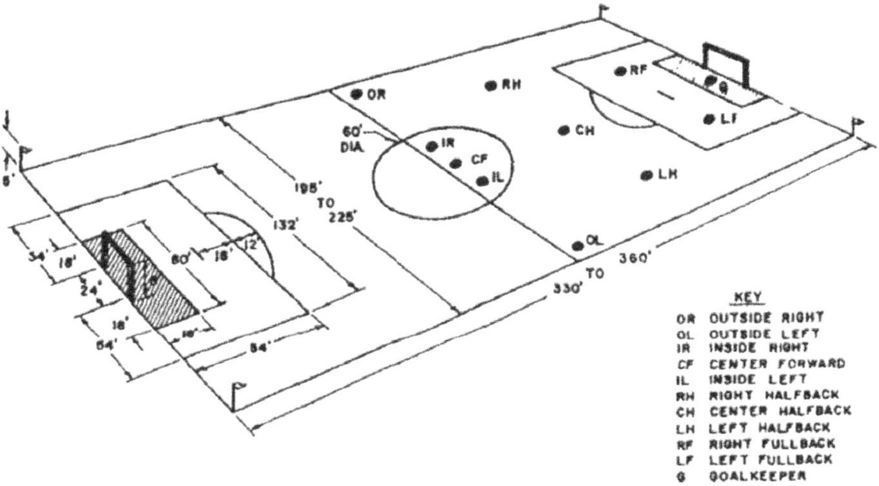

Figure 5-16. Soccer.

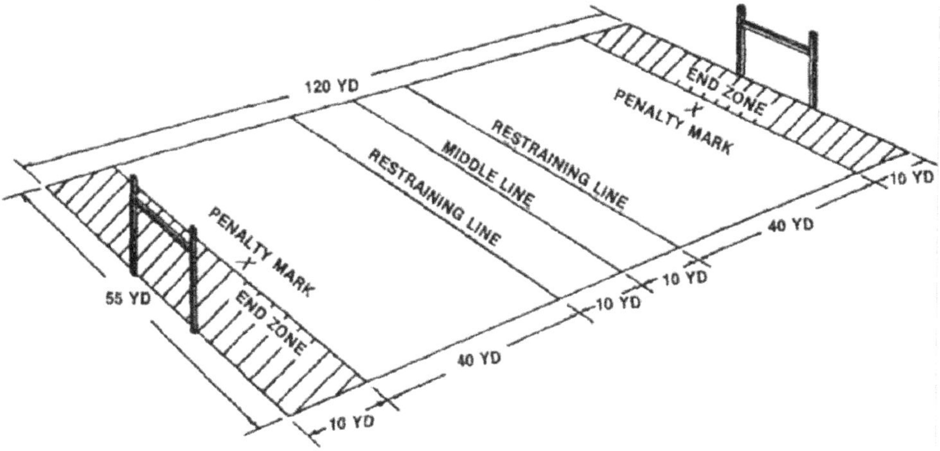

Figure 5-17.Speedball Field.

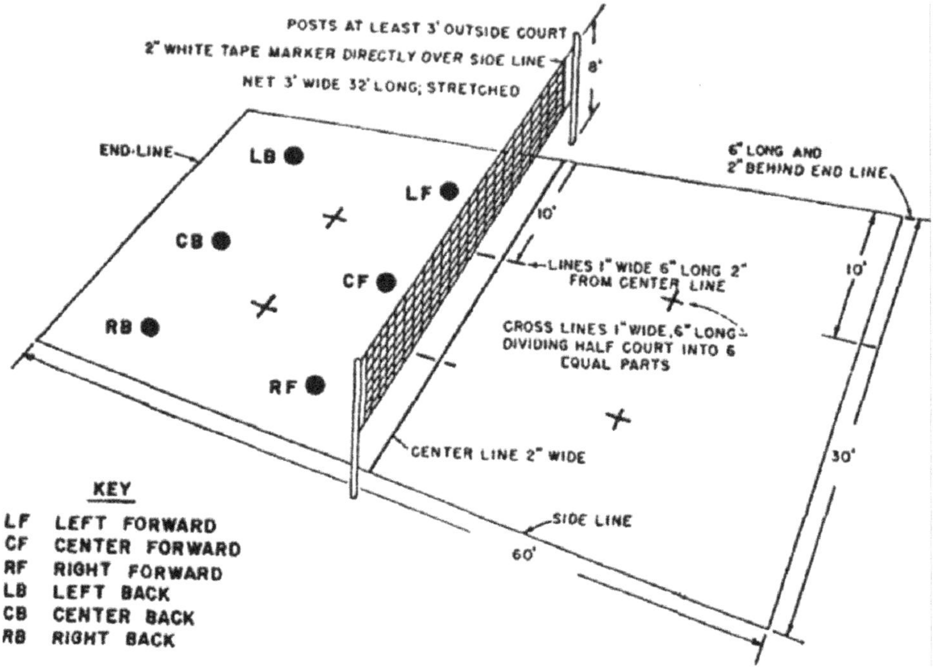

Figure 5-18. Volleyball.

PTS	UPHILL CLIMB	ROPE CLIMB	EVACUATION	FIRE AND MANEUVER	FORCED MARCH
100	1:20-BELOW	0:10-BELOW	0:25-BELOW	1:20-BELOW	26:00-BELOW
99	1:21-1:22	0:11		1:21-1:22	26:01-26:12
98	1:23-1:24	0:12	0:26	1:23-1:24	26:13-26:24
97	1:25-1:26	0:13		1:25-1:26	26:25-26:36
96	1:27-1:28	0:14	0:27	1:27-1:28	26:37-26:48
95	1:29-1:30	0:15		1:29-1:30	26:49-27:00
94	1:31-1:32	0:16	0:28	1:31-1:32	27:01-27:12
93	1:33-1:34	0:17		1:33-1.34	27:13-27:24
92	1:35-1:36	0:18	0:29	1:35-1.36	27:25-27:36
91	1:37-1:38	0:19		1:37-1:38	27:37-27:48
90	1:39-1:40	0:20	0:30	1:39-1:40	27:49-28:00
89	1:41-1:42	0:21	0:31	1:41-1:42	28:01-28:12
83	1:43-1:44	0:22	0:32	1:43-1:44	28:13-28:24
87	1:45-1:46	0:23	0:33	1:45-1:46	28:25-28:36
86	1:47-1:48	0:24	0:34	1.47-1:48	28:37-28:48
85	1:49-1:50	0:25	0:35	1.49-1:50	28:49-29:00
84	1:51-1:52	0:26	0:36	1:51-1:52	29:01-29:12
83	1:53-1:54	0:27	0.37	1:53-1:54	29:13-29:24
82	1:55-1:56	0:28	0:38	1:55-1:56	29:25-29:36
81	1:57-1:58	0:29	0:39	1:58-1:58	29:37-29:48
80	1:59-2.00	0:30	0:40	1:59-2:00	29:49-30:00
79	2.01-2:02	0:31	0:41	2:01-2:02	30:01-30:12
78	2:03-2:04	0:32	0:42	2:03-2.04	30:13-30:24
77	2:05-2:06	0.33	0:43	2:05-2:06	30:25-30:36
76	2:07-2:08	0.34	0.44	2:07-2:08	30:37-30:48
75	2:09-2:10	0:35	0:45	2:09-2:10	30:49-31:00
74	2:11-2:12	0:36	0:46	2:11-2 12	31.01-31:12
73	2:13-2:14	0:37	0.47	2.13-2:14	31:13-31:24
72	2:15-2:16	0:38	0:48	2.15-2:16	31:25-31:36
71	2:17-2:18	0:39	0:49	2:17-2:18	31:37-31:48
70	2:19-2:20	0:40	0:50	2:19-2.20	31:49-32.00
69	2.21-2:22	0:41	0:51	2:21-2:22	32:01-32:12
68	2:23-2:24	0:42	0:52	2:23-2:24	32:13-32:24
67	2:25-2.26	0:43	0:53	2:25-2:26	32:25-32:36
66	2:27-2:28	0:44	0:54	2.27-2:28	32.37-32:48
65	2:29-2:30	0.45	0.55	2.29-2 30	32.49-33:00

Figure 6-1.Physical Readiness Test Scoring Table for Male.

	EXAMPLE 1	EXAMPLE 2	EXAMPLE 3
Event #1	100 pts	100 pts	99 pts
Event #2	100 pts	99 pts	86 pts
Event #3	85 pts	75 pts (F)	80 pts
Event #4	90 pts	79 pts (F)	100 pts
Event #5	79 pts	85 pts	90 pts
Pts	454 pts	438 pts	455 pts
Final Score	79.0% (F)	69.0% (F)	91.0%
(F) = Failure			

Figure 6-2.Scoring Examples.

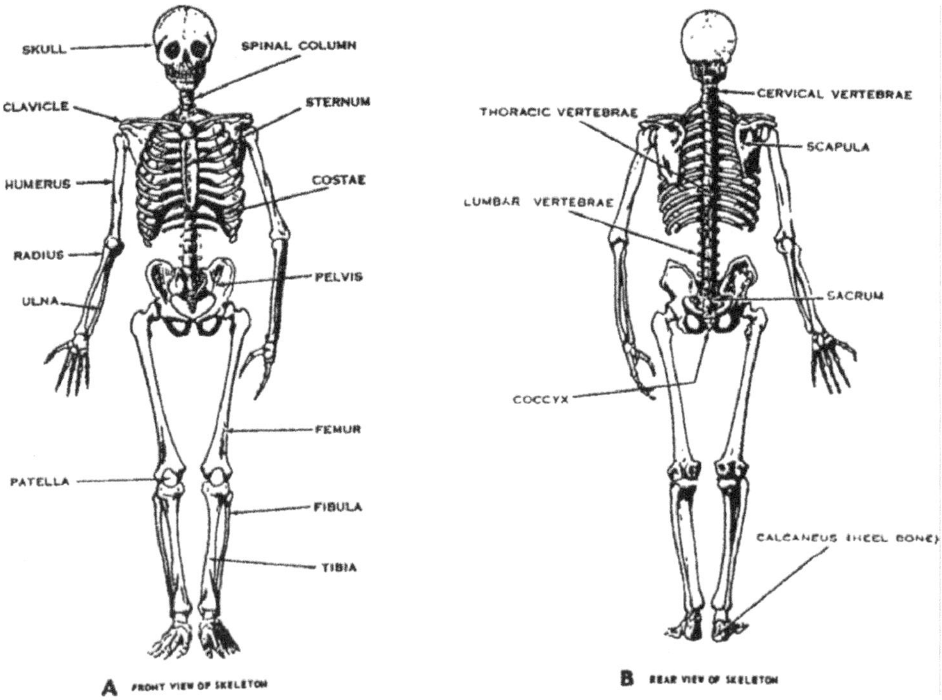

Figure 7-1.Front and Rear Views of Skeleton.

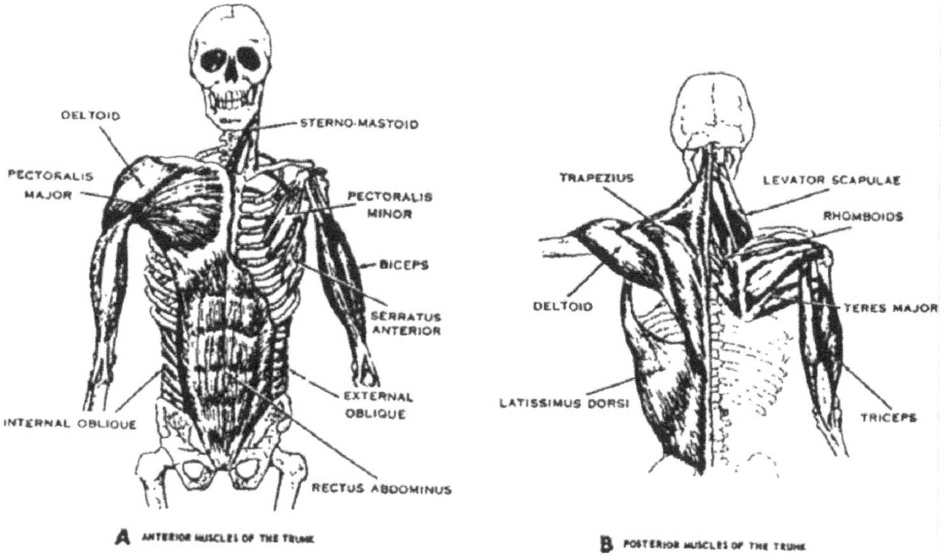

Figure 7-2.Muscles of the Trunk.

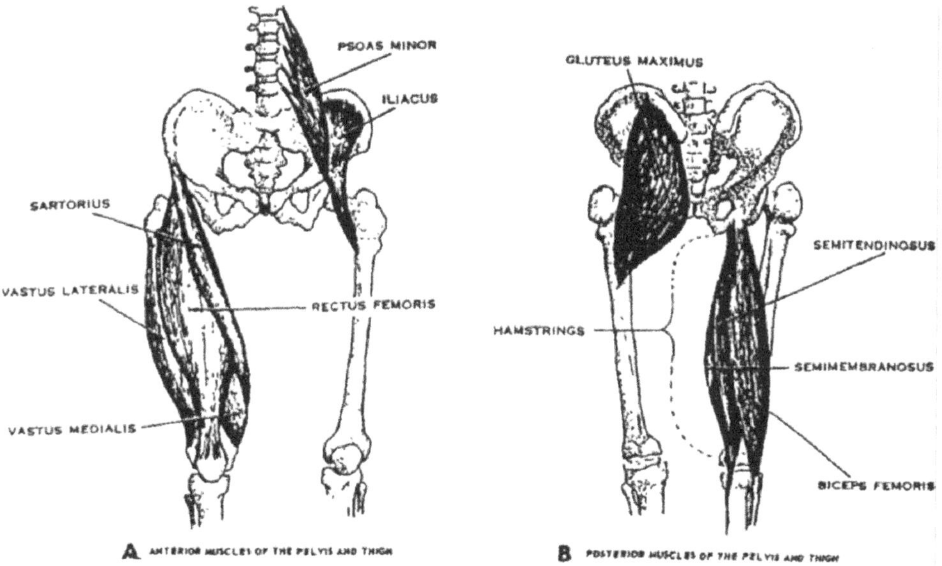

Figure 7-3.Anterior and Posterior Muscles of the Pelvisand Thigh.

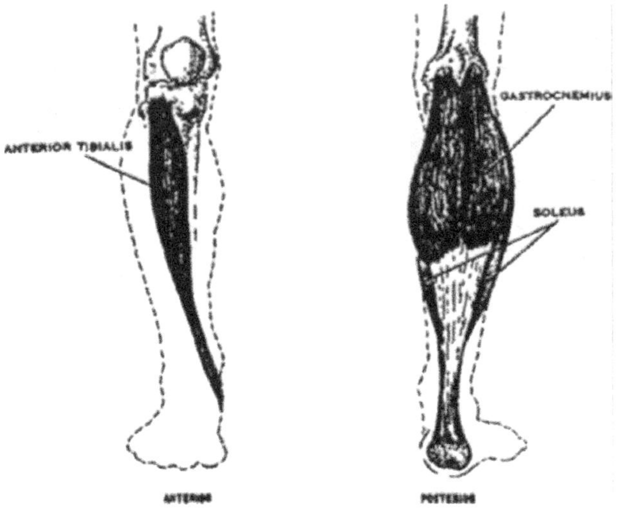

Figure 7-4. Muscles of Lower Leg.

Figure 7-5.Circulation of Blood from Heart to the Body.

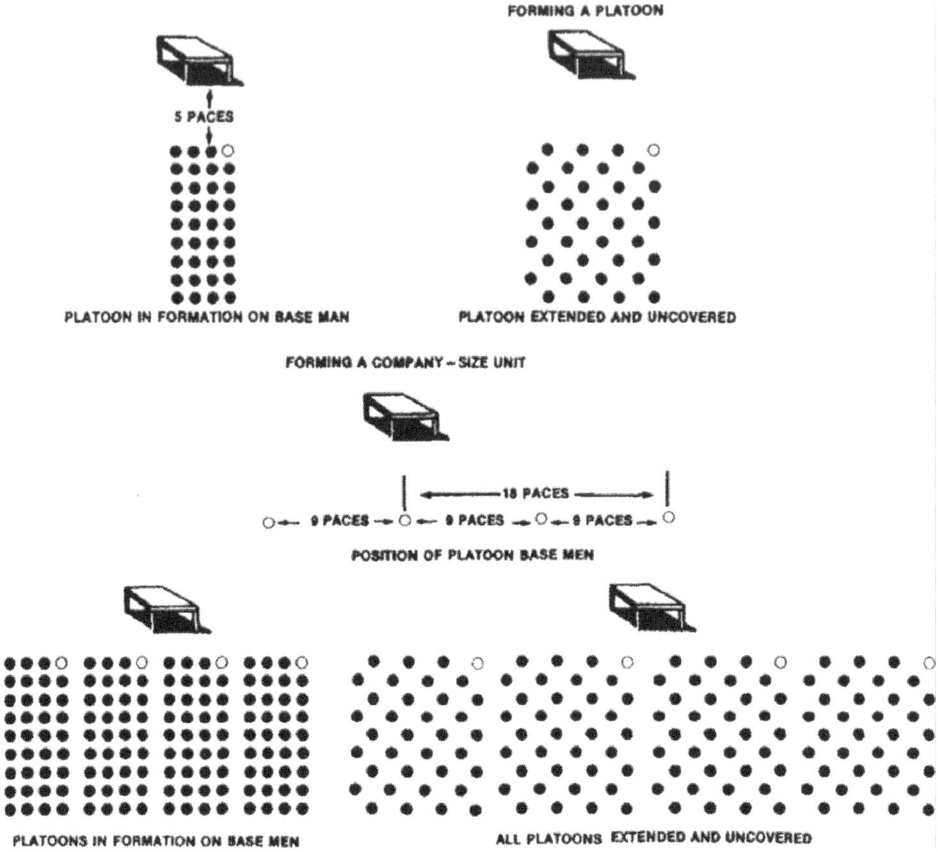

Figure A-1. Forming the Extended Rectangular Formation.

Figure A-2.Circle Formation.

TRAINING CIRCULAR
No. 21-24

HEADQUARTERS
DEPARTMENT OF THE ARMY
WASHINGTON, DC, 10 September 1997

RAPPELLING

CONTENTS

DISTRIBUTION RESTRICTION: Approved for public release; distribution is unlimited.

*This publication supersedes TC 21-24, 24 September 1991.

Page

CHAPTER 3. HELICOPTER RAPPELLING

CHAPTER 4. SPECIAL PATROL INFILTRATION/EXFILTRATION SYSTEM

PREFACE

This circular provides basic rappelling techniques to soldiers and leaders for the conduct of rappelling operations. It serves as the primary reference for both resident and nonresident instruction presented to cadets, officer candidates, and both commissioned and noncommissioned officers. This circular also discusses several advanced techniques dealing with infiltration and exfiltration.

Safety is always the most important consideration when conducting training. This circular provides guidelines for commanders to conduct operations safely. Commanders at all levels must analyze the complete training event to determine the degree of risk involved to men and or equipment (Appendix A). After determining the risks, risk reduction options or controls will be integrated into the training activity. These options or controls may range from a safety briefing or providing additional safety resources, to selecting other means of accomplishing the mission.

Rappelling, FRIES, and SPIES operations are inherently dangerous; therefore, the safety notes and considerations presented in this circular are the minimum acceptable standards.

The proponent of this publication (except for Chapter 5) is the US Army Infantry School. Submit changes for improving this publication on DA Form 2028 and forward it to the Commandant, US Army Infantry School, ATTN: ATSH-RB, Fort Benning, Georgia 31905-5593 (e-mail address: IKNERH@BENNING-EMH2.ARMY.MIL).

The proponent for Chapter 5, Fast-Rope Insertion/Extraction System, is the US Army Special Operations Command. Submit changes or comments for Chapter 5 on DA Form 2028 and forward it to the Commander, US Army Special Operations Command, ATTN: AOOP-TRS, Fort Bragg, North Carolina 28307-5000.

Unless this publication states otherwise, masculine nouns and pronouns do not refer exclusively to men.

CHAPTER 1

TOWER RAPPELLING

The introduction to rappelling is taught on a static tower (Figure 1-1). Using the building block approach to training, soldiers systematically progress to more demanding platforms, including taller static towers, and finally to helicopters. The static tower used may vary in size and height from 34 to 90 feet. The concept of learning the basic rappelling techniques before helicopter operations does not vary. (Units may conduct ground training and wall-side tower rappelling without a rappel master. In this case, the unit commander would be responsible for the safety of the rappelling training.)

Figure 1-1. Rappel tower.

Note: Each static rappel tower should resemble the structure or aircraft being trained to. The Army's standard design for building static rappel towers can be obtained through the Plans and Support Branch of the post Engineers, Building 75, Fort Benning, Georgia 31905 (commercial 706 545-3307 or DSN 835-3307). Strict adherence to the design specifications is required for safety considerations. Rappel towers should be inspected annually by post safety or post engineers.

Section I. PERSONNEL

This section discusses the personnel involved in training rappelling and their duties and responsibilities.

1-1. RAPPEL MASTER

The proponent for accreditation, evaluation, and information for Army rappel master courses is Directorate of Operations and Training, US Army Infantry School, ATTN: ATSH-OTT-F, Fort Benning, Georgia 31905-5593.

a. **Duties and Responsibilities.** The rappel master is responsible for the safety of rappellers. He ensures that all equipment (installation, unit, and personal property) is serviceable. He personally supervises the rappelling operation.

b. **Qualification.** Rappel master qualification is awarded only after the successful completion of a TRADOC accredited rappel master course, which includes the following subjects:

- Duties and responsibilities of a rappel master.
- Safety SOP, regulations, and references.
- Construction of a deployment bag.
- Conduct of an equipment rappel off the rappel tower.
- Conduct of a lock-in.
- Talking a rappeller through completion of a rappel.
- Conduct of ground training.
- Inspecting for proper hook-ups.
- Inspecting and maintaining equipment.
- Inspecting and maintaining snaplinks.
- Inspecting and maintaining rappelling gloves.
- Inspecting and maintaining rappel ropes.
- Identifying the rappel capabilities of aircraft used.
- Controlling rappels from UH-1H or UH-60 aircraft.
- Tying knots (square, bowline, half-hitch, Prusik), safety lines, and rappel seats (Swiss seat, Australian seat).
- Inspecting a rappel seat.
- Aircraft rigging for rappelling operations.
- Aircraft command and control.

c. **Proficiency Maintenance.** To remain current, rappel masters must execute their duties in a tactical or training exercise once every six months. If rappel masters do not execute their duties once every six months, they must take a refresher class taught by a current rappel master. The refresher class includes the subjects listed in the rappel master qualifications paragraph (1-1b).

1-2. RAPPEL SAFETY OFFICER

The RSO is an air assault-qualified or ranger-qualified SFC or above who serves as the OIC for the conduct of rappel operations.

a. The RSO is responsible overall for the safety of all rappellers and ensures that all safety precautions are followed.

b. The RSO briefs VIPs, visitors, and inspecting authorities on training, safety requirements, and layout of training areas.

1-3. RAPPEL LANE NCO

a. **Duties and Responsibilities.** Safety is the rappel lane NCO's number one priority. The rappel lane NCO—

- Ensures proper safety procedures are followed.
- Ensures proper hookup once directed to a rope station.
- Issues commands and maintains eye contact with the rappeller at all times.

b. **Qualifications.** The rappel lane NCO holds the rank of corporal or above, is ranger- or air assault-qualified, and is selected by the commander. Each tower rappel lane must have a qualified rappel lane NCO.

c. **Training.** The rappel lane NCO must also train on the following subjects:

- Responsibilities and safety requirements.
- Inspection and maintenance of equipment.
- Identification of satisfactory anchor points.
- Identification of safe and unsafe hookups.
- Establishment of a rappel point.
- Inspection of a rappel seat.
- Coaching techniques.
- Rappelling procedures.
- Emergency procedures.
- Belay control procedures.

d. **Participation.** The rappel lane NCO must participate in at least seven rappel operations: three as a rappeller; two as an assistant rappel lane NCO; and two performing the duties of a rappel lane NCO under the supervision of a qualified rappel master.

e. **Proficiency Maintenance.** If a rappel lane NCO has not conducted his duties within the last six months, he must complete the training listed in paragraph 1-3c under the supervision of a current rappel master.

1-4. RAPPELLER

Rappel qualification requirements apply to the individual rappeller. Participants in tower rappel training must complete the following requirements under the supervision of a rappel master. The unit commander ensures that personnel successfully complete these requirements before beginning aircraft rappel training.

- Identify all rappelling equipment.
- Demonstrate the construction and attachment of the rappel seat and the rappel rope to the seat.
- Identify unsafe attachments, equipment, rope connections, and seat construction.
- Define terms used in rappelling operations.
- Identify knots used in rappel operations.
- Understand and demonstrate rappel commands.
- Demonstrate rappelling positions.
- Demonstrate belaying procedures.
- Exhibit satisfactory performance from a rappel tower of at least 34 feet in height (two rappels with equipment and weapon, two without equipment and weapon). Two rappels are conducted from the free side of the tower (no wall).
- Demonstrate the ability to lock in.

1-5. BELAYER

Belay requirements are a subtask of basic rappel requirements. Soldiers must know how to belay before conducting rappelling training. The belayer—

- Assumes a position at the base of the lane about one pace away from the tower area.
- Ensures that the rappel ropes are even with the ground during tower rappels.
- Loosely holds the rappel rope with both hands so as not to interfere with the rappeller but still be able to stop the rappeller should he fall.
- Immediately stops the rappeller by pulling downward on the rappel ropes if the rappeller shouts "falling" or loses control of his brake hand during descent.
- Does not wear gloves to ensure a firm grip on the rappelling rope.
- Watches the rappeller at all times, and maintains constant voice or visual contact.
- Wears a helmet to prevent injuries from falling debris.

1-6. BELAY SAFETY

The belay safety must be ranger- or air assault-qualified. He ensures belay personnel are performing their duties properly. Rappel training requires one belay safety for each two rappel stations.

Section II. PREOPERATIONS BRIEFINGS AND SAFETY PROCEDURES

The rappel master ensures participants have a basic understanding of requirements and safety procedures before conducting training.

1-7. SAFETY

The following personnel and equipment must be present during static tower training.

- Two military rappel ropes for each rappel station.
- One safety officer.
- One rappel master for each rappel site.
- One rappel lane NCO per rappel station.
- One medic with medical kit and backboard.
- One safety or medical evacuation vehicle with driver.
- One belayer for each rope station. Rappellers alternate stations.
- One belay safety for each two rappel stations (four ropes).

1-8. SAFETY BRIEFING

As in all training, a safety briefing precedes rappel operations. The rappel master briefs all personnel on safety to include the following instructions.

a. Each rappeller ensures loose clothing and equipment are secured.

b. Rappel seats are tied by the soldier and inspected by the rappel master before climbing the tower. Rappel seats are removed upon completion of every rappel, retied, and reinspected by a qualified rappel master or rappel lane NCO before subsequent rappels.

c. Rappellers climb the tower only when directed by the rappel master or rappel lane NCO.

d. Rappellers stay in the center of the tower until instructed to move to a rappel point.

e. No more than three personnel are behind each rappelling point.

f. If using a troop ladder, only three soldiers are on the ladder at one time. Soldiers do not climb the ladder until told to do so by a rappel master.

g. All rappel masters, rappel lane NCOs, instructors, and anyone else standing near the edge of the top of the tower must wear a restraining strap or safety rope. The strap or rope must be attached to an anchor point.

h. No one should lean or sit on the railing or banisters of the tower.

i. No one is allowed within 3 feet of the edge of the tower without being secured.

j. When attaching the rappel rope to the snaplink, rappellers pull the slack towards the anchor point. The rappel master or rappel lane NCO physically check each hookup.

k. All personnel weighing more than 200 pounds will conduct a standard hookup rappel to determine if they require a friction hookup. A friction hookup is created by placing an additional two ropes in the gate of the snaplink (for a total of six ropes in the snaplink).

l. Combat equipment is positioned on the rappeller so that it does not interfere with the brake hand. The weapon must be slung diagonally across the back with the muzzle pointing down, and on the opposite side of the brake hand.

m. Heavy duty gloves are required for all rappel training.

n. Kevlar helmets with chin straps fastened are worn during tower rappel training.

o. While on the tower, the rappeller maintains eye contact with the rappel master or rappel lane NCO and receives all commands from them.

p. The rappeller ensures that he has a belayer on his rope.

q. The belayer does not wear gloves and keeps both hands on the rope at all times. He also faces the rappeller at all times.

r. All tower rappelling is performed with a double strand of rope.

s. No running is allowed on the tower.

t. No smoking or eating is allowed near the tower.

u. All participants who are unable to rappel, lack confidence, or refuse to rappel are reported to the rappel master or OIC. These participants are immediately removed from the training area.

v. The RSO and rappel master must be aware of overconfidence and carelessness of some rappellers. The rappel master ensures all personnel are tower-qualified before beginning aircraft rappel training.

1-9. TOWER SAFETY AND PREPARATION

The rappel master is in charge of the tower. He conducts a visual and physical inspection of every item of equipment to include the structural lumber and timber, the ladder, the platform floor, and all anchor points.

a. The static tower will not be used during thunderstorms or excessively high winds. If ice is present, or if the platform is slick from rain, rappelling will be delayed until conditions are safe.

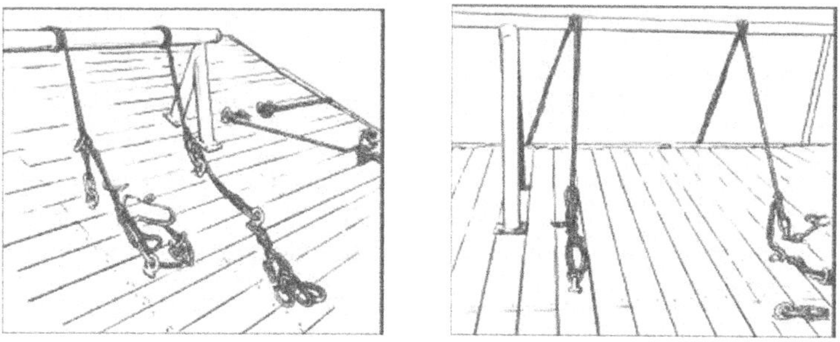

A B

Figure 1-2. Example of rappel tower anchor points.

b. All rope stations are rigged with two anchor points (Figure 1-2A, B, C). The first anchor point is a middle-of-the-rope knot, and the second is an end-of-the-rope anchor knot. The rappel master removes all the slack between the knots to create
equal tension on the anchor points. He ensures that no less than 10 feet of rope is on the ground during static rappelling.

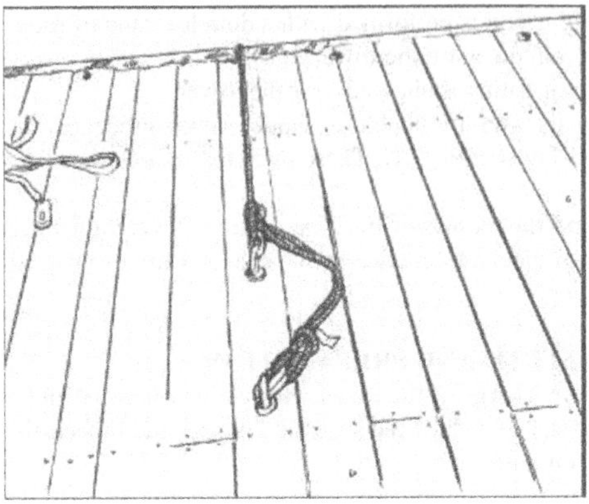

C

Figure 1-2. Example of rappel tower anchor points (continued).

1-10. RAPPELLER PREPARATION

Before conducting a rappel, each rappeller must prepare his individual clothing and equipment.

> **WARNING**
>
> **Failure to properly prepare rappellers could result in bodily injury or damage to equipment.**

a. Secure shirt tails, loose clothing, equipment, straps, and long hair.

b. Wear a helmet during rappelling. Properly fasten all straps and ensure the helmet is in serviceable condition.

c. Wear heavy leather workman's gloves.

d. Wear identification tags.

e. When rappelling with equipment, LBE or LBV should be unfastened in the front or fastened loosely behind the back of the rappeller. The rucksack should be worn high and tight on the back of the rappeller to allow the brake hand to reach the small of the back. Rucksack adjustment straps will be tied across the chest or tucked away.

f. Sling the weapon diagonally across the back with the muzzle down. Ensure the muzzle is on the guide hand side and the stock is towards the brake hand.

Note: Soldiers rappelling with equipment in excess of 50 pounds may want to consider using a friction hookup.

Section III. RAPPELLING PROCEDURES

This section discusses procedures used in tower rappelling.

1-11. SEAT-HIP RAPPEL

When using the seat-hip rappel, friction is created by a snaplink that is inserted in a sling rope seat and fastened to the rappeller. This method provides a faster and more controlled descent than other methods. Wear gloves to prevent rope burns. An alternate technique is to insert a second snaplink into the first snaplink (attached to rappel seat) and run the rope through the second snaplink. This allows easier disengagement from the rappel rope without running the entire rope through the first snaplink. To disengage from the rappel rope using the alternate technique, release the tension from the rope by opening the gate of the first snaplink and removing the second snaplink (with the rope attached).

a. The rappel seat is constructed as follows (Figure 1-3A through 1-3T):

(1) Place the midpoint (center) of the length of the sling rope on the hip opposite the brake hand (the brake hand is the strong hand) (Figure 1-3A, B, C).

(2) Bring the sling rope around the waist above the hip bone. Tie a double overhand knot over the navel (Figure 1-3D, E, F, G, H).

(3) Let the two free ends of the sling rope fall to the ground in front (Figure 1-3I).

(4) Bring the two free ends of the sling rope down between the legs and up over the buttocks. Ensure that the two free ends do not cross (Figure 1-3J).

(5) Pass the ends of the ropes over the rope that is tied around the waist at the two points above the center of the two rear seat pockets (Figure 1-3K).

(6) Grab the free end of the rope that is on the left side of the body with the left hand, and the free end of the rope that is on the right side of the body with the right hand.

(7) Squat down and simultaneously pull on both running ends of the ropes and stand up. This will tighten the seat.

(8) Take the two running ends of the rope down and back over the waist rope from the inside. Bring the running ends back under the ropes that are going across the buttocks (Figure 1-3L).

(9) Tie the two running ends with a square knot and two overhand knots on the hip opposite the brake hand (Figure 1-3M, N).

(10) Place any excess rope in the trouser pocket near the square knot (Figure 1-3O, P).

(11) With the gate down and the hooked end of the snaplink against the navel, place the end of the snaplink through the single rope that is around the waist and the two ropes forming the double-hand knot (Figure 1-3Q).

(12) Rotate the snaplink a half turn so that the gate is facing up and will open away from the body (Figure 1-3R, S, T).

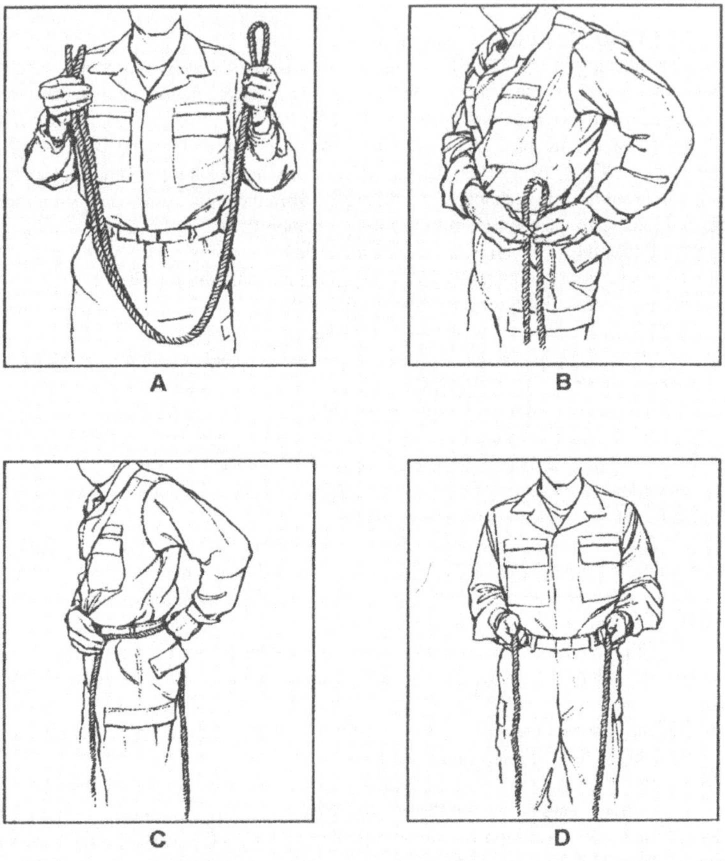

Figure 1-3. Construction of rappel seat.

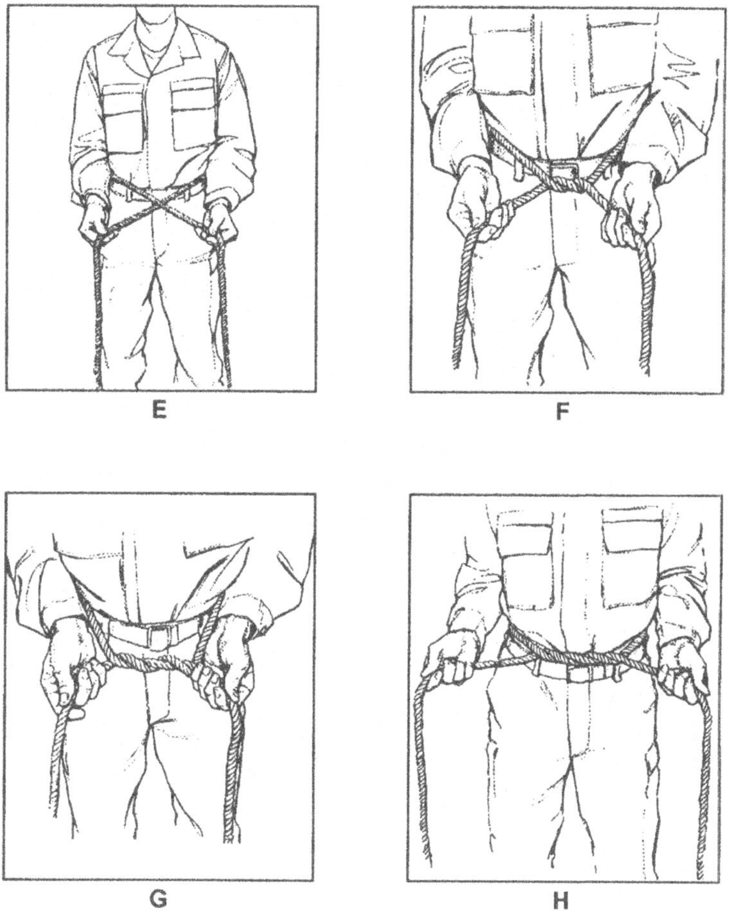

Figure 1-3. Construction of rappel seat.

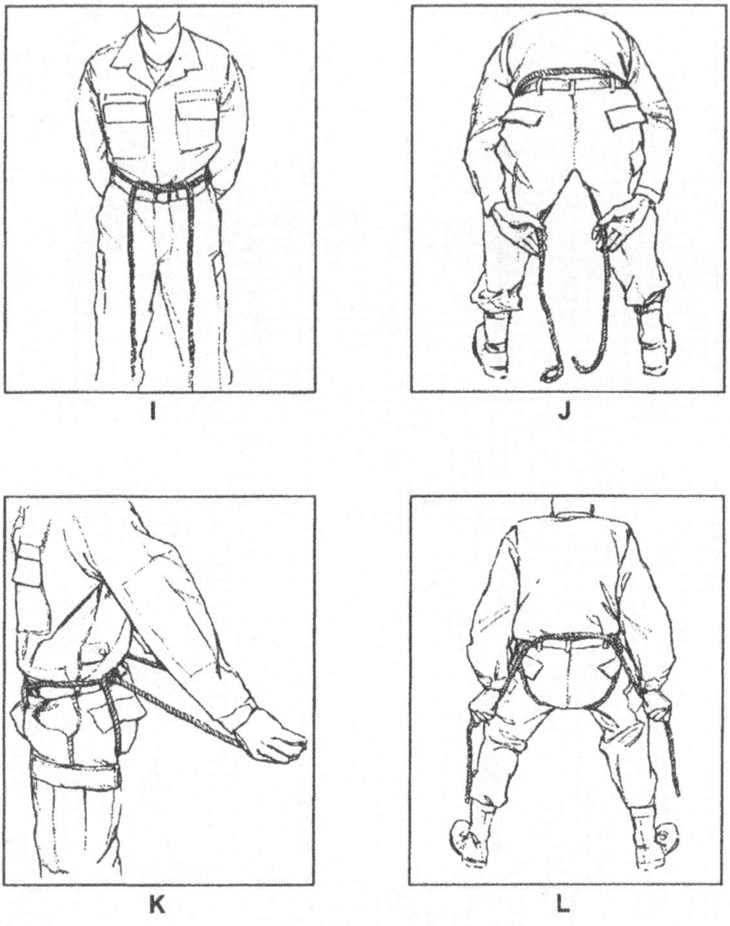

Figure 1-3. Construction of rappel seat.

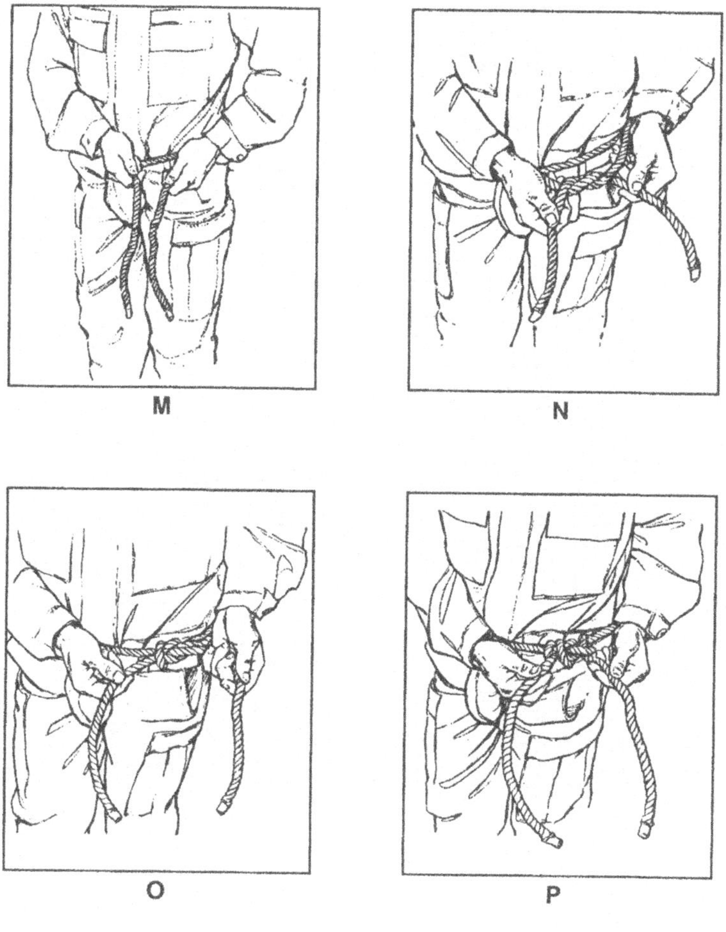

Figure 1-3. Construction of rappel seat.

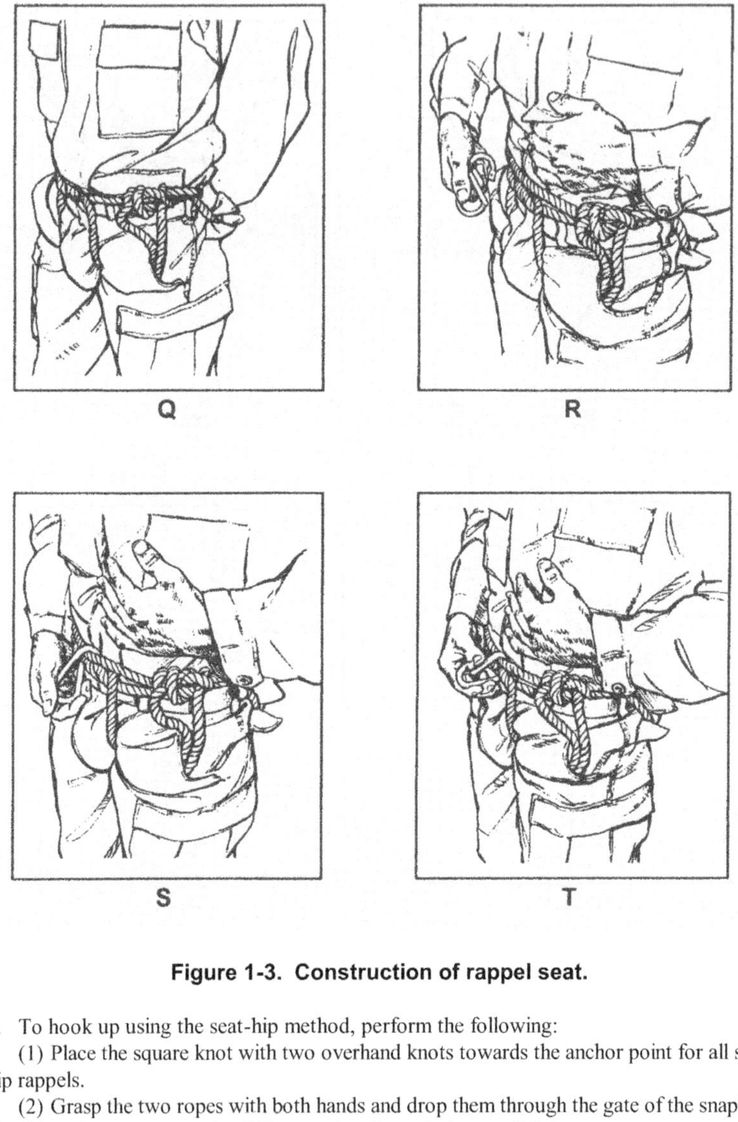

Figure 1-3. Construction of rappel seat.

b. To hook up using the seat-hip method, perform the following:

(1) Place the square knot with two overhand knots towards the anchor point for all seat-hip rappels.

(2) Grasp the two ropes with both hands and drop them through the gate of the snaplink. (At this point, two ropes should be running through the snaplink.)

(3) Using the hand closest to the anchor point, pull the slack towards the anchor point. Rotate the slack under and then over the top of the snaplink.

(4) Drop the two ropes a second time through the gate of the snaplink. (At this point, four ropes should be running through the snaplink.)

(5) Place the guide hand on the rope between the anchor point and the snaplink (palm facing up).

(6) Place the brake hand around the running end of the rope (palm facing down). Place the brake hand with the rope in the small of the back.

1-12. AUSTRALIAN RAPPEL

To hook up for the Australian rappel (Figure 1-4), perform the following:

a. Hook up to the Australian seat.

(1) Grasp the 9-foot sling rope at the midpoint (center) of its length, and double the rope.

(2) Place the doubled sling rope around the back and waist. Ensure that the rope is above the hipbone, but below the ribs.

(3) Tie a *square knot with two overhand knots* over the navel.

(4) Place any excess rope in the trouser pocket nearest the excess.

b. When the Australian seat is donned, face away from the anchor point and to the side of the rappel rope. (Stand to the same side of the rope as the brake hand. Determine left and right of the rope while facing the anchor point.) Place a snaplink onto the seat with the gate facing up, hinge closest to the body. Place it on the hip corresponding to the brake hand. The rappel master grasps the rappel rope and lays the rope into the snaplink. He then places one hand between the snaplink and the anchor point and draws slack toward the anchor point. He rotates the slack down, under, and over the rope and into the snaplink. The rappel master then slides the snaplink directly to the rappellers back.

c. Serve the running end of the rappel rope with the brake hand and prepare to rappel. During descent, brake by drawing the rappel ropes diagonally across the chest with the ropes running from near the waist to the pocket of the opposite shoulder.

Figure 1-4. Australian rappel.

1-13. CLIMBING PROCEDURES

Before climbing the ladder, the rappel master, safety OIC, or rappel lane NCO checks each rappeller's equipment.

a. The rappeller kicks the sand off his boots before climbing.

b. The rappeller grasps the outside of the ladder while climbing, not the rungs (when possible). If stairs are built for the tower, the rappeller grasps the railings as appropriate.

c. Just before climbing up the ladder, the rappeller sounds off, "(name) climbing," and then begins climbing up the ladder. Once at the top and clear of the ladder, the rappeller sounds off, "(name) clear."

d. Once off the ladder, the rappeller waits until the rappel master or lane NCO directs him to proceed to a rope station. At this time, the next rappeller in line may start to climb the ladder.

1-14. TOWER PROCEDURES

After the rappellers climb the tower, the following procedures are adhered to:

a. Once directed to a rope station, the rappel master or lane NCO ensures proper hookup for rappelling.

b. At this time, the rappeller sounds off with "on rappel" and the belayer sounds off with "on belay."

c. While maintaining his brake, the rappeller (on command from the rappel master or lane NCO) steps over the safety rail and faces the anchor point.

d. At this point, the rappel lane NCO sounds off with the following verbal commands and arm-and-hand signals.

(1) *Get Ready*—Rappel master or lane NCO extends both arms to the front with fists clenched and thumbs pointing upward (Figure 1-5). This alerts the rappeller. Each rappeller then looks over his brake hand shoulder to check for the belay man. The rappeller then looks at the rappel master. The rappel master or lane NCO makes his second check of the hookup, rappel seat, snaplink, and equipment.

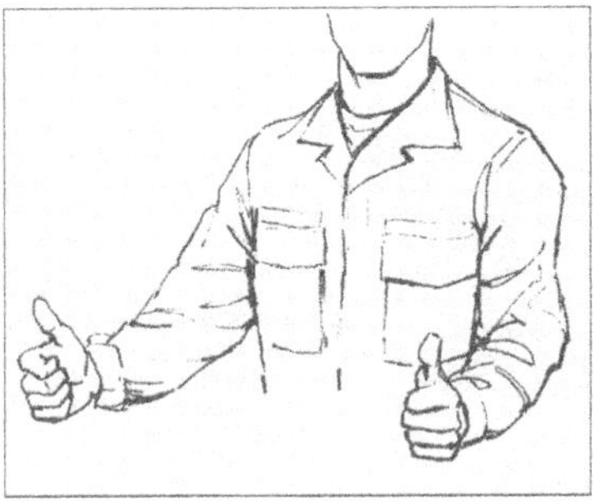

Figure 1-5. Example of arm-and-hand signal GET READY.

(2) *Position*—The lane NCO extends both arms to the front, elbows locked, forearms pointed downward, and fingers extended. He makes a circular motion with both forearms rotating in opposite directions (Figure 1-6). With the brake hand in the small of the back, the rappeller rotates 180 degrees out onto the wall or skid mock up and assumes an L-shaped position. The feet should be shoulder width apart, balls of feet on the wall or skid, knees locked, and body bent at the waist (Figure 1-7).

Figure 1-6. Example of arm and hand signal POSITION.

Figure 1-7. L-shape position.

(3) *Go*—The lane NCO extends his right arm with the elbow locked, fingers extended, thumbs around the index fingers, and points directly at the rappeller (Figure 1 8). This initiates the rappel. The rappeller flexes his knees and jumps vigorously backwards. At the same time, the rappeller throws his brake hand out at a 45-degree angle, letting the rope slide through both the brake hand and the guide hand. The rappeller looks over his brake hand shoulder at all times during descent.

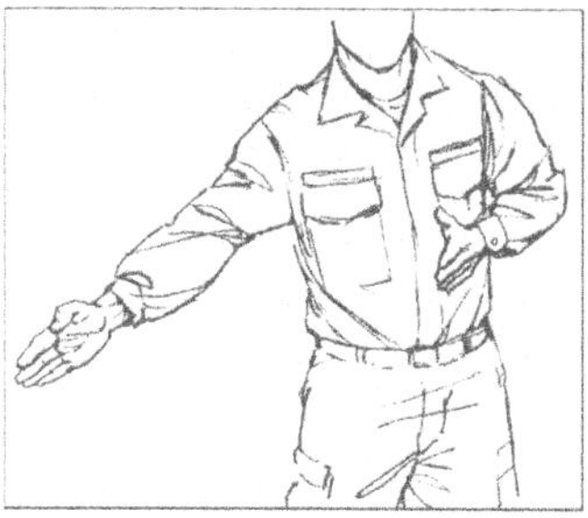

Figure 1-8. Example of arm-and-hand signal GO

e. The rappeller descends in a smooth, controlled manner.

f. The rappeller maintains eye contact with the ground at all times.

g. The rappeller maintains a modified L-shape position during descent with the feet shoulder-width apart, knees flexed, and buttocks parallel to the ground (Figure 1-9).

Figure 1-9. L-shape position while rappelling.

h. When carrying equipment or additional weight, a modified L-shape is used with the legs slightly lower than the buttocks to compensate for the added weight.

i. The rappeller's back is straight. He looks over the brake hand shoulder.

j. The guide hand is extended on the rope with the elbow extended and locked.

k. The rope slides freely through the guide hand, which is used to adjust equipment and to assist in balance during descent.

l. To brake, the rappeller places the brake hand (with rope in hand) in the small of the back and then grasps the rope firmly with the brake hand.

Note: Do not grip the rope firmly with the brake hand while the brake hand and brake arm are extended at the 45-degree angle. If this is done while rappelling, the brake hand and glove may become entangled in the snaplink causing injury to the hand and causing the rappeller to become hung up on the ropes.

m. Releasing tension on the rope and moving the brake hand out to the rear at a 45-degree angle regulates the rate of descent.

n. The rappeller never lets go of the rope with his brake hand until the rappel is completed.

o. After the rope is cleared and the rappeller is off rappel, he acts as the belayer for the next rappeller.

1-15. RAPPEL TOWER TRAINING FOR UH-1H HELICOPTER

Training on the rappel tower for helicopter skid rappelling prepares soldiers to rappel from a UH-1H helicopter (Figure 1-10). The rappeller is hooked up while he sits on the platform just above the helicopter skid. On the rappel master's command GET READY, the rappeller looks over the edge of the tower to ensure the running ends of the ropes are on the ground (Figure 1-10A). On the command SIT IN THE DOOR, the rappeller rotates his feet and legs off the platform and places them on the skid (Figure 1-10B). On the command POSITION, the rappeller turns around and assumes an L-shape position (Figure 1-10C). On the command GO, the rappeller bounds away from the helicopter skid and rappels to the ground. The rappel master is responsible for the proper procedures and safety.

A B C

Figure 1-10. UH-1H helicopter skid rappel training.

1-16. RAPPEL TOWER TRAINING FOR UH-60 BLACKHAWK HELICOPTER

Rappel tower training for the UH-60 Blackhawk is similar to that for the UH-1H with the exception that the UH-60 has no skids on which to stand. Therefore, the edge of the rappel tower is used as a pivot point to assume the L-shape position. All commands are the same except for SIT IN THE DOOR, which does not apply to the UH-60. The stances for each command are also different, as shown in Figure 1-11.

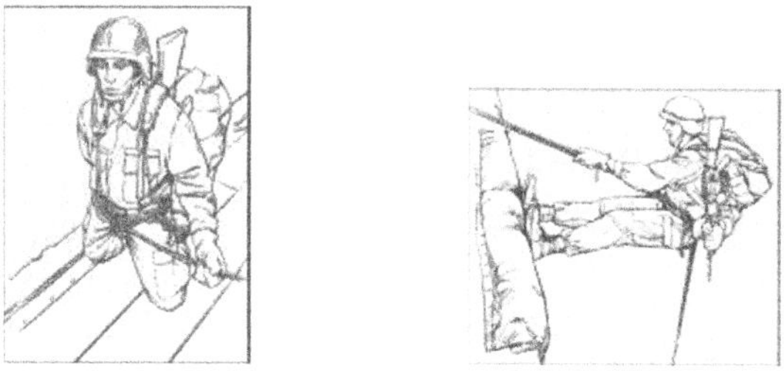

A B

Figure 1-11. UH-60 helicopter rappel training.

1-17. EMERGENCY LOCK-IN PROCEDURES

Mastering the lock-in procedure during tower training is critical before advancing to helicopter rappelling. Using the lock-in procedure allows personnel to hold in position for an extended period of time.

a. If the helicopter gains altitude above the length of the rappel ropes, the rappeller immediately brakes, locks in, and waits for the descent of the aircraft. Procedures for lock-in are as follows:

(1) Place the brake hand in the small of the back and brake to a complete stop.

(2) Release the guide hand from the ropes.

(3) Bring the guide hand around the back and grasp the running end of the two rappel ropes behind the brake hand.

(4) Using the guide hand, bring the two running ends of the rappel ropes around to the front.

(5) Secure these two running ends of the rappel ropes with the two anchor ends of the rappel ropes in the guide hand. This is now the new brake hand.

(6) Take the old brake hand out of the small of the back. Bring it around to the front and grasp the two ropes from the anchor point at a point just above the new brake hand. The old brake hand is now the new guide hand.

(7) Face the rappel master and wait for his command to lower to the ground.

(8) When the command is received from the rappel master to continue the descent, bring the brake hand to a 45-degree angle to the rear. When it is time to brake, bring the new brake hand around to the front diagonally across the chest.

b. If an engine fails or an aircraft emergency occurs during rappelling, the rappellers on the ropes descend as rapidly as possible and move from beneath the aircraft to the sides. Rappellers will maintain control of ropes, if possible.

1-18. COMMUNICATIONS

The rappeller at the top of a rappel point must be able to communicate with those at the bottom. Radios, hand signals, and rope signals are considered during a tactical rappel. For training situations, use the following commands:

COMMAND	GIVEN BY	MEANING
LANE NUMBER, ON RAPPEL	Rappeller	I am ready to begin rappelling.
LANE NUMBER, ON BELAY	Belayer	I am on belay and you may begin your rappel.
LANE NUMBER, OFF RAPPEL	Rappeller	I have completed the rappel, cleared the rappel lane, and am off the rope.
LANE NUMBER, OFF BELAY	Belayer	I am off belay.
LANE NUMBER, FALLING	Rappeller	I am falling. Be alert below—belay man brake.

1-18. DEMONSTRATION

After explaining the procedures to all rappellers, the rappel master should have an assistant demonstrate one complete cycle of rappelling from the static tower. This ensures that the rappellers can hear all the proper commands and see the actions and techniques used on the static tower.

CHAPTER 2

GROUND RAPPELLING

Ground rappelling is a technique that allows soldiers to negotiate mountains and cliffs safely and rapidly. Before rappellers participate in mountain and cliff rappelling they should have completed all the requirements for tower training. To ensure the rappel is safely conducted and meets the mission requirements, rappellers should consider several factors.

2-1. PERSONNEL

The personnel responsibilities for mountain and cliff rappelling are the same as for tower rappelling. As a minimum the following personnel are needed for mountain and cliff rappelling.

a. **Mountaineering Safety Officer (MSO).** The MSO is a SFC or above who is ranger qualified or is a graduate of the Northern Warfare Training Center (NWTC). He has overall responsibility for the safety of all mountaineering participants and ensures that all safety precautions are followed.

b. **Rappel Master**. The rappel master is a qualified officer or NCO in charge of training on ground, tower, and aircraft rappel safety procedures.

c. **Medical Officer or Noncommissioned Officer.** Qualified rappellers and aidmen must know MEDEVAC procedures.

d. **Radiotelephone Operator.** The RATELO monitors the radio and reports to the RSO.

e. **Rappel Lane NCO.** The rappel lane NCO is a qualified officer or NCO in charge of rappelling on a wall or ground rappel lane (Chapter 1, paragraph 1-3).

2-2. SUSTAINMENT TRAINING

Before conducting ground rappel training, the unit or element conducts sustainment training.

a. Review the construction of a rappel seat, seats to be used, and hook-up procedures.

b. Conduct two rappels on the 34-foot tower wall (20-foot tower is acceptable): one with equipment, one without equipment. Conduct two rappels from the free side of a 34-foot or higher rappel tower.

c. Demonstrate the ability to lock in during one of the rappels from the free side of the tower.

2-3. SELECTION OF A RAPPEL POINT

The selection of the rappel point depends on factors such as mission, cover, route, anchor points, and edge composition (loose or jagged rocks). The anchor point should be above the rappeller's departure point. Primary and secondary anchors must be solid (natural anchors are preferred). The rappeller should be sure that the rope reaches the bottom or a place from which he can further rappel or climb. Also, the rappel point should be carefully tested and inspected to ensure the rope will run freely and that the area is clear of obstacles that could be dislodged. If a sling or runner is used for a rappel point, it should be tied twice from two separate loops. Suitable loading and off-loading platforms should be available.

2-4. ESTABLISHMENT OF A RAPPEL POINT

A rappel lane is the area and rope that a rappeller will use during his descent. It should have equal tension between all anchor points by establishing primary and secondary anchor points (Figure 2-1). If one anchor point fails, the rappel rope should not extend. All of the methods discussed below can be performed with a single or double rope. A double rope application should be used for safety.

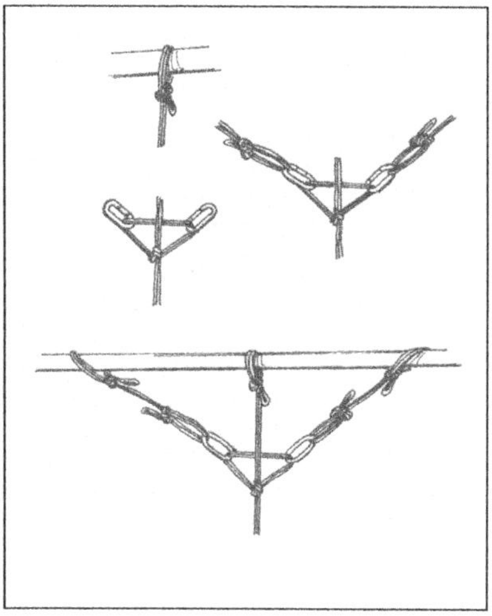

Figure 2-1. Rappel point.

a. If a rappel lane is *less* than half the rope length, the rappeller may apply one of the following techniques:

(1) Double the rope and tie a three-loop bowline around the primary anchor to include the primary anchor inside two loops and enough rope in the third loop to run to the secondary anchor (another three-loop bowline secured with an overhand knot).

(2) Double the rope and tie it around a secure anchor point with a round turn anchor bowline secured with an overhand knot (or any appropriate anchor knot).

(3) Double the rope and establish a self-equalizing anchor system (Figure 2-2) with a bowline-on-a-bight or figure-eight-on-a-bight knot. Tie off on the long standing and with a round turn anchor bowline.

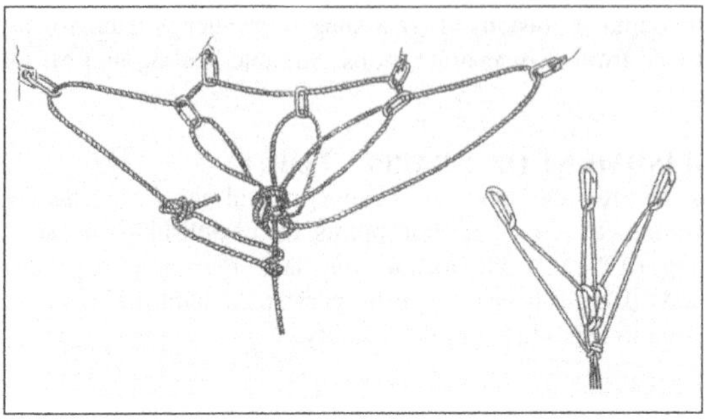

Figure 2-2. Self-equalizing anchor system.

(4) In an emergency, double the rope and place it behind or through a secure anchor point, or tie a runner around an anchor point with a snaplink inserted and place the rope through the snaplink. To preclude a rappeller from sliding off the end of the rappel lane, tie a double figure eight (square knot or double fisherman's knot) at the bottom end of the rope with both ends.

b. If a rappel lane is *greater* than half the rope length, the rappeller may apply one of the following techniques:

(1) Using two ropes, tie a round turn anchor bowline around a primary anchor point. Take the remaining rope (the tail from the primary anchor bowline) and tie another round turn anchor bowline to a secondary anchor point. The secondary anchor point should be in a direct line behind the primary anchor point. The anchor can be either natural or artificial. The ends of the rappel lane ropes should be offset by 15 centimeters (6 inches) so that the rope ends feed freely through the rappeller's snaplink.

(2) Using two ropes, establish a three-piton anchor system using a bowline-on-a-bight knot (or figure-eight-on-a-bight knot) and tie off on the long-standing end with a round turn anchor bowline.

(3) In an emergency, use two ropes and tie the two ends together with a joining knot. Place the joined ropes behind or through an anchor point, or tie a runner around an anchor point with a snaplink inserted and place the joined rope through the snaplink. Offset the joining knot to the left or right of the anchor. Tie off the bottom end of the rope with a joining knot to prevent a rappeller from sliding off the end of the rappel line.

c. Due to the length of the rappel, the rappel rope may not reach the anchor. If the rope is used to tie the knots, it may be too short to accomplish the rappel.

(1) When using a natural anchor—

(a) Tie a sling rope, runner, or another rope around the anchor with a round turn anchor bowline.

(b) Tie a fixed loop (figure eight or butterfly) in one end of the rappel rope, which is attached to the round turn around the anchor through the two snaplinks (opposing gates).

(2) When using an artificial anchor—

(a) Tie off a sling rope, runner, or another rope to form a loop.

(b) Put the loop through the snaplinks that are attached to the artificial anchor point.

(c) Bring the bottom of the loop up and connect it to the snaplinks that are between the artificial anchors (chocks, pitons, or bolts).

(d) Grasp the snaplinks that are between the chocks/pitons and pull them down and together.

(e) Tie a fixed loop (figure eight or butterfly) in the end of the rappel rope and connect this to the snaplinks that have been pulled together.

Notes: 1. Rerouted figure-eight knots can be used instead of bowlines.
2. Runners may be used from one or more anchor points.

d. Backfeed (stack) the rope to ensure it does not snarl when thrown. Take off one wrap at a time and let it fall to the ground, ensuring that no kinks, knots, or twists occur that might hinder the rope from feeding out. When the rope is backfed, anchor off one end of the rope. The two methods used to throw the rope are underhand and overhand. Use the overhand method when trees or shrubs are on or near the rappel point.

e. Set up a retrievable rappel point by applying one of the following techniques:

(1) When the rappel is *less* than half the total length of the rope, double the rope. Place the rope around the primary anchor with the bight formed by the midpoint. Join the tails of the rappel rope and throw the rope over the cliff. Tie a clove hitch knot around a snaplink just below the anchor point ensuring the locking bar inside the snaplink points away from the gate opening end and faces uphill. Snap the opposite standing portion into the snaplink. Upon reaching the bottom of the cliff, pull on the part of the rope to which the snaplink is secured to allow the rope to slide around the anchor point.

(2) When the length of the rappel is *greater* than half the length of the rope used, join two ropes around the anchor point (double fisherman's knot or square knot). Adjust the joining knot so that it is away from the anchor. Tie a clove hitch knot around a snaplink just below the anchor point ensuring the locking bar inside the snaplink points away from the gate opening end and faces uphill. Snap the opposite standing portion into the snaplink. Upon completion of the rappel, pull the rope to which the snaplink is secured to allow the rope to slide around the anchor point.

WARNING

When setting up a retrievable rappel using only a primary point, take care in selecting the point. Ensure soldiers have a safety line when approaching the rappel point, with only the rappeller going near the edge.

2-5. TYPES OF RAPPELS

Types of rappels include body rappel, hasty rappel, seat-hip rappel, seat-shoulder rappel, Australian rappel, and buddy-evacuation. (Seat-hip rappel and Australian rappel are described in Chapter 1, paragraphs 1-10 and 1-11.)

a. **Body Rappel.** Face the anchor point and straddle the rope. Pull the rope from behind and run it around either hip, diagonally across the chest, and back over the opposite shoulder (Figure 2-3). Then, run the rope to the brake hand, which is on the same side of the hip that the rope crosses (for example, the right hip to the left shoulder to the right hand). Lead with the brake hand down and face slightly sideways. The foot corresponding to the brake hand precedes the guide hand at all times. Keep the guide hand on the rope above to guide—not to brake. Lean out at a sharp angle to the rock. Keep the legs spread well apart and relatively straight for lateral stability and the back straight to reduce friction. Turn the BDU collar up to prevent rope burns on the neck. Wear gloves, and use other clothing to pad the shoulders and buttocks. To brake, lean back and face directly toward the rock area so the feet are horizontal to the ground.

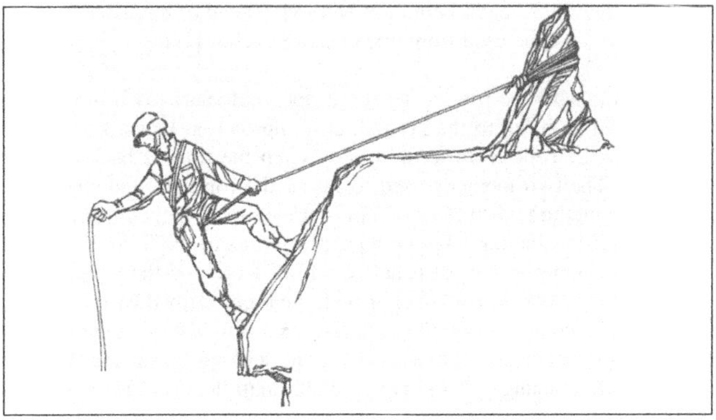

Figure 2-3. Body rappel.

b. **Hasty Rappel.** Facing slightly sideways to the anchor, place the ropes horizontally across the back (Figure 2-4). The hand nearest to the anchor is the guide hand, and the other is the brake hand. Wear gloves to prevent rope burns. To stop, bring the brake hand across in front of the body locking the rope. At the same time, turn to face up toward the anchor point. Use this rappel only on moderate rock pitches. The hasty rappel's main advantage is that it is easier and faster than the other methods, especially when the rope is wet.

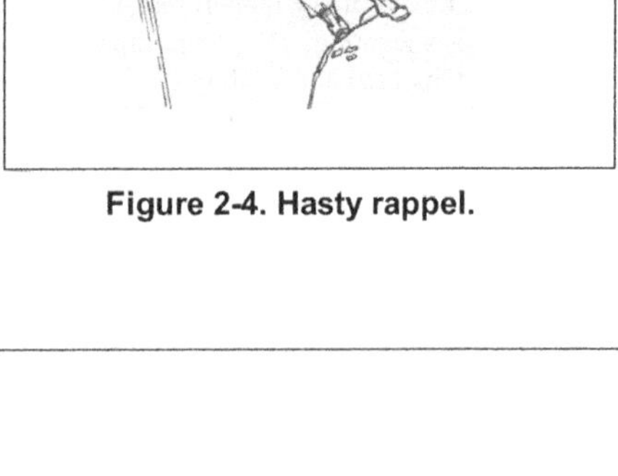

Figure 2-4. Hasty rappel.

Figure 2-5. Seat-shoulder rappel.

c. **Seat-Shoulder Rappel.** To hook up for the seat-shoulder method, face the rappel point (Figure 2-5). Snap into the rope that passes up through the snaplink. Bring the rope over one shoulder and back to the opposite hand (left shoulder to right hand). Use the same technique in the descent as in the body rappel. This method is faster than the body rappel, less frictional, and more efficient rappellers with packs and during night operations.

d. **Buddy-Evacuation Rappel.** Use the buddy-evacuation rappel to evacuate an injured soldier from a cliff or steep terrain. Face the cliff and assume a modified L-shape body position to compensate for the weight of the victim on the back. The victim is top-rope belayed from above, which provides the victim with a point of attachment to a secured rope.

Note: To use this rappel, the victim must be conscious.

(1) The method for securing a victim to a rappeller's back is described below:

(a) To secure the victim to the carrier's back with a rope, the carrier ties a standard rappel seat (brake of choice, depending on the injury) and rests his hands on his knees while the victim straddles his back (Figure 2-6.)

(b) A 4.2-meter (14-foot) sling rope is used. A 45-cm (18-inch) tail of the sling is placed on the victim's left hip. (This method describes the procedure for a seat-hip rappel with right-hand brake.)

(c) The long remaining end of the sling rope is routed under the victim's buttocks and passed over the victim's and carrier's right hip. The rope is run diagonally, from right to left, across the carrier's chest, over his left shoulder, and back under the victim's left armpit.

(d) The rope is then run horizontally, from left to right, across the victim's back. The rope is passed under the victim's right armpit and over the carrier's right shoulder.

Figure 2-6. Patient secured to carrier's back.

(e) The rope is run diagonally, from right to left, across the carrier's chest and back across the carrier's and victim's left hip.

(f) The two rope ends should now meet. The two ends are tied together with a square knot and overhand knots.

(g) The knot is positioned on the victim's left hip. The carrier's shoulders should be padded to prevent cutting by the rope.

(2) An alternate method is to use two pistol belts hooked together and draped over the carrier's shoulders. The victim straddles the carrier, and the belay man secures the loose ends of the pistol belts under the victim's buttocks. Slack in the pistol belt's sling should be avoided, since the carrier is most comfortable when the victim rests high on his back. (See FM 8-35.)

(3) A large rucksack can be slit on the sides near the bottom so that the victim can step into it. The victim is belayed from the top with the carrier conducting a standard rappel. The carrier wears the rucksack with the victim inside.

Note: Rucksacks are not designed to support the weight of a soldier and his gear. Therefore, this technique is used only as a last resort.

(4) A casualty secured to a carrier, as described above, can be rappelled down a steep cliff using a seat-shoulder or seat-hip rappel. The casualty's and rappeller's shoulders should be padded where the sling rope and rappel lines cross if a seat-shoulder rappel is used. The buddy team should be belayed from above with a bowline tied around the victim's chest under his armpits. The belay rope must run over the rappeller's guide hand shoulder (Figure 2-7).

Figure 2-7. One-man carry.

2-6. RAPPELLING PROCEDURES

Proper rappelling procedures must be followed to ensure the safety of personnel.

a. The rappel lane NCO is tied off to a separate anchor point for safety, if possible. He—

(1) Ensures that the anchors are sound and the knots are properly tied.

(2) Ensures that loose rock and debris are cleared from the loading platform.

(3) Allows only one man on the loading platform at a time and ensures that the rappel point is orderly run.

(4) Ensures that each man is properly prepared for the particular rapel: gloves on sleeves down, helmet with chin strap fastened, gear prepared properly, and rappe seat and knots correct (if required). He also ensures that the rappeller is correctly hooked up to the rope and is aware of the proper braking position.

(5) Ensures that the proper signals or commands are used.

(6) Dispatches each man down the rope.

(7) Is the last man down the rope.

(8) Ensures that the rope does not run over sharp edges, and pads the rappelling surfaces as necessary to protect the rope.

(9) Ensures that the rope reaches the bottom or is at a place where additional rappels can be made.

b. The first rappeller down—

(1) Selects a smooth route for the rope that is clear of sharp rocks.

(2) Conducts a self-belay with a prusik sling tied from himself to the rappel rope.

(3) Clears the route and places loose rocks, which the rope may dislodge, far enough back on ledges to be out of the way.

(4) Ensures the rope reaches the bottom or is at a place from which additional rappels can be made.

(5) Ensures that the rope runs freely around the rappel point when pulled from below.

(6) Clears the rappel lane by straightening all twists and tangles from the ropes.

(7) Belays subsequent rappellers down the rope.

(8) Takes charge of personnel as they arrive at the bottom (off-loading platform).

Note: A rappeller is always belayed from the bottom except for the first man down. The first man belays himself down the rope by using a safety line attached to his rappel seat that is hooked to the rappel rope with a prusik knot. As the first man rappels down the rope, he "walks" the prusik knot down with him.

2-7. DUTIES OF THE RAPPELLER

To ensure the safe and efficient conduct of rappelling operations, each rappeller must know the duties of his job.

a. Each rappeller down shouts, "Off rappel" (if the tactical situation permits), untangles the ropes, and ensures the ropes run freely around their anchors. When silence is needed, a planned signal of pulling the rope is substituted for the voice signal. After the rope is cleared and the rappeller is off rappel, he acts as the belayer for the next rappeller.

b. All rappellers inspect ropes often when many soldiers are rappelling.

c. The last rappeller to descend constructs a retrievable rappel point and rappels down. Then he gently pulls the rope to prevent the rising rope end from entangling with the other rope. He stands clear of the falling rope and any rocks that it may dislodge.

Note: Soldiers wear gloves for all types of rappels for protection from rope burns. Also, bounding rappels are discouraged since this stresses the anchor and causes undue wear and friction on the rope.

d. Rappellers descend in a smooth, controlled manner. The body forms an L-shape with the feet shoulder-width apart, legs straight, and buttocks parallel to the ground. When carrying equipment or additional weight, a modified L-shape is used with the legs slightly lower than the buttocks to compensate for the added weight. The rappeller's back is straight. He looks over the brake shoulder. The guide hand is extended on the rope with the elbow extended and locked. The rope slides freely through the guide hand, which is used to adjust equipment and to assist balance during the descent. The rappeller grasps the rope firmly with the brake hand and places it in the small of his back. Releasing tension on the rope and moving the brake hand out to his rear at a 45-degree angle regulates the rate of descent. The rappeller never lets go of the rope with his brake hand until the rappel is completed.

e. During training, the lane number must be understood. In combat, a series of tugs on the rope may be substituted for the oral commands to maintain noise discipline. The number of tugs used to indicate each of the commands is IAW the unit SOP.

2-8. BELAYER

The belayer assumes a position at the base of the lane about one pace away from the rock area. He ensures that the rappel ropes are even with the ground during rock and tower rappels. The belayer loosely holds the rappel ropes with both hands so as not to interfere with the rappeller but still able to stop the rappeller should he fall. If the rappeller shouts "Falling" or loses control of his brake hand or descent, the belayer immediately stops the rappeller by pulling downward on the rappel ropes. To ensure a firm grip on the rappelling rope, the belayer does not wear gloves. (Because no friction exists between the belayer's hands and the rappelling rope, gloves are not required for safety. However, whether or not to wear gloves is decided by the OIC/NCO.) The belayer watches the rappeller at all times and maintains constant voice or visual contact with the rappeller. The belayer wears a helmet to prevent injuries from falling debris. All commands are spoken loudly and clearly.

CHAPTER 3

HELICOPTER RAPPELLING

Helicopter rappelling can provide a means of quick insertion with or without an LZ.

Section I. PERSONNEL

The personnel required for helicopter rappelling are the rappel master, the rappel safety officer, the pilot in command, the rappelers, and the belayers.

3-1. RAPPEL MASTER

A qualified rappel master is aboard each aircraft. Safety is the rappel masters number one priority. The rappel master—

- Ensures internal communication between the pilot and rappel master, and external communication between the aircraft and the ground.
- Inspects all equipment and uses only authorized, serviceable equipment.
- Inspects and tests all anchor points and knots before the mission starts.
- Ensures that all rappellers receive a safety briefing and the pilots and aircrew receive an air mission brief.
- Ensures that the rappelers are rappel qualified before conducting helicopter rappelling to include tactical rappelling.
- Maintains communications with the pilot at all times.

3-2. RAPPEL SAFETY OFFICER

The RSO is a SFC or above who is either air assault or ranger qualified. He has overall responsibility for the safety of all rappellers and ensures that all safety precautions are followed. He maintains communications at all times with the pilot and rappel master through FM radio. He alerts the rappel master and pilot of any unsafe acts.

3-3. PILOT IN COMMAND

The pilot in command (PIC) of the aircraft has the following responsibilities:

a. Ensures that the aircrew and all non-aircrew personnel are briefed and understand their responsibilities during rappelling operations, including aircraft safety and action in the event of an emergency.

b. Ensures that the donut ring anchoring device assembly and or aircraft anchor points have been inspected for completeness and functionality, and that they are installed properly.

c. Emphasizes procedural techniques for clearing, recovery, and or jettison of ropes.

d. Keeps the aircraft centered over the target with corrections from the rappel master as required.

3-4. RAPPELLER

In addition to the tower qualification requirements outlined in Chapter 1, the individual rappeller must complete advance training under the supervision of a qualified rappel master in order to participate in tactical helicopter rappelling operations. He must—

- Satisfactorily complete three rappels from a helicopter from a height of 60 feet (two rappels with combat equipment and weapon).
- Demonstrate confidence and proficiency in the techniques, procedures, and equipment used in rappelling from a helicopter.
- Know the rappelling equipment used in helicopter operations and any special equipment required for helicopter rappelling.

3-5. BELAYER

A belayer is assigned to each rope. He is responsible for walking the rope beneath the helicopter during the descent. (Walking the rope is defined as removing the slack from underneath the helicopter by walking backwards with the rope as the helicopter descends to land.) The belayer makes sure the ropes are not caught on the aircraft skids or tires to ensure a safe landing.

Section II. TRAINING

Training for helicopter rappelling includes sustainment training and refresher training.

3-6. SUSTAINMENT TRAINING

Before conducting helicopter rappel training, the unit or element conducts sustainment training.

a. Review the construction of a rappel seat, equipment to be used, and hook-up procedures.

b. Conduct two rappels on the 34-foot or higher tower wall: one without equipment, one with equipment. Conduct two rappels from the open side of a 34-foot or higher rappel tower.

3-7. REFRESHER TRAINING

Refresher training is routinely conducted to maintain acquired skills. Soldiers who have not performed a helicopter rappel during the past six months will undergo refresher training consisting of three satisfactory rappels from a tower (one with weapon and equipment and one executing a lock in) before executing a helicopter rappel.

Section III. PREOPERATIONAL BRIEFINGS AND SAFETY PROCEDURES

This section discusses safety, medical and communications requirements, and the procedures to follow during unusual conditions (adverse weather/terrain conditions, night operations). Personnel must use sound judgement to determine what action to take depending on the nature and severity of the conditions.

3-8. MEDICAL COVERAGE

A qualified and equipped medic will be present to respond to any mishap. Medical transportation must also be available. Absence of a medic, medical equipment, or transportation will terminate training. If the situation warrants and the installation cannot support a MEDEVAC mission, the rappel aircraft may be used as a last resort MEDEVAC vehicle.

3-9. COMMUNICATION REQUIREMENTS

During helicopter rappel training, the RSO will have radio communications with the aircraft. Voice communications are required before starting aircraft rappelling. Additionally, the RSO will inform the PIC to stop operations if an unsafe condition develops.

3-10. ADVERSE WEATHER/TERRAIN CONDITIONS

Rappel operations will not be conducted under the following conditions

- Ambient temperature is 30 degrees Fahrenheit or less.
- Winds in excess of 30 knots.
- Lightning strikes within 1 nautical mile of rappelling operations.
- Wind chill factors caused by the helicopter's rotor wash or extraction cruise air speeds, which could cause cold weather injuries.
- Water or ice on the rope inhibiting the ability of the rappellers to control their descent.
- The rope is exposed to the elements for a sufficient length of time to freeze, thereby reducing its tensile strength.
- Blowing particles produced by rotor wash causes the aircrew or the rappel master to lose visual contact with the ground.

3-11. NIGHT OPERATION REQUIREMENTS

The following requirements are necessary for night rappelling operations.

a. One chemlight will be attached to the end of the rope and one on each rappeller.

b. One chemlight will be secured to the attachment point of the rope.

c. Night vision goggles (NVG) will not be worn by rappellers during the descent. Aircrew members will wear NVG as required during night operations. NVG lighting criteria will be IAW Army regulations, specific aircraft aircrew training manuals, unit SOPs, or the tactical environment.

3-12. SAFETY BRIEFING

The following safety measures are enforced.

a. Loose clothing and equipment are secured.

b. Helmets are worn with chin straps fastened.

c. Rappellers wear identification tags and earplugs, carry identification cards, and role down their sleeves.

d. Weapons are slung diagonally across the back with the muzzle pointing down on the guide hand side.

e. All seats and rappelling equipment must be inspected by a rappel master before rappelling.

f. No running is allowed within 50 feet of the aircraft.

g. Personnel approach and depart the helicopter from the front and forward of the rear of the cargo doors. When approaching or departing the helicopter, personnel bend their bodies forward at the waist to ensure clearance of the rotor blades. At no time will personnel go near the rear of the aircraft.

h. Upon boarding the aircraft, the rappeller sits or kneels down, hooks up, and applies his brake hand to the small of his back.

i. While in the helicopter, the rappeller maintains eye-to-eye contact with the rappel master and receives commands from him.

j. The rappeller ensures that he has a belayer on his rope at all times when conducting training at a hover site.

k. During the descent, the rappeller maintains eye-to-ground contact.

l. If the rappeller sees his rope coming off the ground or sees that his belayer has lost control of his rope, he immediately brakes and executes a lock-in. He then waits for commands from the rappel master.

m. The rappeller brakes once every 30 feet during descent.

n. The belayer does not wear gloves. He keeps both hands on the rope and his eyes on the rappeller at all times.

o. All rappelling will be conducted using a double strand of rope.

Section IV. DEPLOYMENT OF ROPES

Deployment of the ropes from a helicopter is a critical task. It can cause a planned rappelling operation to fail, or it can increase the time required to conduct the operation. This is due to the likelihood of the ropes becoming entangled. To prevent this, ropes must be deployed using a positive control technique. Two of the techniques that may be used are the *deployment bag* technique and the *log coil* technique.

3-13. DEPLOYMENT BAG TECHNIQUE

The rappeller places the deployment bag (D-bag) (standard) on a flat surface with the stow loop facing upward. If the D-bag still has a static line, the rappeller removes it by cutting the static line where it attaches to the bag. A 12-foot section of the static line can be used as a safety line in the helicopter.

a. Use a sandbag (about one E-tool spade full of sand) as a weight in the deployment system. Roll the sandbag into a rectangular shape, tie it, and then place it in a small plastic bag, rolled and taped. Place a retaining band over the middle of the weight, and place the weight on top of the D-bag. The weight should be about the same width as the D-bag.

b. Lay out and inspect the two ropes. The working ends (closest to the D-bag) should be even. Place a round turn with the two working ends of the ropes on the weight's retaining band. Working on top of the deployment bag, start forming figure eights. The stack should consist of 8 to 10 figure eights, one on top of another—do not exceed the width of the D-bag. Then, starting from either side, center a retaining band over the stack. Ensure that it is over all the figure eights in the stack. Repeat the process each time, placing one stack in front of the other. Continue until about 10 feet of rope remains. If one rope is shorter than the other, the end of the shorter rope should be about 10 feet from the last stack.

c. For a primary anchor point, measure down about 4 feet from the end of the shortest rope. Using both ropes, tie a bowline without a half hitch. This knot is the primary anchor point.

d. For a secondary anchor point, tie an end-of-line bowline with a half hitch toward the end of the ropes. The dressed knot should be 18 to 22 inches from the primary anchor point knot when using the UH-1H helicopter and 22 to 30 inches when using the UH-60 helicopter. If excess rope remains, S-fold and tape it between the two knots. Ensure that 3 to 6 feet of rope remain from the last stack of figure eights.

e. To place the ropes in the D-bag, remove the bag from under the stack of rope. Place the weight into the bottom of the D-bag, and place all the stacks of figure eights, in the order they were made, one on top of the other into the bag.

f. Two left, two right, and two rear loops are in the opening of the D-bag. ***Do not use the two rear loops.*** Close the flap on the D-bag and push the loops through the aligning holes on the flap. Using an 8-inch piece of gutted 550-pound cord, tie the two loops together with a square knot and two hitches. Ensure that two 120-foot ropes are coming out of the center of the flap—not to one side. Repeat the process and tie the right loops together. Wrap the excess rope lengthwise around the bag.

Note: The sizes of the loops of the bowline should be no larger than an average size fist.

3-14. LOG COIL TECHNIQUE

The rappeller lays the running end of the double rope along the length of the coiling log (Figure 3-1). He then coils the double rope around both the running end of the rope and the coiling log. The rope must be coiled evenly and tightly.

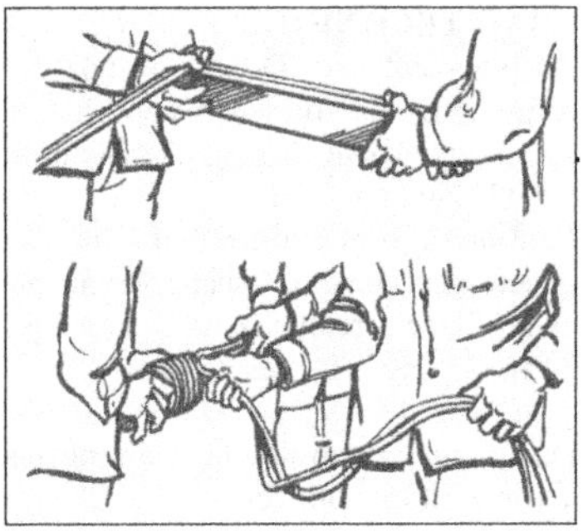

Figure 3-1. Coiling Log

Section V. RAPPELLING OPERATIONS FOR UH-1H IROQUOIS HELICOPTER

The UH-1H helicopter provides a safe, stable aerial platform from which to conduct rappelling operations when landing is not feasible.

3-15. CHARACTERISTICS

The UH-1H is a single-engine, medium-speed, single-main rotor helicopter that can transport eight rappellers, one rappel master, and one three-man crew. It has a lift capacity of 2,300 pounds and is equipped with several floor-mounted tie-down fittings, seven of which are used during rappelling operations.

3-16. RIGGING OF UH-1H HELICOPTER FOR RAPPELLING

The rappel master performs the following to rig the UH-1H helicopter.

a. Removes all seats.

b. Locks the doors in the open position. If no locks are present, removes doors. Also, he removes small cargo doors.

c. Pads and tapes all sharp edges on the floor, door ledge, and all protrusions on the skids. Ensures each door ledge has a scuff pad to protect the rope from contacting the metal door ledge.

d. Secures the donut ring to the center of the floor of the helicopter. The donut ring has six snaphooks numbered clockwise, with 12 o'clock being toward the front of the helicopter. Positions the clamps of the donut ring toward the aft end of the helicopter. Ensures the front two snaplinks and rear two snaplinks are facing outside of the donut ring and hooks them to the tie-down ring with the snaphook facing down. Ensures the center two snaphooks face into the donut ring and are connected to the floor tie-down ring with the gate facing down (Figure 3-2). Attaches the free-floating safety ring to the center floor tie-down ring using a seventh snaphook.

Note: Using this technique has caused equipment failure in many aircraft. The cable clamp nuts must be inspected and certified as airworthy by the pilot or maintenance crew before flying the mission.

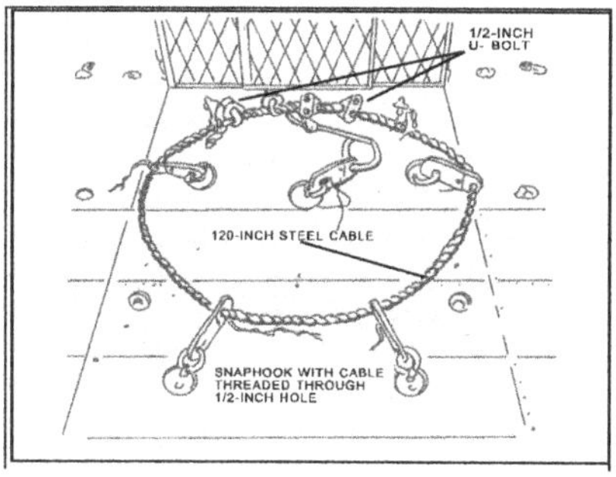

Figure 3-2. Donut ring attached to the floor of the helicopter.

3-17. CONSTRUCTION OF ANCHOR POINTS

In addition to the usual equipment requirements for rappelling operations, an anchor assembly is fabricated. This anchor assembly is commonly known as a *donut ring*. The secondary anchor point is a *floating safety ring*.

a. **Donut Ring.** This primary (or No. 1) anchor point for the rappelling ropes is constructed from a 1/2-inch steel cable with a steel wire core. The cable is 120 inches long, consists of 6 strands (18 wires per strand), and has a tensile strength of 21,000 pounds (Figure 3-2). The completed donut ring has a tensile strength of 3,000 pounds.

(Figure 3-3 shows the donut ring and rappel rope connection in a helicopter.)

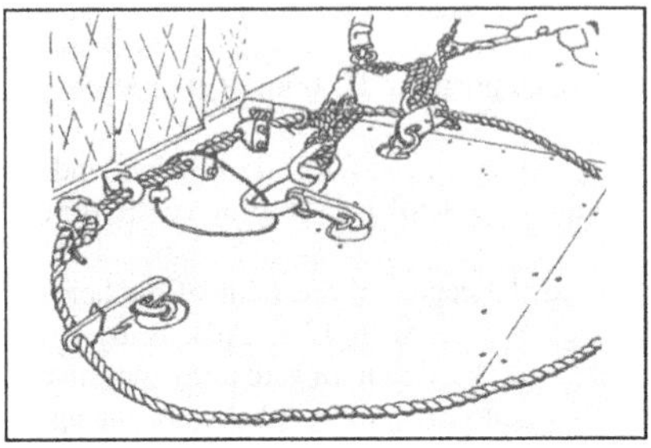

Figure 3-3. Donut ring and rappel rope connection in a helicopter.

(1) Thread six parachute static line snaphooks onto the 120-inch steel cable so that four snaphooks are facing out with the gates down, and the center two snaphooks are facing in with the gates down. Drill the end portion of each snaphook to make a 5/8-inch diameter hole, and thread the cable through the holes in the snaphooks.

(2) Overlap the ends of the 120-inch cable 20 inches to form a circle. Secure the ends with four 1/2-inch U-bolts placed at 2- to 3-inch intervals.

(3) Attach two U-bolts to each dead end of the cable so that the bolts engage the dead end.

(a) Before torquing the U-bolts, position a 12-inch length of chain or 1/8-inch diameter cable on the center of the overlapped 120-inch steel cable so that it remains in position between the two center U-bolts.

(b) Tighten each nut of the U-bolts with a torque wrench (if possible) to 40-foot pounds (480-inch pounds).

(c) After the U-bolt clamps have been attached and tightened, fasten a steel plate (drill to fit) over the open end of the U-bolt studs and spot weld in place to prevent loosening.

b. **Floating Safety Ring**. The floating safety ring is referred to as the secondary (or No. 2) anchor point for the rappelling ropes (Figure 3-4). The snaplink at the end of the rappelling rope is hooked to this connection. Either of the following two types of floating safety rings may be used:

(1) *First method.* Thread an elliptical rappelling ring through the free end of the keeper chain (cable). It is constructed of cold-rolled steel that is 1/4-inch in diameter, with inside dimensions of 2 1/4 inches (minor axis) and 4 inches (major axis). Thread a seventh parachute static line snaphook onto the ring before welding. Weld this ring together so that it withstands a force up to 3,000 pounds.

(2) *Second method.* Attach two snaplinks to the aircraft tie-down ring in the center of the donut ring. Insert the first snaplink through the free end of the keeper chain (cable) and the tie-down ring with the gate down. Insert the second snaplink through the free end of the keeper chain (cable) and the tie-down ring with the gate up. Tape the snaplink

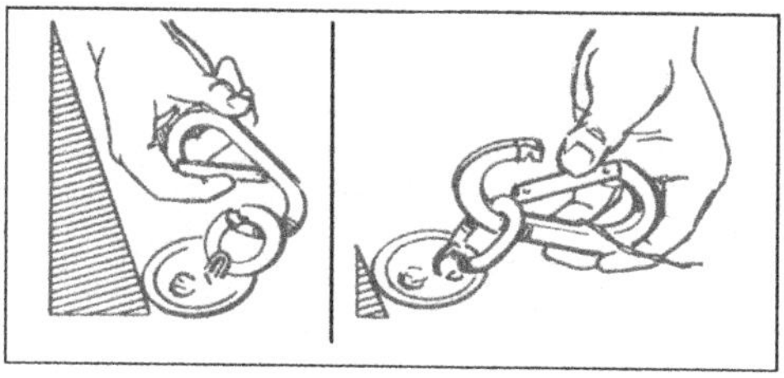

gates closed with masking tape. Then, tape the snaplinks together to ensure that the snaplink gates are on opposite sides of each other.

Figure 3-4. Floating safety ring formed with two snaplinks.

c. **Rappel Rope Anchor Points.** The rappelling rope is connected to the floating safety ring and the donut ring as follows:

(1) *Number 1 anchor point (donut ring).* Using a bight about 5 feet from the end of the standing part of the rope, make one turn through the snaplink forming a round turn. The bowline is the preferred method of attaching rappel ropes to snaplinks or anchor systems. The round turn with two half hitches is a reliable means of attaching ropes to anchor systems. Secure the round turn to the snaplink with two half hitches. Make the connection to the donut ring by attaching the snaplink gate upward. Ensure the gate faces upward with the opening away from the knot.

(2) *Number 2 anchor point (floating ring).* Attach the No. 2 snaplink the same as the first with some exceptions. Using a bight about 2 feet from the end of the standing part of the rope, connect the snaplink to the rope the same as the first connection. Tape the end of the standing part of the rope and the knots with masking tape (or green engineer tape) to secure them in place. Make the connection to the floating safety ring the same as the connection to the donut ring (Figure 3-5). Four ropes can be connected to the floating safety ring using two snaplinks (Figure 3-6).

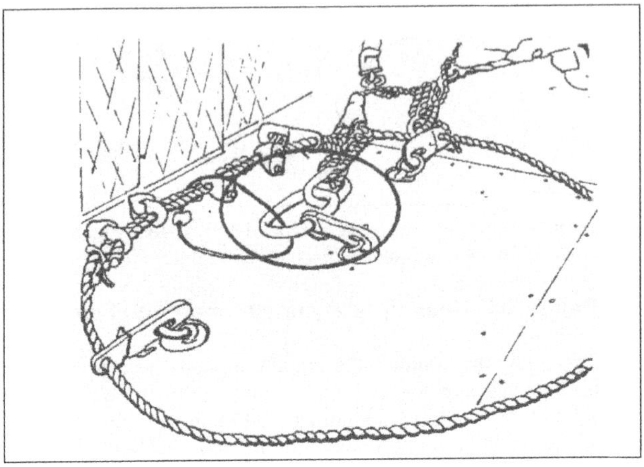

Figure 3-5. Rappel rope connection using two snaplinks for the floating safety ring.

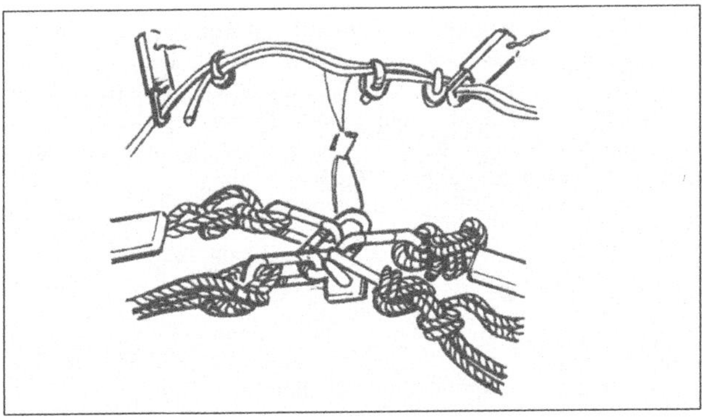

Figure 3-6. Four rappel ropes connected to the floating safety ring (two snaplinks).

3-18. SEATING ARRANGEMENTS AND LOADING TECHNIQUES

Rappelling operations are executed quickly and safely to ensure the success of the operation in a combat or training environment. All rappellers must know the seating arrangements and the procedures for loading (Figure 3-7). They must be rehearsed under the supervision of the rappel master before conducting the operation.

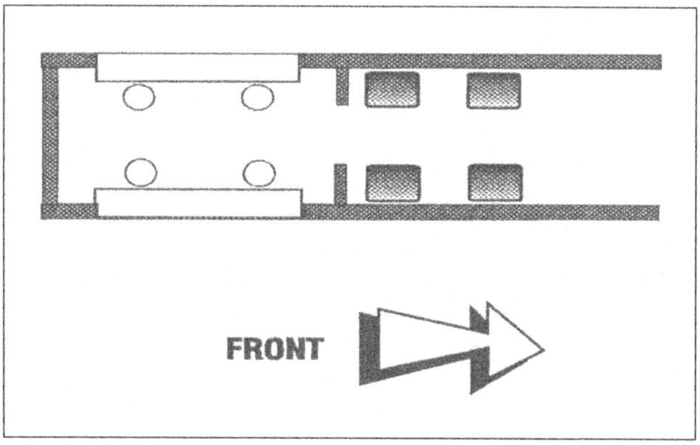

Figure 3-7. Seating arrangement.

a. The first four rappellers are hooked up to the donut ring and use safety belts while in flight. The remaining rappellers use seat belts or safety straps. For safety emphasis during flight, each rappeller's brake hand is in position with no slack in the rope between the brake hand and the donut ring. The rappeller ensures that his rappelling rope is not tangled on any part of the interior of the helicopter or his equipment. He ensures that the coiling log or D-bag is located in such a position that he drops the rope with ease, using his guide hand. The coiling log or D-bag is secured with half of it under the rappeller's inside leg and the other half on top of his outside leg.

b. The rappellers should board the aircraft and sit along the leading edge of each door. They hook up with the rappel seat square knot on the inboard side of the aircraft and the brake hand on the outboard side. The rappel master sits in the center to aid visual inspection of personnel and equipment.

c. For a flight time of less than five minutes and for training and operations where speed is a factor, the rappel master directs the rappellers to be seated in the door facing outward with their feet on the skid. In a tactical situation and before the helicopter comes to a hover, the rappellers assume an upright position facing inside the helicopter.

d. The aircraft is cleared as fast as possible. This reduces the time that the aircraft is in the high-hover attitude to a minimum.

3-19. RAPPELLING PROCEDURES

The rappel master ensures all personnel follow the rappelling procedures for the safe and efficient execution of the operation. These procedures are rehearsed under the supervision of the rappel master.

a. Rappellers (eight maximum) are identified, and rappel masters tell them which door of the aircraft to exit. Rappellers approach the aircraft once it is firmly on the ground, and the crew chief signals when it is safe to approach. The approach to the aircraft should be about 45 degrees to the nose of the aircraft allowing the pilot to see the rappeller.

b. If the rappel master is not already on the aircraft, he boards first and secures himself to the aircraft with a safety harness. He dons a headset to maintain communications with the pilot and positions himself behind the crew chief's chair.

c. As the rappellers approach the aircraft, they hand their anchor attaching points (snaplinks connected to their ropes) to the rappel master. The rappel master connects one set of snaplinks to the floating safety ring and one set of snaplinks (the primary) to the donut ring.

d. All rappellers use safety belts. For safety emphasis during flight, each rappeller's brake hand is in position with no slack in the rope between the brake hand and the donut ring. Rappellers ensure that their rappelling ropes are not tangled on any part of the interior of the helicopter or their equipment. They ensure that they can drop their D-bag or log coil using the guide hand. The D-bag or log coil is placed so it does not fall out of the aircraft. Once over the target area, the rappel master issues the rappel commands and ensures the rappel ropes are on the ground (about 20 feet of rope should be on the ground).

e. On the command GO, the rappeller flexes at the knees and vigorously pushes away from the skid gear, allowing the rope to pass through his brake hand and guide hand. The descent is accomplished smoothly at a rate of about 8 feet per second, avoiding jerky stops. The rappeller initiates his braking action slowly when he is about halfway to the ground. The safety officer or NCO ensures that at least a five-second delay is maintained between each rappeller. Another technique used to deploy rappellers is to issue the command GO to the rappellers diagonally opposite each other. In the early phase of performance, the rappeller's feet are together (particularly if over woods or jungle) and his guide hand is used (not the brake hand) to remove any entanglements.

f. Upon reaching the ground, the rappeller clears the rappel rope through the snaplink (or rappel ring) until the rope is free. If other rappellers are to follow on the same rope, the rappeller on the ground separates or untwists the ropes and becomes the belayer for each subsequent rappeller. If ropes are to be released from the aircraft following the last rappeller, untwisting and separating the ropes is not necessary. A two-second interval is maintained between exit groups (two men exiting at the same time).

g. The rappel master releases the rope from the donut ring after he confirms (by visual inspection) that the rappeller is off rappel. He then drops the rope away from the helicopter.

3-20. RAPPELLING COMMANDS

Helicopter rappelling is conducted in a noise-filled environment. Each rappeller is trained to know the rappel commands and to understand the actions required of him during the execution of each command. Due to the noise created by the helicopter, all rappel commands should be accompanied by arm-and-hand signals. All commands are issued by the rappel master and confirmed orally by the rappellers.

a. The helicopter rappelling commands are as follows:

(1) GET READY. The rappeller checks his combat equipment, looks toward the donut ring, and pulls the rope to check the anchor point connection. He then checks his

snaplink to ensure the rope is properly seated in the snaplink on his rappel seat.

(2) THROW ROPE. The rappeller drops his deployment bag out and away from the helicopter with his guide hand. He ensures that the rope does *not* fall between the side of the helicopter and the skid. The rappel master ensures the rope is touching the ground and is free of tangles and knots. The rappeller then replies, "Rope okay."

(3) SIT IN THE DOOR. The rappeller swings his legs to the outside of the helicopter and takes up a sitting position. (This command applies only to the UH-1H helicopter.)

(4) POSITION. The rappeller pivots 180 degrees on the skid. His feet are shoulder-width apart, the balls of his feet are on the skid, his knees are locked, his body is bent forward in an L-shape position, and his brake hand is on his buttock. The rappel master makes a visual inspection of the snaplink and rappel ring.

(5) GO. The rappel master points at the rappeller to exit. One rappeller from each side rappels at the same time. The rappel master gives the command GO to the rappellers that are diagonal to each other—for example, the front left rappeller and the rear right rappeller. The rappeller flexes his knees and jumps backward, letting the rope run through the brake hand and guide hand.

b. The arm-and-hand signals that are used with rappelling commands are as follows:

(1) GET READY (Figure 3-8). The rappel master extends both arms to the front with fists clenched and thumbs pointing upward.

Figure 3-8. Arm-and-hand signal for GET READY.

(2) SIT IN THE DOOR (Figure 3-9). The rappel master extends both arms to the front with elbows locked, fingers extended, thumbs running along the index fingers, and palms facing downward. He bends slightly at the waist so that the arms are below the waist. Then, he moves both arms in a crisscross waving motion, alternating left over right and right over left.

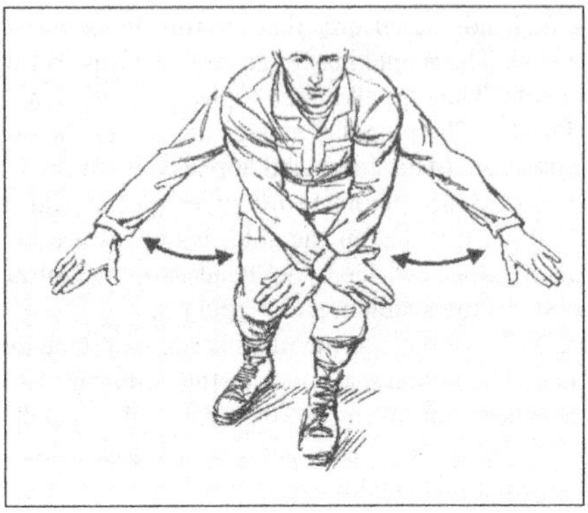

Figure 3-9. Arm-and-hand signal for SIT IN THE DOOR.

(3) THROW ROPES (Figure 3-10). The rappel master extends both arms to the front with elbows locked, fist clenched and index fingers extended. He points at respective rappellers, brings the forearms to an upright position and then down to the elbow-locked position.

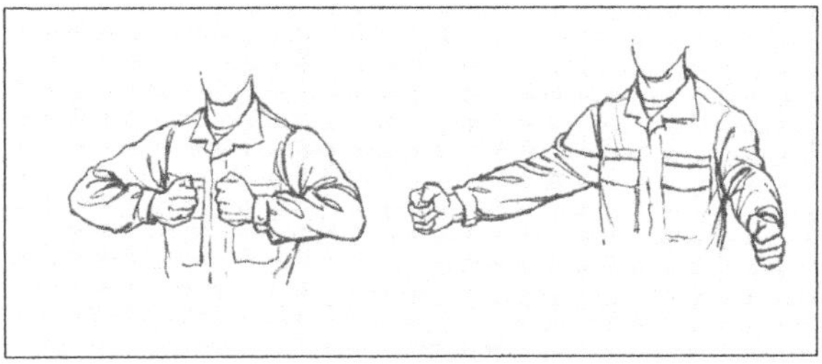

Figure 3-10. Arm-and-hand signal for THROW ROPES.

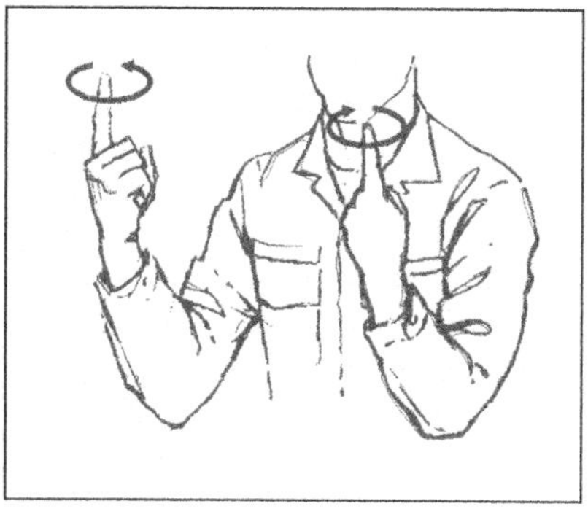

Figure 3-11. Primary arm-and-hand signal for POSITION.

(4) POSITION. The two arm-and-hand signals for POSITION are:

(a) Primary signal (Figure 3-11). The rappel master extends both arms to the front, elbows bent, forearms pointed upward, fists clinched, and index fingers pointing upward. He makes a circular motion with both forearms rotating in opposite directions.

(b) Alternate signal (Figure 3-12). The rappel master extends both arms to the front with elbows locked, fists clenched, and index fingers extended. He bends at the waist so that his arms are below his waist and makes a circular motion with arms rotating in opposite directions.

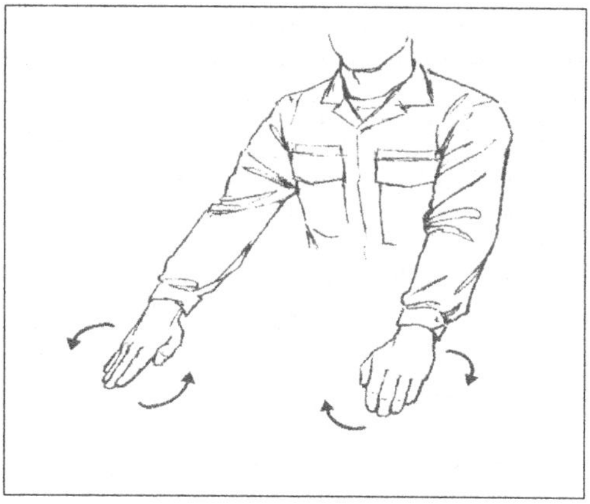

Figure 3-12. Alternate arm-and-hand signal for POSITION.

(5) GO (Figure 3-13). The rappel master extends an arm with elbow locked, fingers extended, and thumb along the index finger, and he points directly at the rappeller.

Figure 3-13. Arm-and-hand signal for GO.

3-21. INSPECTION AND SAFETY CONSIDERATIONS

The rappel master and pilot (or his representative) conduct a joint inspection of the aircraft to ensure the safety of all personnel and serviceability of equipment.

a. Cargo doors are locked in the open position or cleared for closing, depending on the mission.

b. All loose objects in the cargo compartment are removed or secured forward.

c. Sharp edges or protrusions on the cargo floor and door ledges are taped.

d. The donut ring and floating safety ring are serviceable and properly attached to tie-down fittings.

e. A headset and intercom jack for the rappel master are available and operational.

f. Serviceable safety harnesses are available for the rappel master and crew chief.

g. The recovery rope is installed and properly stowed.

h. Unused floor rings are taped down.

Section VI. RAPPELLING OPERATIONS FOR UH-60 BLACKHAWK HELICOPTER

The techniques used by the rappeller when rappelling from different aircraft are similar. However, positioning, seating, and the tie-down anchor point are different. Each rappeller is well trained on each aircraft before conducting a rappel.

3-22. CHARACTERISTICS

The UH-60 is a twin-engine, medium-speed, single-main rotor helicopter that can transport 10 rappellers, 1 rappel master, and 1 four-man crew when the center row of troop seats is removed. It is equipped with four 4,000-pound (load limit) cabin ceiling tie-down fittings or rings that are located in the ceiling of the troop/cargo compartment. These fittings are used as the primary rappelling rope anchoring points. The UH-60 is also equipped with eight 3,500-pound (load limit) cargo restraint net rings. Four of the eight rings are located in the ceiling of the troop/cargo compartment. These rings are used as the secondary rappelling rope anchoring points.

3-23. RIGGING OF UH-60 FOR RAPPELLING

The rappel master rigs the UH-60 helicopter by performing the following:

a. Locks both cargo doors in the open position.

Note: For arctic or other cold-weather operations, or during flights of long duration, the cargo doors are closed and locked until the time specified for opening them.

b. Removes the center row of troop seats.

c. Tapes any sharp edges or protrusions on the cargo floor and door ledges that may come in contact with the rappeller or the rappelling rope.

d. Stows loose equipment forward in the cargo compartment.

e. Extends the rappel master's intercom cord to the rear over the aft utility drain line and tapes the cord to the overhead troop seat support tube.

f. Installs the floor restraint provisions for rappellers No. 1 through 6.

g. Rigs and connects rappelling ropes to the aircraft's primary and secondary anchoring points.

(1) *Primary anchor points.* Ties a bowline with a half hitch about 4 feet from the standing end of the rope. Attaches the two primary snaplinks to the respective cabin tie-down fitting ring with gates facing in the opposite directions (Figures 3-14 and 3-16)

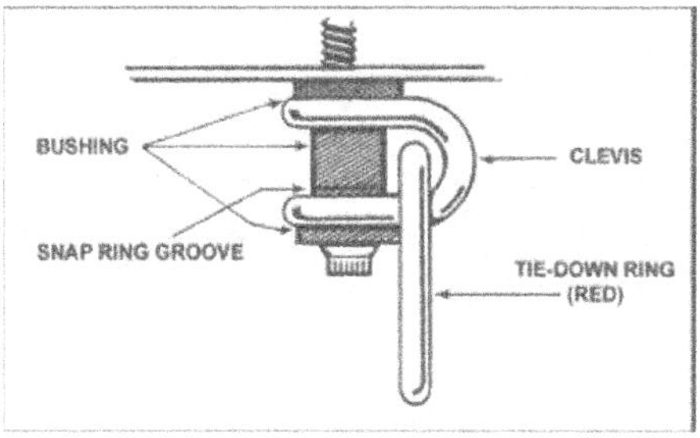

Figure 3-14. Primary snaplink attaching point.

(2) *Secondary anchor points.* Ties a bowline with a half hitch about 1 1/2 feet from the standing end of the rope. Attaches the secondary snaplink to the adjacent overhead cargo restraint net ring (Figures 3-15 and 3-16).

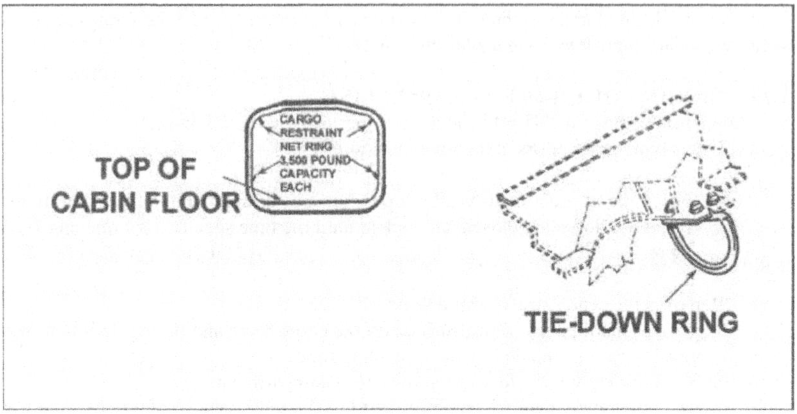

Figure 3-15. Secondary snaplink attaching point.

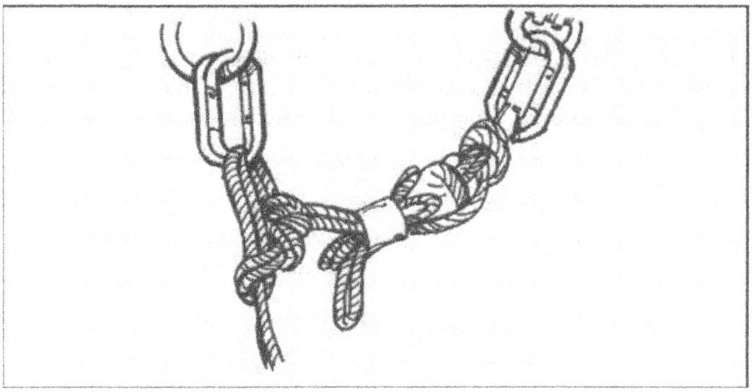

Figure 3-16. Primary and secondary snaplink attaching points.

h. Removes and secures the cargo hook access door and deploys the cargo hook in the DOWN position.

i. Installs the recovery rope for endangered rappellers.

j. Tapes the unused floor rings.

Note: There must be a minimum of 22 inches and a maximum of 30 inches between the anchor knots.

3-24. SEATING ARRANGEMENTS AND LOADING TECHNIQUES

Rappelling operations are executed quickly and safely to ensure the success of the operation in a combat or training environment. All rappellers must know the seating arrangements and the procedures for loading. They must be rehearsed under the supervision of the rappel master before conducting the operation.

a. A maximum of 10 rappellers, with and without combat equipment, and 1 static rappel master is seated and restrained aboard the UH-60 helicopter when the center row of troop seats are removed (Figure 3-17).

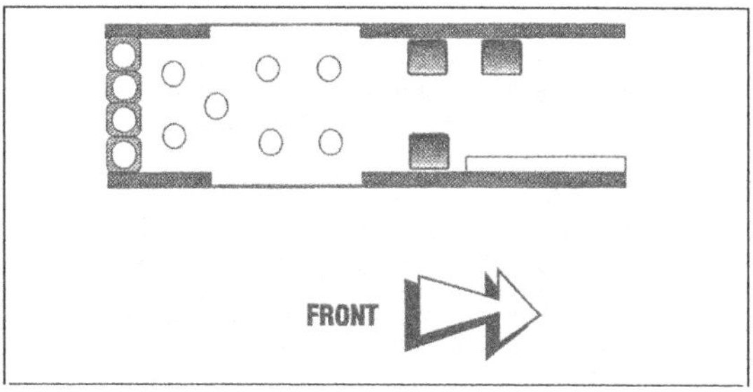

Figure 3-17. Rappelling personnel seating arrangement.

(1) Rappellers No. 1 through 6 and the rappel master are seated on the cargo floor. The rappellers are restrained using 3 1/2 feet of 1-inch tubular-nylon webbing (or equivalent) that is run through the cargo tie-down rings attached to the floor of the aircraft. The webbing is attached to the rappeller by means of a snaplink that is hooked to a Swiss seat. The rappel master is secured by his safety harness.

(2) Rappellers No. 7 through 10 are seated on the troop seats across the aft end of the cargo compartment. They are secured by the seat belts.

b. Rappellers are organized into 10-man sticks. They approach the aircraft from the left or right side in reverse order. They are divided into two groups and approach and enter the aircraft at the same time, depending on the situation. Rappellers No. 9, 7, 5, 3, and 1 enter the aircraft through the left door; they are seated and restrained. Rappellers No. 10, 8, 6, 4, and 2 enter the aircraft through the right door; they are seated and restrained. The static rappel master is restrained in the center of the cargo compartment using a C-3A troop-type safety belt or a safety harness during take-off and landing.

DANGER

BECAUSE THE LOWEST ARC OF THE ROTOR BLADE OCCURS AT THE DIRECT FRONT OF THE AIRCRAFT, APPROACHING THE AIRCRAFT AT THIS POINT COULD RESULT IN PERSONNEL INJURY OR DEATH.

3-25. RAPPELLING PROCEDURES

The rappel master ensures all personnel follow the rappelling procedures for the safe and efficient execution of the operation. These procedures are rehearsed under the supervision of the rappel master.

a. The rappel master is equipped with a headset and maintains communications with the pilot at all times. He wears a safety harness and stations himself in the center of the cargo floor to maintain control of the rappellers, their ropes, and all anchor attaching points.

Note: If all rappellers are rappelling into the same area, individual rappel ropes are not needed. The ropes used by rappellers No. 1 through 4 are used by the succeeding rappellers.

b. After all rappellers are in the aircraft, rappellers No. 1 through 4 hand the rappel master their primary and secondary anchor point snaplinks. The rappel master hooks the primary anchor point snaplink into the respective cabin tie-down fittings and the secondary anchor point snaplink into the respective cargo restraint net rings. After the primary and secondary anchor point snaplinks are secured, the first four rappellers hook onto their respective rappel rope. The rappel master checks the first four rappellers for the correct hookup.

c. After the first four rappellers exit and clear their ropes, the rappel master hands the next four rappellers their rappel ropes to hook up. Left-handed rappellers take extra precaution to ensure the correct hookup. Problems can be avoided by placing left-handed rappellers in the positions where they initially have the rope to their nonbrake side.

WARNING

Extreme care must be taken in hooking up and positioning personnel that require repositioning of their rope to their proper brake-hand side. This could result in bodily injury or damage to Army property.

d. After checking the rappellers for the correct hookup, the rappel master moves the next four rappellers into position on the floor of the aircraft. After the four rappellers exit and clear their ropes, the rappel master hands the last two rappellers their rappel ropes to hook up.

e. After checking the rappellers for correct hookup, the rappel master moves them into position on the floor of the aircraft. After the last rappeller is off the rope, the rappel master releases the ropes.

f. If the rappel master rappels, the crew chief is instructed to release or retrieve the ropes.

3-26. RAPPELLING COMMANDS

The UH-60 rappelling is conducted similar to the UH-1H rappelling with the exception that no skids are on the UH-60 on which to stand. Therefore, the edge of the floor, along the door of the helicopter, is used as a pivot point to assume the L-shaped rappelling position.

a. GET READY. This alerts the rappeller. The rappeller and rappel master perform final checks of the hookup, rappel seat, snaplink, and equipment.

b. THROW ROPES. The rappeller looks down below the aircraft to ensure that no one is under the aircraft. Keeping his brake hand in the small of his back, he tosses the deployment bag with the ropes *out and away* from the helicopter with his guide hand. The rappeller observes that the ropes are touching the ground and are not knotted or entangled.

c. POSITION. From a kneeling position with the brake hand in the small of the back, the rappeller rotates 90 degrees so that he is facing the inside of the aircraft and the rappel master. The rappeller then places his heels on the edge of the floor of the helicopter doorway and leans out into an L-shaped position. His feet are shoulder-width apart, the balls of his feet are on the edge of the helicopter doorway, his knees are locked, and his body is bent at the waist toward the helicopter.

d. GO. The rappel master initiates the rappel with this command. The rappeller flexes his knees and jumps backwards. At the same time, the rappeller throws his brake hand out at a 45-degree angle, letting the running ends of the ropes slide through both the brake hand and guide hand.

3-27. INSPECTION AND SAFETY CONSIDERATIONS

The rappel master and pilot (or his representative) conduct a joint inspection of the aircraft to ensure the safety of all personnel and serviceability of equipment.

a. Cargo doors are locked in the open position or cleared for closing, depending on the mission.

b. All loose objects in the cargo compartment are removed or secured forward.

c. Sharp edges or protrusions on the cargo floor and door ledges that may come in contact with the rappeller or his rappelling rope are taped.

Note: Do not tape the door latches or handles; this can interfere with door operations.

d. Primary and secondary rappelling anchor points are serviceable and securely attached to the aircraft structure.

e. A headset/helmet and intercom jack for the rappel master are available and operational, and the intercom extension cord is secured overhead.

f. Serviceable safety harnesses are available for the rappel master and crew chief.

g. The cabin ceiling tie-downs must have safety wire installed to ensure they do not come undone or unraveled, and the bolt head must be stamped H.

Section VII. RAPPELLING OPERATIONS FOR MH-53 HELICOPTER

The MH-53 helicopter is a highly versatile aircraft. It is used for a variety of missions, usually in conjunction with special operations.

3-28. CHARACTERISTICS

The MH-53 is a two-engine, single-rotor, heavy-lift helicopter. It has a crew of six and can be refueled while in flight. It has a precision navigational and communications package, which makes it excellent for conducting deep infiltration and exfiltration missions. The MH-53 helicopter can fly at night in all weather and terrain conditions. It can follow down to 100 feet and operate from unprepared sites. It is fitted with three gun stations for a mix of 7.62-mm miniguns and caliber .50 machine guns.

3-29. RIGGING OF MH-53 HELICOPTER FOR RAPPELLING

When rappelling from the crew entrance door, rappellers connect the ropes with locking snaplinks to the two 10,000-pound tie-down rings under the left scanner's window. When rappelling from the ramp, rappellers connect the ropes with locking snaplinks to the two 10,000-pound tie-down fittings on the left and right sides of the aircraft at station 522. Any sharp edges that could damage the ropes should be padded or taped. A length of fire hose may also be used over the portion of the rope that comes in contacts with the door or ramp edge.

3-30. SEATING ARRANGEMENTS AND LOADING TECHNIQUES

The MH-53 is a large, versatile helicopter. The seating arrangements and loading techniques are numerous. The using unit develops a seating and loading SOP that ensures the safety of all personnel and permits smooth, efficient execution of the mission.

3-31. RAPPELLING PROCEDURES

The rappel master or his assistant are at the two exit points (ramp and door). The rappel master maintains communications with the helicopter commander and relays all commands and time warnings. The commander issues time warnings at 20-minute, 10-minute, 5-minute, and 1-minute intervals. During limited visibility, the rappel master may use NVDs to observe the safety/belayer. Chemical lights are attached to the top or bottom of the rappel rope. Arm-and-hand signals should be IAW The Air Assault School Handbook and FC 57-2.

Note: Night vision goggles cause limited depth perception and a tunnel vision effect.

a. The rappel master is secured by a harness and ensures the proper hook up of each rappeller. Once hooked up, the rappellers release their safety belts. At the command of the rappel master, the rappellers position themselves to ease immediate deployment.

> **WARNING**
>
> **Before conducting deployments off the ramp, the deploying team members must be briefed on the importance of maintaining separation between members. About 24 to 27 inches are desired. This separation helps maintain aircraft center of gravity limits.**

b. The rappel master deploys ropes only after the helicopter is in a stable hover over the target area and he has given the command ROPES. He ensures the ropes are on the ground before giving the command GO. After the last rappeller is off the rope, the rappel master retrieves or cuts the ropes away.

3-32. RAPPELLING COMMANDS

The helicopter rappelling commands for the MH-53 are as follows:

a. GET READY. This command is given as the helicopter approaches the rappel site. It alerts the rappellers of the approach to the site and signals them for a final equipment check.

b. POSITION. This command clears the first rappellers into position for deployment.

c. THROW ROPE. This command is given once the helicopter reaches a stable hover.

d. GO. The rappellers exit.

3-33. INSPECTION AND SAFETY CONSIDERATIONS

The rappel master briefs all personnel participating in the operation. He conducts hands-on inspection to ensure the safety of all personnel and serviceability of equipment.

a. Tie-down fittings are serviceable.

b. All sharp or protruding edges that may come in contact with the rappelling ropes are padded or taped.

c. All ropes are retrieved or cut away before forward movement of the helicopter.

d. Only three rappellers are deployed at a time—two from the ramp and one from the personnel door.

CHAPTER 4

SPECIAL PATROL INFILTRATION/EXFILTRATION SYSTEM

The special patrol infiltration/exfiltration system (SPIES) provides an excellent form of exfiltration over short distances. It is not recommended for infiltration because team members are exposed the entire time. Due to the nature of SPIES operations, a thorough briefing is required for all participants before the operation. Personnel being extracted must train extensively in the SPIES extraction before infiltration. For the other personnel involved, a complete preoperations briefing is held before the operation. This is crucial when additional assets are involved other than the extraction helicopter such as gunships, aerial observers, or artillery support.

Section I. SPIES MASTER DUTIES AND QUALIFICATIONS

The SPIES master holds the rank of sergeant or above (may be waived by the commanding officer). The commanding officer ensures that this qualification is entered on the soldier's training record.

4-1. QUALIFICATIONS

To qualify to be a SPIES master, the sergeant must participate in at least three SPIES operations (observe twice and execute SPIES master duties once under observation). For example, he hooks up the helicopter, assists in the preparation of an operation, and conducts a successful operation under the supervision of a qualified SPIES master. A SPIES master is conscientious and thorough, and he knows all aspects of a SPIES operation. He can give an effective pilot's brief, use the aircraft communications equipment, and understand aviation terminology.

4-2. DUTIES

The SPIES master is responsible for the safe conduct of the SPIES operation.

a. **Preflight Duties.**

(1) Inventories and inspects all SPIES equipment.

(2) Briefs the pilot and other concerned personnel about details of the operation, especially the extraction and dismounting procedures.

(3) Ensures that an ICS helmet and gunner's belt are available for the SPIES master's use. Uses a sling rope if no belt is available. Connects and checks the operation of the ICS to be used. (ICS communications is established between the SPIES master and pilots on all SPIES operations.)

(4) Attaches the SPIES rope to the helicopter IAW the guidance in this chapter.

(5) Ensures that nothing is adrift in the aircraft that may fall on a team member.

(6) Checks the location of the emergency axe. Places the axe where readily available, yet secure enough so as not to endanger the men on the SPIES rope. (Inspects the axe to ensure that it is sharp.)

b. **Extraction Duties.**

(1) On arrival at the team's estimated position, assists the pilot in determining the exact location of the team members.

(2) As the aircraft approaches the team's location, aids the pilot (using the clock system) in placing the aircraft directly above the team.

(3) Requests permission from the pilot to drop the SPIES rope when the aircraft is hovering above the team.

(4) Drops the rope, taking care to avoid striking team members on the ground.

(5) Notifies the pilot when the rope is down, and reports all altitude corrections to ensure that team members reach all SPIES attachment points.

(6) Watches for the "thumbs-up" signal from all team members.

(7) On receipt of the "thumbs-up" signal, advises the pilot that the team is ready for extraction and requests a vertical lift-off.

(8) Advises the pilot of the team's position, the location of any potential obstacles, and the avoidance of horizontal movement.

(9) If a team member becomes entangled with an obstacle during the extraction, notifies the pilot and requests that the vertical lift be stopped. If the situation is critical, prepares to cut the SPIES rope (the anchor point or cargo straps) after the team members are secured to the obstacle or on the ground.

(10) When positive that all obstructions are clear, advises the pilot to obtain a safe altitude (about 350 feet AGL for training purposes or as the situation dictates in combat) or to transition into forward flight.

(11) At frequent intervals during the flight, advises the pilot on the safety status of all team members. Maintains a constant visual watch on the team and checks security of the SPIES attachments often.

c. **Dismounting Duties.**

(1) On arrival at the dismounting area, informs the pilot as to the approximate distance of the lower rope end from the ground.

(2) Once the pilot starts the vertical descent, continually informs him as to the approximate distance of the lower rope end from the ground.

(3) Informs the pilot of any horizontal drift that occurs and any obstructions near the SPIES rope. Also informs the pilot of any oscillation that may occur.

(4) Informs the pilot when the rope is about 25 feet above the ground and again when it is 10 feet above the ground. Ensures that the rate of descent is slow enough to enable the team members to land and get out from under team members safely.

(5) Reports when the first man initially touches down, when the last team member starts to safely move away from under the helicopter, and when all team members are disconnected.

(6) On order of the pilot, either retrieves the SPIES rope into the helicopter or disconnects the SPIES rope and drops it to the ground. When using the UH-1H/UH-60 helicopter, the only way to retrieve the SPIES rope while in the air is by having an arranged recovery rope attached with a 16-foot sling rope. In some cases, the SPIES master joins two 12-foot-long sling ropes to haul the SPIES rope aboard and attaches the rope about 5 or 6 feet below the cargo hook or cargo strap hookup point. The type of knot used to connect the sling (or recovery) rope to the SPIES rope is self-tightening in nature (for example, the prusik knot). The SPIES master fastens the standing end of the sling rope to the deck tie-down or uses a snaplink. Although it is important to keep the line out of the way, the primary consideration is its length. The rope must be long enough for any oscillation in the SPIES during flight.

Section II. PREOPERATIONS BRIEFINGS AND PROCEDURES

Before conducting a SPIES training mission, the participants must have a basic understanding of the requirements. The SPIES master conducts briefings to ensure the soldier extracted and the pilot know the procedures. As SOPs are developed and units train together, the SPIES master simply refers to the SOP. He always gives a safety briefing and conducts an equipment inspection.

Note: SPIES has not been approved for Army-wide use. Approval must be obtained from division level or higher commander before the conduct of any training.

4-3. SAFETY

As in all training, a safety briefing must precede operations using the SPIES. The briefing should consist of, but not be restricted to, a review of the following:

- Characteristics of equipment associated with the SPIES.
- Equipment inspection.
- Proper donning of the harness.
- Methods of extraction and insertion used.
- Emergency signals.

4-4. COMMUNICATIONS

Due to the noise level associated in all helicopter operations, radios are used for communication. They are also used before the arrival of the helicopter. Precise arm-and-hand signals are established in case of radio failure or poor communications due to static or noise overriding the audio output of the radio. During the first part of the operations, the SPIES master either observes (daytime) or knows that a definite procedure is taking place (night or jungle) while the teams are hooking up to the SPIES rope for extraction. Headsets and voice suppressors replace the handset for better radio procedures. This allows the RATELO on the ground to use both hands while the helicopter hovers for a faster and safer hookup. If radio communications are hampered, special procedures are used along with hand or light signals.

4-5. EXTRACTION PROCEDURES

After the team has been located, the SPIES master assists the pilot in directing the helicopter to the proper distance over the team.

a. The team leader positions himself so that he can move and approach the rope as it is dropped by the SPIES master. Once the rope is clear of any obstacles, the team leader signals the team to their assigned positions along the 10 hookup points. Using the primary or harness snaplink, each team member hooks to the D-ring on his side of the line. This is the primary hookup. Once the team is hooked up, the team leader hooks into the alternate or second hookup point using the safety line and snaplink. He then faces forward along the line so that he is heading in the direction he is traveling when the aircraft starts its ascent.

b. The SPIES rope is held up and routed over the shoulder closest to the rope. With the other hand, the team members give thumbs-up signals to allow both the team leader and the SPIES master to see that they are ready to go. The team leader physically inspects (if time and situation permits) or hooks himself in on the lowest point along with the RATELO to ensure the running end is clear of all obstacles. He gives the thumbs-up signal to the SPIES master. This thumbs-up signal (at night, an arranged light signal), continues until the team is lifted off the extraction surface. The team leader also signals (arms held straight to the sides) when the team is at safe altitude (approximate 10 feet above the tallest obstacle at the extraction site). The helicopter starts a transition in a horizontal direction on its return flight.

4-6. EMERGENCY PROCEDURES

During the flight, from extraction until the team is safely and quickly detached from the SPIES rope, each team member should be aware of any problems that arise from above or below. The man above checks the man below. At the first sign of danger or if an emergency arises, the team leader or a team member places his free hand on his head. Upon observing anyone on the SPIES rope with his hand on his head, the SPIES master tells the pilot to make an emergency landing in the nearest and safest area.

4-7. DISMOUNTING PROCEDURES

During the familiarization training phase, the SPIES master ensures all members know they should immediately head in the direction of the nose (12 o'clock) of the aircraft. This allows the pilot to see that the team is out from under the aircraft. When an emergency situation arises, the pilot can make a better appraisal of the situation if he can see all the members of the team. When the helicopter is making a scheduled landing, the team ensures that the SPIES rope does not interfere with the aircraft and that the aircraft does not land on the rope.

4-8. INSPECTION OF EQUIPMENT

The SPIES is inspected by a certified SPIES master at six-month intervals and whenever the SPIES master questions serviceability. Outdated, spliced, abraded, or cut rope is removed from service. However, loose or broken stitching in excess of three stitches is not repaired.

CAUTION

Acid contamination, cuts, or fraying of harness or sling webbing constitute nonrepairable damage.

a. The certified SPIES master inspects the harness and suspension sling webbing for signs of contamination from oil, grease, acid, or rust. He also checks for cuts, twists, fading, excessive wear, fusing (indicated by unusual hardening or softening of webbing fibers), fraying, burns, abrasions, and loose or broken stitching. The damaged harness or suspension sling is removed from service and returned to supply for appropriate disposition.

b. The SPIES master inspects all hardware for signs of corrosion, pitting, ease of operation, security of attachment, bends, dents, nicks, burrs, and sharp edges.

Replacement of hardware (except chest strap adapter) that requires unstitching of webbing renders the harness unserviceable.

c. The SPIES master replaces the V-ring by cutting the strap above the stitching. He folds and stitches a new end section of the leg strap. If the leg strap is damaged, the SPIES master returns the harness or suspension sling to supply for appropriate disposition.

d. The SPIES master checks ropes, harnesses, and suspension slings for expiration of service or total life, which is 7 years of service (opening manufacturer's package) or 15 years from date of manufacture, whichever occurs first. The SPIES master also ensures the rope is free of splices. He ensures the eye loop at the end of the SPIES rope is not broken, frayed, or loose.

e. The SPIES master inspects the rope surface for cuts, excessive abrasions, and snags. Cuts on the rope are excessive when there are four or more cut strands in any 5-inch length. The two-to-one braided rope has 12 pairs of strands (24 individual strands) around the circumference. Abrasion is extensive when torn yarns are equivalent to that of four strands of any 5-inch length. Rope that has been subjected to heavy loads may display glazed areas where it has worked against hard surfaces. This condition may be caused by paint or fused fibers. After long use, the rope may become fuzzy on the surface (although this should be minimized with the surface coating). However, the effect on the rope's strength is negligible.

f. The SPIES master inspects the rope for signs of contamination by acid, alkaline compounds, salt water, fire extinguishing solutions, and petroleum-based solvents. Although the ropes gradually change color, such changes do not indicate decrease in strength unless the change is due to contact with strong chemicals. Changes in color caused by chemicals are usually spotty; changes caused by use are uniform throughout the length of the rope.

4-9. RIGGING OF UH-1H HELICOPTER FOR SPIES OPERATION

Personnel who are not familiar with SPIES are encouraged to observe or take part in the rigging of the helicopter. This builds confidence in the equipment and assists in a more comprehensive training of new SPIES masters.

a. **Equipment.**

- One 120-foot SPIES rope with deployment bag.
- Two 11-foot, 3- or 4-loop cargo slings or two 9-foot, 3- or 4-loop cargo slings.
- Two Type IV connector links (four if aircraft does not have cargo hooks).
- Heavy duty tape (100-mph tape).
- One 12-foot sling rope.
- Five oval snaplinks (nine if the aircraft does not have cargo hooks).

NOTE: If the aircraft does not have cargo hooks, use four 11-foot or four 9-foot, 3- or 4-loop cargo slings.

b. **Installation.**

(1) The primary attachment point for the SPIES rope is the cargo hook. The end of the SPIES rope having a polyurethane encapsulated eye is attached to the cargo hook.

The two 9- or 11-foot long cargo suspension slings are joined to form one continuous sling, using a Type IV link. The sling is stretched out on the helicopter deck. One end is taken under the helicopter and through the eye of the SPIES rope. It is then connected on the other end of the sling using a Type IV link assembly. The sling must pass between the helicopter skids and the fuselage.

(2) Once the the SPIES rope and cargo straps are in place, the straps running across the deck of the helicopter are secured in place by at least four and as many as eight snaplinks. The snaplinks are evenly spaced across the deck and alternated from one side of the strap to the other and top and bottom, so that the first snaplink is to the rear of the strap and going around the bottom two straps. The next snaplink is in the front of the cargo strap and around the top two sections of the strap. This process continues until at least four points are established.

(3) If eight snaplinks are available, then each tie-down has two reversed. If no hook is available or is not working properly, the SPIES can be used safely by doubling the cargo slings and Type IV links. Thus, two cargo straps are side by side for a total of four slings and four Type IV links. Padding is used around the edge of the fuselage to protect the sling from damage.

(4) Care must be taken when using the UH-1H helicopter because of the ways in which it can be outfitted. Some helicopters have a step attached at the doorway. This is an added obstruction not only during installation but also during the operation. Others may have rocket pods or machine guns mounted. Not all of the UH-1 helicopters are hooked up the same way each time.

4-10. RIGGING OF UH-60 HELICOPTER FOR SPIES OPERATION

The SPIES master inspects all materials required to rig the helicopter.

a. **Equipment.**

- One 120-foot SPIES rope with deployment bag.
- Two 11-foot, 3- or 4-loop cargo slings or two 9-foot, 3- or 4-loop cargo slings.
- Two Type IV connector links (four if aircraft does not have cargo hooks).
- Heavy duty tape (100-mph tape).
- One 12-foot sling rope.
- Five oval snaplinks (nine if the aircraft does not have cargo hooks).

Note: If the aircraft does not have cargo hooks, use four 11-foot or four 9-foot, 3- or 4-loop cargo slings.

b. **Installation.**

(1) The primary attachment point for the SPIES rope is the cargo hook. The end of the SPIES rope has a polyurethane encapsulated eye that is attached to the cargo hook. The two 9- or 11-foot cargo suspension slings are joined to form one continuous sling using a Type IV link. The sling is stretched out on the helicopter deck. One end is taken under the helicopter and through the eye of the SPIES rope. It is then connected on the other end of the sling using a Type IV link assembly. The sling passes between the helicopter skids and the fuselage. Padding is used around the edge of the cargo hatch to protect the sling from damage.

(2) Once the SPIES rope and cargo straps are in place, the straps running across the deck of the helicopter are secured in place by at least four and as many as eight snaplinks. The snaplinks are spaced evenly across the deck and alternated from one side of the strap to the other and top and bottom. Thus, the first snaplink can be to the rear of the strap, wrapping around the bottom two straps. The next snaplink is in the front of the cargo strap and around the top two sections of the strap. This process continues until at least four points are established.

(3) If eight snaplinks are available, each tie-down has two snaplinks connecting the same spot, and the swing gates are reversed. If no hook is available or is not working properly, the SPIES can be used safely by doubling the cargo slings and Type IV links. Two cargo straps are side by side with a total of four slings and four Type IV links.

4-11. RIGGING OF CH-46/CH-47 HELICOPTER FOR SPIES OPERATION

The SPIES master inspects all the materials required to rig the helicopter.

a. **Equipment.**

- Two 11-foot, 3-loop slings.
- Two 9-foot, 3-loop cargo slings.
- Four Type IV connectors.
- One 13-foot sling rope.

b. **Installation.**

(1) The SPIES rope is attached using two 9- or 11-foot cargo suspension slings and four Type IV links. The cargo slings are passed through the encapsulated eye of the SPIES rope and attached to the outboard cargo tie-down rings on the aircraft floor. Two tie-down rings are used for each sling. Padding is used around the edge of the cargo hatch to protect slings from damage.

(2) Not all of the tie-down rings will be in the exact same position on the helicopters. This is one of the main considerations in deviating from the installation procedures. When possible, the cargo straps are placed to form two U shapes. One strap is placed forward of the cargo hole in the center of the aircraft floor and one aft or toward the rear of the helicopter. Installed properly, the cargo straps hold the SPIES rope comfortably in the center of and slightly below the opening of the cargo hatch.

(3) The use of snaplinks attached close to all four faulty tie-down points not only ensures a backup in case of a faulty tie-down ring but also reduces the amount of movement in the cargo suspension straps. A total of eight snaplinks are used for added security at each point with the swing gates reversed.

Section III. LAND AND WATER EXTRACTION

If the situation, mission, or terrain suggests the possibility of a SPIES extraction, the leader includes the SPIES harness in each individual's equipment list. If the mission or insertion precludes the wearing of the harness, it will be carried on the top and just inside the flap of the pack being used. Once the extraction helicopter has been requested, the harness is retrieved and donned.

4-12. LAND EXTRACTION PROCEDURES

The SPIES is used only when the team requires immediate extraction or is unable to move to a clear (open) position suitable for helicopter landing.

a. The extraction helicopter proceeds to the area and establishes radio or visual contact with the team. The backup helicopter equipped with the SPIES remains aloft and away from the area. The helicopter maintains visual contact with the LZ and monitors radio communications.

b. After the team has indicated readiness for pickup and the tactical situation has stabilized, the extraction helicopter moves to the LZ by the safest route. When the helicopter is above the team's location, the SPIES master drops the SPIES rope on order of the pilot after the aircraft has obtained a stable hover at slightly above treetop height.

c. The team hooks up the same as in the familiarization procedures. Team members sling individual weapons over the shoulders, securing weapons and equipment to withstand the wind. A safety line is attached to prevent lost weapons during SPIES operations. When the team is ready, the team leader gives a thumbs-up signal.

d. During extraction, the team RATELO maintains communications with the extraction helicopter. He gives a verbal backup to the thumbs-up signal and relays all information during the flight. His location is near or at the bottom hookup point so he can assist in giving accurate information about the extraction, clearing of obstacles, and descent.

e. Lift-off of the extraction helicopter is vertical until the SPIES rope has cleared all obstacles. With the barrel directed downward at a 45-degree angle and outward, the team members fire their individual weapons using the hip position.

f. Once the SPIES rope is clear of all obstacles, the extraction helicopter changes to horizontal flight and departs the area by the safest route. Airspeed is limited to 70 knots in moderate climates and 50 knots in cold climates while team members are attached to the SPIES rope.

g. When the extraction helicopter has reached a safe dismount area, a transaction is made into hover flight at an altitude of 250 feet AGL. A vertical descent is started with the SPIES master continuously providing information to the pilot on the distance from the ground to the lower end of the SPIES rope. The vertical descent rate of the aircraft (at touchdown) is less than 5 feet per second.

h. When the team reaches the ground, they immediately move out toward the front of the helicopter. All team members rapidly unhook themselves and teammates that need assistance. Once this is completed, they either move away from the area and set up security or assist in clearing the rope if the helicopter is going to land.

4-13. WATER EXTRACTION PROCEDURES

The SPIES is suitable for extracting teams from the water. For this procedure, three inflatable life vests or any type of floatation device is tied to the SPIES rope to provide buoyancy for the rope while in the water. One floatation device is tied at each end of the attachment points, and one floatation device is tied in the middle of the attachment point area just above the middle two sets of D-rings. The team members to be extracted wear their SPIES harnesses under their individual life vest. They can also wear swimming fins, masks, and snorkels (amphibious operations) to ease hooking up to the SPIES rope within the spray area beneath the hovering helicopter.

a. After the extraction helicopter has attained a stable hover above the team member's location, the SPIES master drops the SPIES rope (with floatation attached) on order from the pilot. When the team members complete the hookup to the SPIES rope, the team leader signals the SPIES master to start lift-off. Lift-off of the extraction helicopter is vertical until all team members and the bottom end of the rope have cleared the water. During the initial lift-off, team members know that they are to be dragged through the water. The team members prepare to roll on their backs until clear of the water.

b. Flight speed and altitude are the same as over land. The dismounting procedures remain the same, except when landing on a ship. Once on board, the team members take their orders from personnel in charge of the deck.

Section IV. AFTER-OPERATIONS PROCEDURES

After-operations procedures include repairing, cleaning, and proper storage of the rappelling equipment.

4-14. REPAIRING AND CLEANING OF EQUIPMENT

The contaminated ropes are washed with a mild detergent, such as liquid dish soap, and cold water followed by a rinse in clean, fresh water. Ropes are dried at a temperature not to exceed 140 degrees Fahrenheit. Stubborn oil, grease, hydraulic fluid, and other petroleum stains can be removed with the cleaning agent xylene (Grade A or B, TT-X 916).

4-15. STORAGE OF EQUIPMENT

To avoid ultraviolet deterioration, the nylon materials should be protected from direct sunlight. The SPIES rope is stowed in an aviator's kit bag for protection when not in use. Bins or similar facilities are used for storage of SPIES equipment. Shelves used for storage should be at least 4 inches from the walls and 12 inches from the floor. Areas used for storage should be well ventilated and free of oil, acid, cleaning compounds, and other contaminants. Equipment must *not* be stowed above or near hot water pipes or heating apparatus.

CHAPTER 5

FAST-ROPE INSERTION AND EXTRACTION SYSTEM

The fast-rope insertion and extraction system (FRIES) is employed by small units to make rapid infiltration and exfiltration using rotary-wing aircraft in confined areas. The use of FRIES is normally restricted to situations where aircraft cannot land, and when mission requirements preclude the need for large amounts of equipment, and ammunition or heavy crew-served weapons.

Note: Headquarters, Department of the Army (HQDA) policy specifies that FRIES is not approved for Army-wide use and names the Commanding General, United States Army Special Operations Command (CG USASOC) as the executive agent for FRIES doctrine. The use of FRIES is restricted to special operations forces, pathfinders, long-range surveillance units, and HQDA-approved schools with a USASOC-approved FRIES program of instruction. Requests for other units to conduct FRIES operations should be submitted to Commander, USASOC, ATTN: AOOP-TRS, Fort Bragg, NC 28307-5000.

Section I. GENERAL

FRIES information is subject to change. Units that recognize the need to update this chapter should submit recommendations to Commander, USASOC, ATTN: AOOP-TRS, Fort Bragg, NC 28307-5000.

5-1. OBJECTIVES

The objectives of this chapter are to—

- Prescribe the requirements for FRIES qualification and training, including proficiency sustainment, safety, and administrative procedures.
- Establish the scope of the duties and responsibilities of key personnel who conduct FRIES operations.

5-2. GUIDANCE FOR COMMANDERS

Units that have HQDA approval to perform FRIES operations are authorized to conduct initial FRIES qualification and FRIES master (FRM) qualification training. Training requirements are established in this publication, USASOC Regulation 350-6, and applicable SOAR policies. Aircraft material requirements are specified by the US Army Technology Applications Program Office (TAPO) .

a. FRIES training and operations are inherently risky; therefore, safety is paramount. All training and operations require in-depth attention to detail from planning through preparation and execution.

b. The following basic guidance applies to FRIES training and operations:

(1) Unit commanders must personally approve FRIES training sites.

(2) A risk analysis and or assessment must be completed before FRIES operations.

(3) Night FRIES operations are medium risk or higher.

(4) Units authorized to conduct FRIES operations may incorporate FRIES activities in field training exercises only after personnel have become FRIES qualified.

(5) Field exercises using FRIES activities are subject to the limitations imposed by applicable directives, including mandatory support requirements.

(6) Requests for waiver of procedures or restrictions prescribed herein must be submitted in writing with endorsement of the first general officer in the requesting unit=s chain of command to HQ USASOC.

5-3. TRAINING PREREQUISITES

> **CAUTION**
> **Trainers must ensure all personnel preparing to participate in FRIES training are tested *with equipment* to ensure they have sufficient upper body strength to safely perform the full scope of roping duties.**

Before allowing soldiers to participate in FRIES training and operations, the unit commander must make sure the soldiers are physically fit. Minimum screening standards that apply to each training participant include—

- Having passed the Army physical fitness test within the past six months.
- Having passed a medical examination within the past two years.
- Being free of any injury or physical condition that could cause a potential safety hazard during FRIES training.
- Demonstration of ability to perform a controlled descent from a height of 15 feet while carrying a 40-pound load (not including the basic duty uniform and combat boots).
- Demonstration of ability to hold a static position on a FRIES rope for 20 seconds, using hands and feet to lock in while carrying a 40-pound load.
- Being qualified and current to fly FRIES activities IAW HQ USASOC and the Special Operations Aviation Regiment (SOAR) Aircrew Training Program and applicable policies (for aircrews).

Note: The peacetime maximum soldier load will not exceed 50 pounds. This includes helmet, weapon, vest, web gear, and rucksack. Rucksack weight will not exceed 35 pounds.

5-4. PERSONAL EQUIPMENT REQUIRED

The minimum personal equipment required for FRIES training or operations includes—

- Heavy leather gloves.

- A helmet with a chin strap.
- Protective goggles.
- A long-sleeve shirt or jacket, long pants, and boots.
- Hearing protection and identification tags for helicopter operations.

Section II. FRIES QUALIFICATION TRAINING

Each major unit (commanded by an O5 or higher) is responsible for conducting its own FRIES qualification and sustainment program. FRIES trainers will be FRIES master (FRM) qualified and current. Unit training should be tailored to unit and situation needs, however proponent requisites will not be reduced without written approval of the CG USASOC.

5-5. INITIAL FRIES QUALIFICATION TRAINING

Before participating in fast-rope operations, personnel are briefed on the FRIES and its purpose, capabilities, and limitations. The briefing also covers the duties and responsibilities of the pilot in command (PIC), safety, FRIES master (FRM), the assistant FRIES master (AFRM), and any ground assistants. Once the FRIES briefing is conducted, the remainder of the initial training is hands-on practice of the proper FRIES operational techniques.

a. Individuals are shown the proper techniques for boarding the aircraft, moving to the door, grasping and descending, locking-in, and clearing the rope. Individuals are also shown the proper wear, hook-up, and use of the STABO extraction harness.

b. After the demonstration, all ropers participate in a practice exercise in which they properly perform the following tasks using a tower and subsequently an aircraft. All ropers must demonstrate a minimum of six properly executed FRIES descents (three without equipment and three with equipment) and at least one successful lock-in.

(1) *Perform FRIES descent and lock-in from a tower at the 34-foot level.* Training should be progressive, starting from a tower at the 34-foot level without equipment and then with equipment. All ropers must complete a successful lock-in at this level. Ropers will not progress above the 34-foot level until after demonstrating the ability to stop descent, lock-in, and hold a stationary position for 20 seconds with equipment.

(2) *Perform FRIES descent and lock-in from a tower at the 60-foot level.* The recommended succession will be no less than three successful FRIES descents on the 60-foot tower, again starting without equipment, completing a lock-in, and finishing with equipment.

(3) *Perform FRIES descent from a helicopter at 60 feet or higher.* The recommended completion should be at least two helicopter descents from 60 feet or higher (one without equipment and one with equipment).

WARNING

During initial qualification, students will not perform rapid exits or have more than three soldiers on a rope at any time. During advanced training, the limit is no more than five personnel with equipment on any rope at any time. In all cases (insertion or extraction), the maximum load allowed on any fast-rope system will not exceed 1,500 pounds.

CAUTION

All ropers should be limited to no more than a total of 10 roping events in a 24-hour period; no more than 6 events should be with equipment.

5-6. FRIES PROFICIENCY SUSTAINMENT TRAINING

Commanders must make sure soldiers participating in FRIES operations receive sustainment training on equipment and procedures within 24 hours before the FRIES operation. Ropers who do not attend FRIES sustainment training will not be allowed to participate in FRIES operations. As a minimum, training will include a review of the following:

- Arm-and-hand signals.
- Individual equipment riggings.
- Aircraft familiarization.
- Safety procedures.
- Any rehearsals the FRM or commander deems necessary.

Section III. FRIES MASTER SELECTION AND QUALIFICATION TRAINING

Selection and qualification of FRIES masters (FRMs) is a unit prerogative. Units will maintain records of qualification and proficiency. These records will accompany SM to new units when a permanent change of station occurs.

5-7. FRIES MASTER SELECTION

Units may train and certify soldiers as FRIES-qualified FRMs. Soldiers selected by units to qualify as FRMs should be the rank of E5 or higher and should have demonstrated leadership abilities with emphasis on maturity and decisiveness.

5-8. FRIES MASTER PREREQUISITES

Soldiers performing FRM, AFRM, and safety duties must have completed initial FRIES qualification training, be experienced in FRIES operations, have completed FRM training, and have executed at least 10 FRIES descents and 2 extractions. An FRM candidate's experience must include no less than five descents with combat load and at least one night descent with equipment from 60 feet or higher.

5-9. FRIES MASTER TRAINING AND CERTIFICATION

During FRM training and certification, FRM candidates must participate in three FRIES helicopter operations (twice observing another FRM and once executing FRM duties under observation of a current FRM) and be certified by either the parent unit or an authorized school. FRMs must be proficient in FRIES operations and must demonstrate proficiency at the following:

- Inspecting, preparing, and rigging all FRIES gear.
- Inspecting and preparing aircraft for FRIES operations.
- Controlling and coordinating actions of AFRMs, safeties, and ropers.
- Preparing and conducting both insertion and extraction operations.
- Conducting troop and pilot briefings.
- Giving arm-and-hand signals.
- Giving time warnings and associated commands.
- Deploying and retrieving FRIES ropes.
- Rigging and lowering equipment.

5-10. FRIES MASTER REFRESHER TRAINING

Units will conduct refresher training to maintain acquired skills. FRMs who have not participated in FRIES operations during the past six months will receive refresher training by a current FRM and serve as an AFRM before performing FRM duties. Refresher training for ropers consists of an FRM briefing and participation in FRIES training.

Section IV. KEY PERSONNEL DUTIES AND RESPONSIBILITIES

The following personnel duties and responsibilities provide baseline requirements for the safe conduct of FRIES operations. Unit standing operating procedures (SOPs) may increase (but will not reduce) training safety requirements.

5-11. AIR MISSION COMMANDER

The air mission commander (AMC) provides overall command and control when conducting large, complex missions. The AMC is selected or appointed by the headquarters responsible for the overall mission. He ensures that all operators, aircrew members, and support elements synchronize their actions during the conduct of the mission and the FRIES operation. An AMC is not usually needed during small-unit qualification or proficiency training events when only one helicopter is used.

5-12. FRIES TRAINING OFFICER

Units that conduct FRIES training appoint a training officer in charge (OIC) or a noncommissioned officer in charge (NCOIC) to manage the training. The OIC is responsible for planning and coordinating various aspects of training to include—

- Coordinating all support activities, such as procuring FRIES equipment, aircraft, training areas, medical support, and communication.

- Assigning qualified personnel to perform FRM, AFRM, ground safety, and medic duties.
- Adhering to procedures for planning, preparing, and executing the operation IAW this TC, USASOC Regulation 350-6, and the participating unit=s SOPs.
- Ensuring the FRMs, AFRMs, PICs, aircrews, safeties, and medics are briefed on the operation.
- Ensuring FRIES training events are not conducted on or over unsafe ground. (The ground must be free of loose material that could become airborne and cause an unsafe condition or obscure vision for aviators or ropers.)
- Ensuring a medic, an aid bag, a backboard, and a dedicated vehicle and driver are on site during all training.
- Ensuring positive communication with aircraft is maintained throughout training.

5-13. PILOT IN COMMAND

The aviation unit providing helicopter support for the FRIES training and operation appoints the PIC. The PIC is responsible for all aspects of the flight and ensures—

- Aircrew members are current and qualified to conduct FRIES operations
- Aircrew members are informed of and understand their responsibilities in fast-rope operations.
- Procedures for planning, preparation and execution are adhered to IAW this TC, USASOC Regulation 350-6, and applicable SOPs and policies.
- Aircraft is properly configured to perform the mission.
- Aircraft is at the proper altitude, airspeed, and location as briefed.
- The command to deploy ropes is not given until the aircraft is at a stabilized hover.
- The aircraft position is maintained to keep ropes in contact with the surface until all descending ropers are on the ground (or extracting ropers are securely attached for extraction).
- Ropes are fully recovered inside the aircraft (or jettisoned) before the aircraft departs the stabilized hover position at the infiltration site.
- All personnel are briefed on in-flight emergency and safety procedures.
- Ropes are never deployed with anything other than night illumination attached to the free end.
- All personnel are briefed on fire support to be provided by the aircraft including:
  - — The nature of fire support the aircraft can provide.
  - — The time the fire support starts, shifts, and stops.
  - — The primary and alternate commands and signals that start, stop, lift, or shift fires.

5-14. FRIES MASTER

Units conducting FRIES operations will designate one overall FRM to organize, coordinate, and supervise the activities of the day, and AFRMs as needed. AFRMs are additional FRM-qualified soldiers who assist the primary FRM. An AFRM will be designated for each roping point. AFRMs are responsible for all preparation, inspection, and command and control of all roping activities on their points. FRMs also must—

- Ensure all FRIES roping personnel, including FRMs, AFRMs, and safeties, are qualified and current and understand the proper roping procedures for FRIES operations.
- Perform safety and serviceability checks on all FRIES equipment and rigging.
- Ensure the attachment bar or points are serviceable and free of any defects or contamination and that the quick-release mechanisms and safety pins are present and serviceable and operate correctly.
- Inspect the rope to make sure it has no contamination, damage, or defects that could make it unsafe.
- Check the rope to make sure it is the correct type and length for the operation (smooth rope for infiltration or looped for exfiltration).
- Ensure the rope is properly attached with safety pin in place and back coiled.
- Ensure the extraction harnesses are serviceable and properly worn.
- Ensure the rope chemlights are correctly rigged and illuminated when needed (two at the mount, two at the end, and two 15 feet from the end).
- Confirm the proper seating arrangement, including seat belts or a safety strap for ropers.
- Coordinate with the PIC and brief the aircrew, safeties, and ropers on the following:
  - — The aircraft flight, approach, flare, and position over the target site.
  - — The height of the aircraft above the target.
  - — A description of the FRIES site (rooftop, small clearing) and size, as well as the actions of the ropers upon arrival at the site.
  - — The signal to deploy the rope, who will deploy it, and how it will be done.
  - — Time warnings and actions at each time warning.
  - — Arm-and-hand signals.
  - — The use of radios and intercoms.
- Brief the PIC and crew chief on verbal commands and final adjustments over the target.
- Confirm the correct target and inform the pilot before the deployment of ropes or ropers.

WARNING

The FRM must ensure that nothing, other than rope illumination, is ever attached to the free end of the FRIES rope during deployment and that equipment is never attached and dropped.

- Deploy the ropes when the PIC gives the command DEPLOY ROPES. (This action may be delegated to the aircrew or safety.)
- Ensure safeties and AFRMS understand the time and activity sequence and all arm-and-hand signals.
- Ensure safeties know the proper procedures to—
- Keep the aircraft on target.
  - — Clear the ropes.
  - — Recover the ropes.
  - — Jettison the ropes.
- Observe the target and pass to the PIC, through the crew chief, final aircraft position adjustments over the target to make sure conditions are safe.
- Deploy the ropes when the PIC gives the command DEPLOY ROPES. (This action may be delegated to the safety.)
- Ensure all ropers know how to hook up for extraction and know the correct arm-and-hand signals.
- Direct the aircrew or safety to inform the pilot when the aircraft is on target (the PIC replies with the command DEPLOY ROPES).

CAUTION

When equipment is to be lowered, it should be lowered before the FRIES is deployed or personnel begin roping. When simultaneous, equipment and ropers will exit from opposite doors of the aircraft.

- — The safety repeats, " DEPLOY ROPES." The FRM signals the AFRMs to deploy ropes and makes a final check, ensuring all ropes from his aircraft reach the surface and are safe for descent (at least 15 feet of rope must be on the surface).
- — When conditions are safe, the FRM commands "GO" to the ropers and points to the rope or leads the stick out.

Note: If the FRM leads the stick out, the AFRM automatically assumes control. He completes marshaling the remaining ropers and is the last man to exit.

- Coordinate actions with the crew chief for the positioning or storage of seat belts, safety straps, rucksacks, door bundles, and communication cords.

Note: During night operations, upon completing descent, the AFRM observes the the ropes of his aircraft. When he sees that all ropes are clear, he issues the prearranged ALL CLEAR hand signal to the aircrew safety. This hand signal tells the aircrew that they are free to jettison or recover the ropes, and fly away.

5-15. FRIES AIRCREW MEMBER AND SAFETY

FRIES safeties are aircrew members who are FRM-qualified and perform the additional duties of the safety during helicopter roping operations. Safeties assist FRMs and pilots as needed. Aircrewmen performing safety duties differ from FRMs and AFRMs in that safeties are members of the aircrew (crew chief or gunner) whose orientation is the flying portion of the operation. Conversely, AFRMs are FRM-qualified members of the ground unit who are performing FRM duties as assistants to the primary FRM and who exit with the ground unit. Safeties remain in the aircraft and operate the doors, extend and retract the rope mount bars, and wear night vision goggles (NVGs) when appropriate. Safeties must have completed FRM training IAW this TC and the aircrew training program (ATP).

a. During roping operations, the safety's four major areas of responsibility are:

(1) Observe all activities in the aircraft and assist the pilot and FRM or AFRM as needed.

(2) Serve as the communication link between the PIC and the FRM. The safety monitors aircraft communication and relays information between the PIC and the FRM, constantly keeping both informed.

(3) Assist the FRMs and AFRMs in lowering cargo and clearing the aircraft of ropers.

(4) Clear and jettison the ropes, when appropriate.

b. The safety performs the following duties:

(1) Relays time warnings, aircraft positioning commands, and other information to assist the FRM.

(2) Conducts aircraft equipment rigging and serviceability inspections, making sure mounts, doors, and ropes are ready and safe for operations both pre-flight and en route.

(3) Verifies that the aircraft is at a stabilized hover with ropes in contact with the surface and remains so (or stops stick until corrections are made).

(4) Checks the rope to make sure 15 feet of rope are on the target surface.

(5) Tells the PIC, "ROPES DEPLOYED", when appropriate.

(6) Tells the PIC, "ROPERS OUT", when the first roper has exited.

(7) Helps the AFRM in controlling and spacing each roper's exit.

(8) Monitors the rope and ropers to make sure they do not get hung up, dragged, or dropped off site.

(9) Observes the exit of the last roper until the roper reaches the surface, moves clear of the rope, and gives the ALL CLEAR signal.

(10) Tells the PIC, "ROPERS DOWN", (station name: aft, forward, left, right) when all ropers are on the surface and conditions are safe to jettison the ropes.

(11) Passes the signals as required to position the aircraft, including HOLD, MOVE LEFT, RIGHT, FORWARD, BACK, UP, or DOWN.

(12) Acts as belay man during cargo lowering when needed.

(13) Jettisons ropes upon receiving the command JETTISON ROPES from the PIC. Tells the PIC, "Ropes are jettisoned," when the ropes have been recovered or jettisoned.

(14) Tells the PIC when ALL CLEAR.

CAUTION

This command tells the pilot the aircraft is cleared for flight.

(15) Deploys the rope for extraction upon receiving the PIC command DEPLOY ROPES.

(16) Observes the extracting rope and personnel and informs or guides the PIC through aircraft positioning and throughout the recovery flight.

5-16. FRIES ROPER

Ropers are responsible for notifying the FRM, AFRM, safety, or pilot if they observe any unsafe act or condition. Ropers may halt or call for a halt of roping for safety at any time. During FRIES insertion training, the number of ropers on the fast rope at one time is limited to three. During extraction training, the total weight per extraction bar or rope will not exceed a total of 1,500 pounds. When rucksacks are worn by ropers, the rucksacks must not exceed a weight of 35 pounds (total roper equipment weight will not exceed 50 pounds). Ropers will—

- Keep hands at head level.
- Maintain visual contact with lower ropers during their descent.
- Maintain a minimum of a 1-second interval on exit to avoid collisions.
- Keep at least two points of contact on the rope (both hands) at all times.
- Use their feet for additional breaking any time needed.
- Execute descents at a safe speed.
- Slow the rate of descent halfway down the rope to avoid landing on each other.
- Move quickly away from the rope upon arrival on the surface.
- Know correct wear of extraction (STABO) harness and extraction procedures.

WARNING

The FRM, AFRM, and safety must make sure all personnel are on the surface and have moved clear of the ropes before aircraft fly away or ropes are jettisoned.

Section V. FRIES RIGGING OF AIRCRAFT

The aviation unit is responsible for preattachment inspection and installation of FRIES anchor point mount bars and associated equipment. All aircraft used to conduct FRIES operations with US Army personnel must be rigged and equipped IAW current parent service directives--for Army aircraft, USASOC, SOAR, and TAPO directives apply. Rigging of the aircraft includes installation and removal of the seats and seat belts (or personnel restraints), and attaching and detaching the ropes.

Figure 5-1. UH-60 helicopter

> **CAUTION**
> **Army aircraft that have the FRIES mounts installed must have an airworthiness release for FRIES operations from the Aviation Troop Support Command (ATCOM) TAPO.**

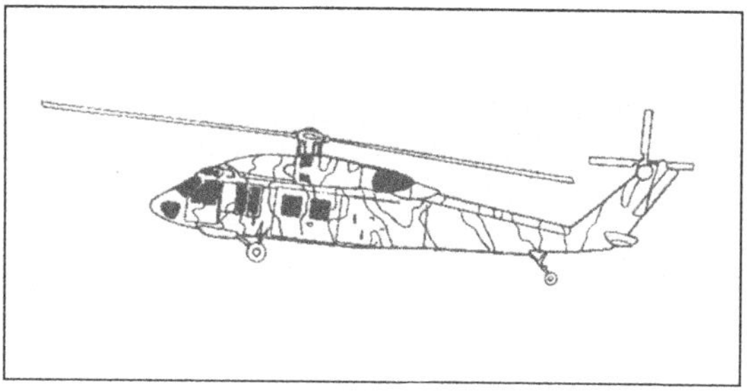

5-17. RIGGING OF FRIES IN UH/MH-60

The UH-60 is a twin-turbine, medium-speed, single-main-rotor helicopter (Figure 5-1).

The aviation aircrew or safety is responsible for—

- Inspecting the FRIES mount bar assembly, attachment hardware, and associated equipment to make sure they are serviceable and ready to use.
- Ensuring the bar assembly is correctly mounted to the aircraft (Figure 5-2).

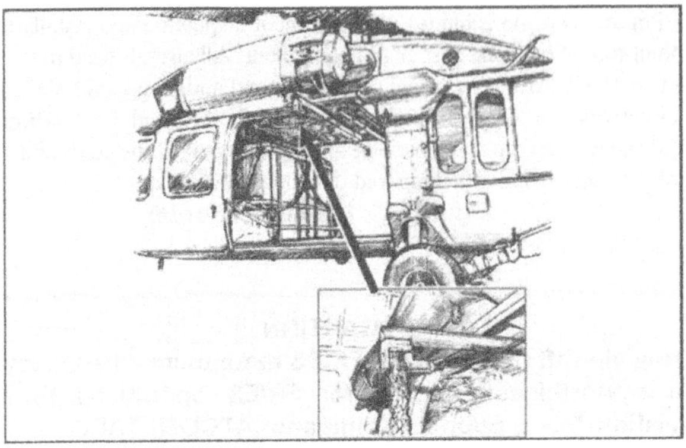

Figure 5-2. Bar assembly correctly mounted to aircraft.

- Ensuring the mount is complete, functional, and free of any defect.
- Ensuring the mount and the aircraft floor are clean and free of any petroleum, oils, or lubricants (POL) or other contaminants that could weaken the rope or cause personnel to slip or fall.
- Ensuring the mount and aircraft are free of sharp edges or burrs that could cut the rope.
- Ensuring no metal fatigue or structural weakness exists that could cause the system to fail.
- Extending locking and retracting mount bars as needed.
- Lowering and belaying equipment as needed.
- Opening, locking, and closing cargo doors as needed.

Note: For cold-weather operations or during flights of long duration, the cargo doors are closed and latched until the time specified for opening. The aircrew will operate doors or directly supervises those who do. In good weather, doors should remain locked open.

- Removing or reconfiguring troop seats.
- Providing seat belts or floor-mounted personnel restraints for ropers' use during flight.
- Attaching and detaching ropes as needed.

Note: Clusters of rucksacks or small door bundles (weight limited to 250 pounds) may be lowered (belayed) on rappelling ropes. Cargo lowered this way must never be free dropped (the belay must start at the time the bundle is pushed [tipped] out the door).

5-18. RIGGING OF FRIES IN CH/MH-47

The CH-47 is a tandem-rotor, medium-transport helicopter (Figure 5-3).

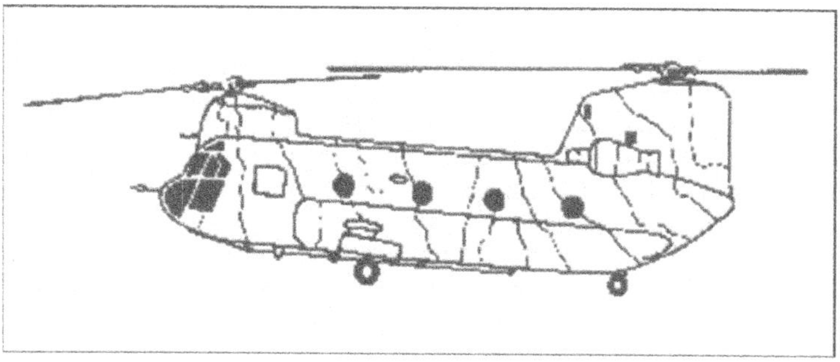

Figure 5-3. CH-47 helicopter.

The aviation aircrew or safety is responsible for:

- Inspecting the FRIES mount bar assembly, attachment hardware, and associated equipment to make sure they are serviceable and ready to use.
- Ensuring the bar mounts are correctly mounted to the aircraft (Figure 5-4, rear; Figure 5-5, winch).

Figure 5-4. Bar mounts correctly mounted, rear.

Figure 5-5. Bar mounts correctly mounted, winch.

- Ensuring the mount is complete, functional, and free of any defect.

- Ensuring the mount and the aircraft floor are clean and free of any POL or other contaminants that could weaken the rope or cause personnel to slip and fall.
- Ensuring the mount and aircraft are free of sharp edges or burrs that could cut the rope.
- Ensuring no metal fatigue or structural weakness exists that could cause the system to fail.
- Extending locking and retracting mount bars as needed.
- Lowering and belaying equipment as needed.
- Opening, locking, and closing cabin doors as needed.
- Removing or reconfiguring troop seats.
- Providing personnel restraints for ropers' use during flight.
- Taping and padding any sharp edges that can come in contact with ropes or personnel
- Attaching and detaching ropes as needed.

5-19. RIGGING OF FRIES IN MH-53

The MH-53 is a USASOC twin-engine, single-rotor, medium-transport helicopter used by the Army (Figure 5-6).

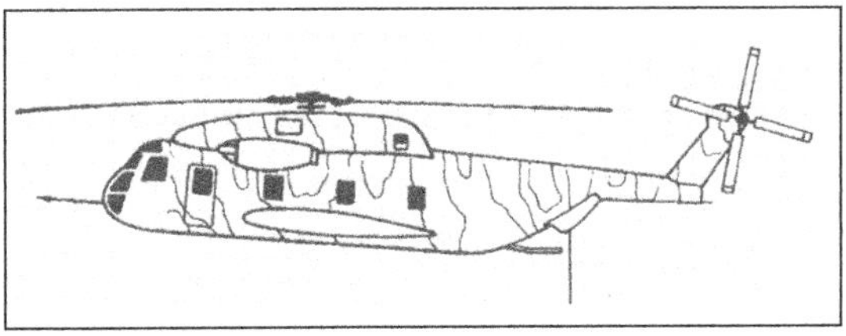

Figure 5-6. MH-53 helicopter.

The MH-53 primary FRIES mount bar assembly is located overhead, just aft of the rear ramp. The ramp mount bar assembly spans the cabin width and has a two-pin release mechanism on each end. The ramp position mount assembly provides two points; each will accommodate one FRIES rope or one cargo lowering rope. The ropes are carried back-coiled on the ramp.

a. The alternate position for fast-rope descent from the MH-53 is the crew entrance in the right side of the fuselage immediately behind the pilot's position. In this position, the fast-rope mount must be connected to the external hoist point. The FRIES master is responsible for inspecting the MH-53 mount system for its presence, security, and operability.

b. The aviation unit is responsible for—

- Installation and maintenance of the fast-rope mount assemblies at each position to be used.

- Configuring the aircraft to accommodate the number of soldiers and type of mission planned.
- Ensuring seat belts or restraint straps are present to secure ropers during flight.

Note: The MH-53 aircrew can improvise handrails to assist roper exits. When used, improvised handrails will be rigged IAW USAF special operations SOPs.

- Configuring the MH-53 for fast-roping when equipped with external 650-gallon fuel tanks.
  - Remove the ramp door to install the cross bar.
  - Place the bar in the brackets with the two quick-release paddles facing forward.
- Configuring aircraft armament (50-caliber machine gun and the GAU-2B ramp minigun) for FRIES operations.
  - When using the aft right roping position, point the 50-caliber machine gun to the 6 o'clock position and secure the gun with bungee cords.
  - When using the left aft rope, disconnect the minigun ammunition chute and stow it in the ammunition can, or disconnect it at the ammunition can and securely wrap it around the minigun. Use bungee cords to secure the ramp minigun in the firing position facing the 6 o'clock position.

Note: Ideally, only the right aft rope is used when the GAU-2B ramp minigun is installed. Although not recommended, the left aft rope may be used, if required.

Section VI. EQUIPMENT MAINTENANCE AND INSPECTION

Before conducting a FRIES operation, the FRM must inspect all FRIES equipment (ropes, STABO harnesses, and mount bars) for serviceability and readiness for use.

> **WARNING**
>
> **FRIES ropes and harnesses are critical life support equipment and must be stored, cared for, and maintained as life support items.**

5-20. ROPES AND HARNESSES

FRIES equipment should be maintained in the same manner as a parachute. The unit rigger section can provide detailed guidance on appropriate inspection, care, and maintenance of FRIES equipment.

a. FRIES ropes and harnesses must always be stored in a clean, cool, dry space out of direct sunlight and free of chemicals or chemical vapors. Equipment that becomes wet with fresh water should be hung up to dry (indoors) on hardwood pegs. Equipment that is exposed to salt water or becomes imbedded with dirt or mud should be washed and rinsed in fresh water (within 72 hours) and hung up to dry (indoors) on hardwood pegs (out of direct sunlight).

b. Before conducting a fast-rope operation, the FRM—

- Inspects the fast rope thoroughly and carefully. Checks the rope length to ensure it is the correct rope for the operation planned.
- Checks the woven loop on the mount end for excessive wear or chemical contamination. Checks the rope along its entire length for fraying, cuts, and chemical contamination.
  - Do not use a rope that is severely frayed. (Light fraying on the rope from normal use does not weaken the rope.)
  - Do not use a rope when any single strand is cut halfway through or has two or more cuts that penetrate one-third or more through any strand's thickness within 1 foot of the running length of the FRIES.
- Inspects the rope for contamination of acid, alkaline compounds, saltwater, fire extinguisher solutions, or petroleum-based solvents. Changes in color caused by chemicals are usually blotchy and have an unusual odor. Although used ropes gradually change color, such changes do not indicate a decrease in strength unless the change is due to contact with strong chemicals. Changes occurring because of use are usually uniform throughout the length of the rope.

WARNING

Any rope known or suspected to be chemically contaminated, cut beyond allowable limits, excessively frayed, or sun damaged must be taken out of service, marked for destruction, and turned in to the rigger section.

- Inspects the extraction loops to the same standard as the main rope. Ensures the woven attachment loops are secure.
- Inspects the harness to ensure—
  - Ropers are wearing the harness under all load-carrying equipment.
  - Ropers have properly fastened all connectors.
  - Harness material and stitching are not cut, torn, or contaminated, and all hardware is free of corrosion and is in operable condition.

Note: Serviceability inspection and maintenance of the extraction harness are the responsibility of the rigger-qualified personnel.

5-21. FRIES MOUNT BARS

The aviation unit is responsible for the installation, removal, storage, and maintenance of FRIES mount bars. The US Army Special Operations Aviation Regiment, Ft Campbell, KY, prepares and provides checklists for these procedures. Aviation units participating in FRIES operations must have and comply with current SOAR guidance.

Aviation units:

- Install and maintain the FRIES mount bar; check mount bolt torque and witness marks.
- Check the FRIES mount bar for security and operability.
- Check the condition of the mount bar; inspect for bends, dents, distortion, cracks, corrosion, and contaminants.
- Ensure the fast-rope quick-release pins are installed and operable (two each side).
- Check the security and condition of the rope-release system, as follows:
  - The mounting bolts are installed and properly torqued.
  - The safety pins are installed, and lanyards are present.
  - The system locks and unlocks as designed.
- Ensure the release cables are attached at both ends and are not broken or kinked.
- Check the release handles to ensure they are securely mounted and can move freely.
- Check the latch as follows:
  - Latch or close the release mechanism and apply a downward pressure on the release arm. Attempt to pull the release handles the release should not open.
  - Relax pressure on the release mechanism and pull the handles the release should open.
  - With no pressure on the release mechanism and with the safety pin installed, attempt to pull the release handles--the release should not open.

Section VII. OPERATIONAL REQUIREMENTS AND LIMITATIONS

This section establishes medical coverage and communication support requirements for FRIES operations, including during adverse weather conditions and night operations.

5-22. MEDICAL SUPPORT

Dedicated medical support is required for FRIES training and demonstrations. A qualified medic with assistant, aid bags, two backboards, two stretchers, and a dedicated vehicle with a driver must be present to respond to any mishaps. The nearest medical evacuation (MEDEVAC) unit will be notified of where and when FRIES training is scheduled. MEDEVAC response will be coordinated as on call. Training or demonstrations will not be conducted without medical support present. The ground force commander may direct the FRIES aircraft to transport life or death (urgent) patients when no other MEDEVAC unit is available, providing the medical team is trained and properly equipped with the appropriate equipment for aerial transport of patients. This must be included in the air mission briefing prior to training and coordinated with the medical team providing coverage. FRIES operations that are evaluated as high risk or that occur at night must have a certified combat lifesaver on location and medics (equipped as stated above) in the immediate vicinity.

5-23. COMMUNICATIONS REQUIREMENTS

During peacetime training and operations, FRIES lift aircraft and ground elements will maintain continuous radio communications. Aircrew personnel clearing ropes at each FRIES station must be on the intercom system, monitoring aircraft internal communications. They must also be proficient in the use of arm-and-hand signals as precoordinated between FRIES users and aircraft crews. The designated aircrew member will relay time warnings between the PIC and FRM. Routine time warnings are 10-, 6- and 1-minute warnings to the FRM. Ground-to-air communications must be continuously maintained throughout FRIES training.

5-24. ADVERSE WEATHER OR TERRAIN CONDITIONS

FRIES training and operations will ***not*** be conducted under the following conditions:

- Anytime temperatures are below 32 degrees Fahrenheit or wind-chill factors that could cause cold-weather injuries exist including those caused by helicopter rotor wash or extraction cruise speed.
- Anytime water or ice on the rope prevents personnel from controlling their descent.
- Anytime the rope is exposed to freezing conditions for a sufficient length of time to freeze, thereby reducing its tensile strength.
- Anytime dust out or white out conditions restrict the aviators or the ropers vision.
- Anytime trees or vegetation on the LZ is too dense for safe training.
- Anytime the LZ is too steep or too slippery for safe training.
- Anytime winds exceed 30 knots or are too gusty for safe operations.

5-25. NIGHT OPERATION REQUIREMENTS

Night operations require the following procedures.

a. Attach two chemlights at the free end of the rope and two above the extraction attachment loops (15 feet from the end of the rope) to aid in determining when adequate rope is on the surface.

b. Attach two chemlights (one primary and one backup) to the loop of the rope, or to the rope mount, to guide personnel exiting the aircraft to prevent glare from obscuring vision.

c. Attach one chemlight or luminous tape to the top of each roper's helmet. If a chemlight is used, select a color different from the colors used on the ropes.

CAUTION

Ropers will not wear NVGs during descent. NVGs limit depth

perception and cause a tunnel-vision effect. Static FRM and Safeties may wear NVGs to help confirm that fast ropes are fully deployed. The aircrew will wear NVGs as required by flight mission needs.

Section VIII. FRIES PROCEDURES

The risks of personal injury and equipment damage during FRIES operations can be reduced by detailed planning, thorough training, and strict adherence to safe procedures. Conducting FRIES operations with heavy loads requires personnel to be proficient in these operations. FRIES users must limit ropers to no more than 6 descents with combat load and no more than 10 total descents in any 24-hour period. The total weight of equipment soldiers may carry during training is limited to never more than 50 pounds. This section provides standardized procedures for FRMs, PICs, and ropers. These procedures are common to all units and aircraft, except as noted.

5-26. FRIES MASTER

The FRM performs the following series of actions.

- Briefs members of his team and aircrew.
- Inspects team members and equipment.
- Installs the FRIES rope in the aircraft and conducts safety checks.
- Relays time warnings to team members. (Time warnings are a tool to help keep aircrew and ropers' actions synchronized and can be modified according to user needs; however the 1-minute warning should always be used.)
- Breaks chemlights, if required. Two chemlights or luminous tape will be attached at the anchor point to identify the handhold point.

WARNING

Equipment must never be attached to the FRIES and dropped. As little as 10 pounds of weight can exceed the load limit of the FRIES mounts when the load stops at the end of a 60-foot drop. The only items to be attached to the fast rope during deployment are the position markers.

- Ensures the rope is properly configured for deployment (back coiled to prevent tangles).
- Ensures the team members are in order of exit no later than the 1-minute warning.
- Confirms target on final approach.
- Deploys the rope and ensures it reaches the ground. During night

operations, the rope is marked with six chemlights: two at the mount, two on the end, and two 15 feet from the end.

- Deploys personnel.
- Accounts for personnel, and signals aircrew.

> **WARNING**
> **Because the FRIES mount in the CH-47 helicopter is narrow, special care must be taken by ropers to avoid fouling weapons or equipment on the frame during exit.**

5-27. PILOT IN COMMAND

The PIC briefs the FRM, safety, and roping personnel before FRIES operations.

a. The PIC briefing includes:

- Approaching, loading, unloading, and departing the aircraft.
- Actions in the aircraft.
- Flight route and checkpoints en route.
- Altitude or height for roping.
- Time warnings to FRM as needed (recommend 10 minutes, 6 minutes, and 1 minute).

b. When the PIC is over the intended objective he calls, "DEPLOY ROPES" Fast-rope warnings and expected aircrew coordination calls are as follows:

(1) The aircrew or FRM provides any final position adjustments of the aircraft with the commands of MOVE (left, right, forward, back, up, down, or hold) as needed to place the ropes ON TARGET. (When the pilot is to hold the position, an explanation will be provided as soon as possible.)

(2) When the FRM deploys the ropes, the aircrew member calls, "ROPES DEPLOYED."

(3) When the first roper exits the aircraft, the aircrew member calls, "ROPERS OUT."

(4) The safety watches the last roper until he is on the ground and gives the prearranged signal. The safety then tells the PIC, "ALL ROPERS AWAY," and identifies his station. This alerts the PIC that the safety is going to either jettison or recover the ropes.

(5) The safety will then either recover or jettison the rope as planned, clear the aircraft all around for obstructions, and call, "ROPES CLEAR (station)" This informs the PIC that the aircraft is clear for flight.

c. The PIC also briefs on emergencies that could occur in the following situations:

- En route (when ropers are inside the aircraft).
- Before deployment of ropes.

- When fast-rope personnel are on the ropes.
- After the ropers are on the surface.

5-28. ROPERS

Ropers perform the following actions:

a. At the command STAND BY (given at 1-minute warning), ropers make final check of themselves and take the prepared-to-exit position.

b. At the command GO, the first roper exits the aircraft, rotates his body 90 to 180 degrees to make sure his equipment clears the aircraft, places the fast rope between the arches of his feet, and begins descent. The roper uses his hands and feet to slow his descent approximately two-thirds of the way down.

c. Subsequent ropers exit at 1-second intervals using the same procedure, but begin slow descent approximately half-way down to avoid landing on other ropers.

d. During descent, ropers keep a lookout and break as necessary to avoid landing on obstructions or on fellow ropers.

e. Ropers prepare to land just before reaching the ground by spreading their legs about shoulder-width apart with their knees slightly bent.

f. At landing, ropers quickly move clear of the ropes to avoid collisions with descending ropers.

CAUTION

According to AR 95-3, anytime operations occur over water, all personnel will wear an approved personal flotation device as identified in FM 57-220 or USASOC Regulation 350-2.

5-29. EQUIPMENT-LOWERING PROCEDURES

Anytime a roper's load exceeds a total of 50 pounds or his rucksack weight exceeds 35 pounds, equipment will be lowered (as clusters or bundles) on separate ropes. Lowering equipment from heights great enough to cause equipment damage necessitates belaying the equipment during descent to avert damage. Belaying devices used in mountaineering allow ropers or safeties to control equipment descent.

Figure 5-7. Rucksacks upright on cargo deck.

CAUTION

When equipment is to be lowered, it should be lowered before the FRIES is deployed or personnel begin roping. When done

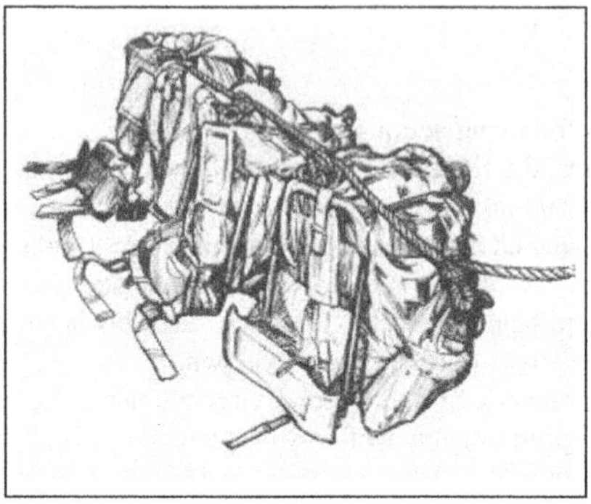

simultaneously, equipment and ropers will exit from opposite doors of the aircraft.

a. Rigging equipment to be lowered requires rigging rucksacks in clusters. Clusters of up to five rucksacks may be lowered (smaller clusters work better). FRMs will ensure the total weight of any cluster does not exceed 250 pounds.

(1) Place the rucksacks upright on the cargo deck of the helicopter (Figure 5-7) .

(2) Lay a sling rope across the frames between the shoulder strap attachment points.

(3) Tie a bowline knot at each end of the sling and a figure eight knot at each rucksack.

(4) Attach a snaplink to the frame of each rucksack, to the loops in the sling rope, and to the end of the lowering rope.

(5) For Lowe rucks, install the issue H-harness on the rucks and join them into clusters with a sling rope tied to the quick-release attachment buckles.

b. Rigging of the lowering rope to belay the rucksacks requires the following actions.

(1) Backcoil a rappel rope on the helicopter floor at the gunner=s position.

(2) Attach a snaplink and belay device (either a figure eight [Figure 5-8] or an SMC ladder [Figure 5-9]) to the floor tiedown (with a snaplink) inboard of the ceiling rappelling anchor point.

(3) Attach a snaplink to the ceiling rappelling ring.

(4) Route the belay rope from the backcoil, through the belay device, then up through the ceiling snaplink, and down to the cluster snaplink.

(5) Tie the end of the belay rope into a bowline with halfhitch and snap it into the cluster snaplink.

(6) The FRM identifies two ropers to push the cluster or bundle out of the aircraft. He also identifies an AFRM or a safety to belay the cluster or bundle. The FRM briefs these personnel on the time to lower the bundle, the signals to be used, and the procedure to clear the belay system before deploying the FRIES. In performing his

belaying duties, the AFRM or safety must The aviation aircrew or safety is responsible for:

- Provide sufficient slack to allow the pushers to eject the cluster from the aircraft.
- Control the descent until the cluster reaches the surface.

Figure 5-8. Belay Device figure eight SMC

Figure 5-9. Belay device ladder

- When the cluster is on the surface, unsnap both the upper and lower snaplinks, drop the remainder of the rope from the aircraft, and immediately look out to make sure conditions are safe to deploy the FRIES.
- Inform the FRM and pilot when conditions are safe.

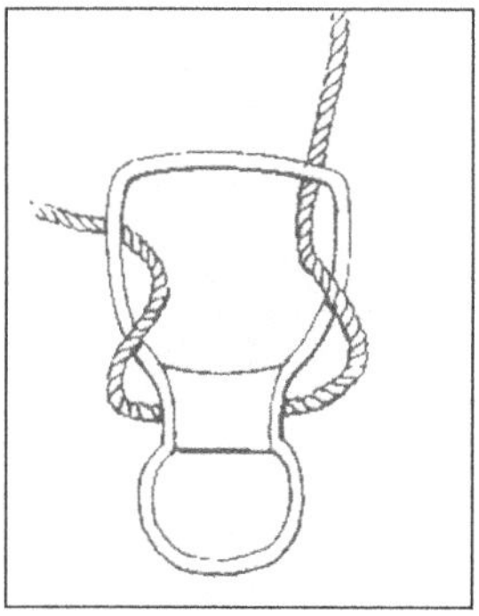

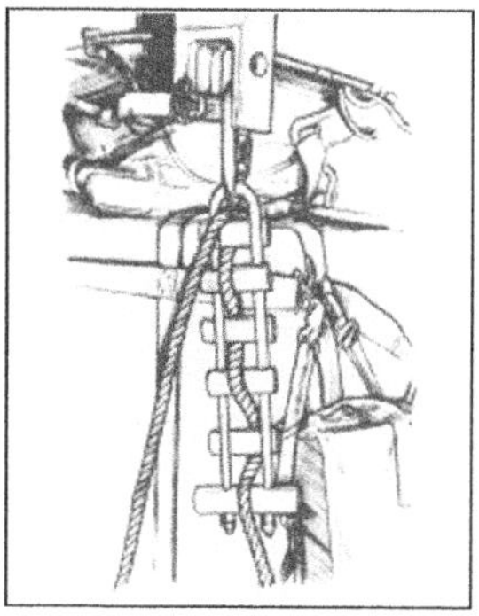

- Deploy FRIES on command

Note: Expect to reposition aircraft before ropers exit to avoid ropers landing on equipment.

5-30. HELICOPTER OPERATIONS FRM CHECKLIST

Table 5-1 shows a checklist of duties of key personnel in fast-rope operations.

PRE-ROPING ACTIONS
Receive a briefing from the OIC or the AMC.
Coordinate and brief all participants.
Rig aircraft and conduct a joint inspection.
Brief ropers, AFRMs, and safeties.
Rig and inspect ropers.

Conduct a static rehearsal.
AIRCRAFT LOADING Position equipment and personnel. Make sure all personnel have straps or seat belts. **ACTIONS IN FLIGHT** Monitor the command net. Monitor the aircrew net. Monitor the flight route. **ACTIONS AT 10-MINUTE WARNING** (applies to long infiltrations) Issue the 10-minute time warning and GET READY. Check equipment and belay system hookup. Check fast-rope hookup. Secure fast-rope bar in position. Make sure rope is back coiled and markers are attached. **ACTIONS AT 6-MINUTE WARNING** Issue the 6-minute time warning. Remove personnel restraints or seat belts. Position personnel and equipment. Break chemlights (night operations). Open aircraft doors, if required. **ACTIONS AT 1-MINUTE WARNING** Issue the 1-minute time warning and STAND BY. Position ropers in stick formation. **ACTIONS AT FLARE** Identify the target area. Deploy bundles/equipment (Safeties) and clear ropes. Deploy FRIES ropes (FRM and AFRM). Check ropes to make sure 15 feet of rope are on the surface (FRM).

Table 5-1. FRM helicopter fast-rope operations checklist

ACTIONS FOR DESCENT
~ FRM, AFRM, or safety positions the number 1 man at the rope. (The FRM may exit first or last.) ~ FRM issues the command GO. AFRM echoes GO command, and the ropers exit the aircraft. Safety informs PIC, "ROPERS OUT."

AFRM or safety controls the rate of exit of the ropers. AFRM exits last. Aircrew or safeties observe the last roper. Safety tells the PIC, ALL ROPERS AWAY, after the last roper is on the surface and signals. PIC issues command JETTISON or RECOVER ROPES. Aircrew or safety jettisons or recovers ropes, and issues ROPES CLEAR report to the pilot.

Table 5-1. FRM helicopter fast-rope operations checklist (continued).

5-31. FRIES COMMANDS AND SIGNALS

The following standardized arm-and-hand signals are provided for FRIES operations. Ground-to-aircraft arm-and-hand signals are military standard as found in the Pathfinder field manual 57-38.

a. Time warnings (10-, 6-, and 1-minute) are given by holding up the appropriate number of fingers and verbally sounding off.

b. The signal for GET READY is the same as the standard Airborne signal—arms extended horizontally, fingers extended vertically and joined.

c. The signal for STAND-BY is arms extended downward at a 45-degree angle from the body with hands closed except for index fingers, which are extended, pointing to the floor

d. To execute the signal for DEPLOY ROPES, the FRM or safety forms both hands into fists at mid-thigh level and sweeps arms horizontally outward. At full extension, he opens his hands as if dropping the ropes out of the helicopter and sounds off with DEPLOY ROPES.

e. The signal for GO is arms raised to horizontal, hands closed with index fingers extended pointing to the ropes.

f. The signal for STOP or ABORT is hands closed into fists with arms raised across the forehead .

g. The signal for SIT-DOWN and DON SEAT BELTS is pointing to the floor, followed by moving both fists around waist, as if donning a belt.

h. The signal for OK is hands raised to eye level, index finger and thumb forming a circle with other fingers extended.

Section IX. EMERGENCY ACTIONS

Procedures for correcting an emergency condition that could reasonably be encountered are described in this section. Multiple emergencies, adverse weather, or other unusual conditions may require modification of these procedures. The nature and severity of the emergency dictate the response necessary; therefore, personnel must use sound judgment in determining the correct action to be taken.

5-32. EMERGENCIES BEFORE ROPING STARTS

All personnel sit down, don their seat belts, and take further instructions from the pilot or crew chief.

5-33. EMERGENCIES AFTER ROPING STARTS

Emergency procedures to be followed after roping starts are discussed in this paragraph.

a. **Unsafe Drift or Premature Liftoff.** Anyone who detects the aircraft having drifted off the site must immediately stop training and inform the FRM, AFRM , safety, or PIC.

(1) The FRM, AFRM, safety, or roper stops stick.

(2) The ropers stop descent and lock in.

(3) The FRM or crewman informs the PIC, and guides him in moving the aircraft back on target.

(4) The unit continues operations.

b. **Rope Hung or Snagged (On Ground, Tree, Building).** Anyone who detects the rope being snagged or hung must immediately stop training and inform the FRM, AFRM, safety, crew chief, or PIC.

(1) The safety ensures ropers are off the rope and are clear.

(2) The aircraft descends or lands as needed.

(3) The safety frees the rope and informs the pilot.

(4) The unit resumes the operation.

c. **Lost Communications.** All training and operations must be planned to include the use of the intercom between the PIC or crewmen and the FRM. If the intercom fails, the following arm-and-hand signals can be used until the rope can be cleared and the intercom is restored.

- Wrists crossed diagonally above the forehead with clenched fists to signal stop stick.
- Finger pointed toward the exit and a verbal GO to signal ropers to exit.
- Open palm moved and faced in the required direction to signal aircraft movement.
- Clenched fist to stop aircraft movement.

Note: Training will not be conducted without radio communications between the pilots or aircrew and the FRM and the ground.

Section X. SAFETY

Safety is everyone's responsibility. All personnel involved in FRIES operations are responsible for identifying hazardous situations and preventing injuries to personnel. Anyone who observes an unsafe condition or act is authorized to halt the operation and inform the FRM or the PIC.

5-34. PREFLIGHT AND INFLIGHT

Anyone who detects an inflight emergency will immediately notify the PIC, crew chief, or safety and follow aircrew instructions.

a. Before a FRIES operation, the FRM will coordinate with the PIC both routine and emergency procedures as well as any unique conditions or potential problems. He also discusses the procedures to be used during flight, during roping, and before the aircraft's departure from the roping site. The FRM then discusses the information with all the AFRMs, safeties, and crew chiefs as well. He reviews the time warnings (including arm-and-hand signals) and appropriate actions such as who will deploy ropes, verify rope status, and perform rope-clearing procedures.

b. The mission prebrief covers any aircraft equipment that, if disturbed, could cause an aircraft or fast-rope problem. The problem areas should be made known to all personnel concerned and preventive measures taken to neutralize or eliminate them where possible.

c. The FRM gives the crew chief the final position adjustment over the target, using hand signals.

d. All safety straps and communications cords that are not in use are securely stowed. All safety straps and communication cords in use should be positioned so they do not present a hazard to personnel movement in the aircraft.

e. After the PIC gives the command DEPLOY ROPES, the FRM is responsible for deploying the rope and making sure it is in contact with the surface before ropers exit.

Note: The safety is responsible for observing rope and ropers and for making sure the pilot maintains aircraft position until all ropers are on the surface.

CAUTION

According to AR 95-3, anytime operations occur over water, all personnel will wear an approved personal flotation device as identified in FM 57-220 or USASOC Regulation 350-2.

5-35. DURING ROPING

The following safety procedures apply during roping.

a. During the actual descent, fast-ropers will not use NVGs because they limit depth perception and create a tunnel vision effect.

b. During descent, ropers must watch for obstructions and lower ropers, and must break as necessary.

c. Individual ropers will lock in during emergencies by wrapping one leg around the rope once or twice and standing on the fast rope with the other foot.

d. Ropers must maintain sufficient intervals on the rope to prevent contact with lower ropers. Individual ropers exit when the head of the man in front of them disappears below the ramp.

e. Ropers execute descents at speeds commensurate with their experience and proficiency in FRIES infiltration operations. (For example, less-experienced ropers should descend slowly until they become more proficient.)

f. The number of ropers on each rope at one time must not exceed the safe limit of the rope mount point in use, as follows:

(1) Do not exceed 1,500 pounds per rope on the MH/UH-60.
(2) Do not exceed 1,500 pounds per rope on the MH/UH-47.
(3) Do not exceed 1,500 pounds per rope on the MH/UH-53.
(4) Do not exceed 1,300 pounds per rope on the MH-60G.

5-36. SAFETY BRIEFING

All units must present a safety briefing before conducting FRIES training or operations. The safety briefing must cover the following (as a minimum):

- The planned event (day or night, tower or helicopter, high or low hover, insertion or extraction).
- Actions on the aircraft (or tower).
- Operation (or training) location and any hazards within 500 meters.
- Identification of key personnel, their location, and their duties.
- FRIES equipment and its characteristics.
- FRIES equipment rigging or inspection.
- Roping personnel preparation (shirttails tucked in, chin strap fastened, loose straps taped, gloves and goggles on).
- Arm-and-hand signals.
- Emergencies and emergency procedures.
- Medical procedures (medic with aid bags, backboards, litters, and dedicated vehicle with driver on site).
- Communications requirements (frequencies, call signs, required reports).
- Any other information essential to safety.

5-37. SAFETY REMINDERS

The following safety reminders will be reviewed by all key personnel before conducting operations.

a. Ropers must wear long-sleeve shirts or jackets, full-length pants, laced boots, helmets, protective goggles, and heavy leather gloves.

b. MILES harnesses must not be worn during fast-roping.

c. Ropers will not wear NVGs.

d. Commanders must personally approve the rope site selected for night operations. The FRIES site must be large enough to permit all ropers, upon reaching the surface, to move clear of the rope and for units to consolidate or re-form.

e. All night rope operations must be assessed as medium risk or higher.

f. FRIES descent training is not authorized with a fast rope longer than 90 feet.

g. Ropers must not carry more than 50 pounds of equipment during FRIES descents. The maximum weight of rucksacks is limited to 35 pounds. When rucksacks exceed 35 pounds (or total load exceeds 50 pounds), gear must be lowered (belayed) on a separate system, preferably from the opposite door.

h. Ropers will not carry bulky equipment that could interfere with FRIES operations. Equipment that cannot be carried by the soldier during roping (due to its weight or bulk) can be lowered (belayed) during insertion. During extractions, rucksacks must be attached to the extraction loops of the FRIES rope using snaplinks.

i. Ropers must not make more than 6 combat equipment descents or a total of 10 descents in a 24-hour period.

j. FRIES training must not be conducted if—

- The aircraft is unstable.
- The fast rope is not fully deployed (15 feet of the rope must be on the surface).
- Any obstacle is present that could interfere with the descent of the ropers.
- The rope is frozen or slippery
- Anyone identifies any unsafe condition.

k. Safety personnel must—

- Position themselves so they can observe active FRIES points with out obstructing ropers.
- Conduct final rigging checks of the mount, equipment, and ropes at the 10-minute warning.
- Ensure the aircraft is over the target area and the ropes reach the ground.
- Constantly keep both the PIC and FRM informed—
  - — When the aircraft is in position for deployment.
  - — When anything is being deployed from the helicopter (ropes, bundles or personnel).
  - — When 15 feet of rope are on the surface.
  - — When the first roper exits on the fast rope.
  - — When all ropers on the surface are clear.
  - — When the ropes are recovered or jettisoned.
  - — When the aircraft is cleared for flight.
  - — When 15 feet of the extraction rope is on the surface.
  - — When extraction personnel are attached to the rope and ready for lift off.
  - — When extracted personnel are safely lowered to the ground.
  - — When extracted personnel are out from under the aircraft.
  - — Any other conditions that could impact safe operations.

Note: If the FRM and the safety exit, the crew chief performs rope clearing and or jettisoning duties.

WARNING

Using the word CLEAR tells pilots the aircraft is ready to fly. Before any use of the word CLEAR, FRMs and safeties must ensure that all personnel are well away from the ropes and the ropes have been fully recovered or released (during insertion), or all personnel are securely attached to the ropes (during extraction).

l. FRIES training must not be conducted in densely wooded areas or areas prone to blowing dust, sand, or snow, which can produce a hazard for ropers or the aircraft.

m. FRIES training and operations must not be conducted during severe weather such as rain, sleet, ice, or extreme cold.

n. The possibility of personnel injury during FRIES training and operations is always present. Every soldier involved in FRIES activities must maintain constant vigilance to prevent personnel injury and equipment damage.

CHAPTER 6

EQUIPMENT

This chapter discusses the basic equipment used for rappelling operations (Appendix B). It covers the types of ropes, the procedures for rope selection, and the care for each rope type. It also discusses the two methods of coiling and the different snaplinks used to join the equipment and rope.

Section I. ROPES

Ropes are the most important items of equipment for the rappeller. They provide access down the obstacle or aerial platform, while ensuring individual safety.

6-1. TYPES OF ROPES

Two types of ropes are available to the rappeller—*static* and *dynamic*.

a. **Static Rope.** This rope allows for minimal stretch of the rope. The static rope stretches about 5 to 15 percent at the point of failure and about 2 percent under a working load. The minimum tensile strength for an 11-mm (7/16-inch) static rope for military use is 4,500 pounds. *Kernmantle* static ropes range from 3 millimeters to 11 millimeters in diameter and are constructed similar to the 550-parachute cord. (*Kern* means core and *mantle* means sheath.) The internal core of the rope is constructed of a continuous multifilament-nylon yarn, which is spun into continuous parallel strands. This internal core is then covered with a nylon-braided outer sheath. Due to the internal parallel strand construction of the static rope, there is less spinning and kinking than with a dynamic rope. Static ropes can be used for slings, harnesses, etriers (stirrups), rappelling, rope installations, hauling lines, and where stretch of the rope is not desired.

b. **Dynamic Rope.** This rope allows for stretch within the fibers of the rope, which can be a disadvantage in rappelling, prusik climbing, and other applications. Dynamic ropes are more susceptible to abrasion and wear. They have about a 5 to 10 percent working stretch. The minimum tensile strength for an 11-mm (7/16-inch) rope for military use is 4,500 pounds. Two types of dynamic ropes are nylon-laid and kernmantle:

(1) *Nylon-laid ropes.* Synthetic fibers are the best climbing and rappelling material. Nylon has become the standard material for climbing ropes and has replaced manila, flax, hemp, and sisal. The military mountaineering critical application rope is constructed in a *laid* fashion with continuous multifilament-nylon fiber yarn twisted into strands. Three strands are twisted into a climbing rope of a specific diameter (Figure 6-1). Nylon-laid ropes are easy to inspect for serviceability by twisting the fibers. Laid ropes tend to untwist slightly when under a load, causing kinking and spinning. They are also susceptible to abrasion. The following specifications pertain to the standard military nylon-laid climbing/rappelling rope:

- 36 1/2 meters (120 feet) long.
- 11 millimeters (7/16 inch) wide.
- At least a 4,500-pound tensile strength.
- Right-hand lay.
- One-third stretch factor (at point of failure).

- May lose as much as 15 percent of its strength when wet.
- Weighs 6 pounds when dry.

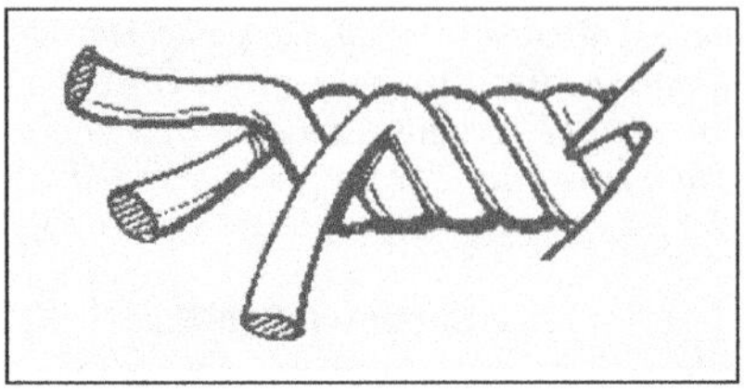

Figure 6-1. Nylon-laid rope.

(2) *Kernmantle ropes.* These ropes are similar to their static counterparts in that they consist of an inner core and an outer sheath. The core is constructed of continuous twisted nylon filaments, which are laid or braided together and enclosed in a tightly braided outer sheath. These ropes are well suited for climbing on rock, snow, or ice where a brief elastic stretch of the rope occurs during a fall. They have a stretch of about 40 percent at the point of failure. The breaking strength is high, and there are no exposed strands for rock crystals to work between to damage the rope (Figure 6-2). There is less sliding friction through the snaplink and over other surfaces, since the outer sheath is smooth. Kernmantle ropes are available in many sizes, lengths, stretch factors, tensile strengths, and fall ratings.

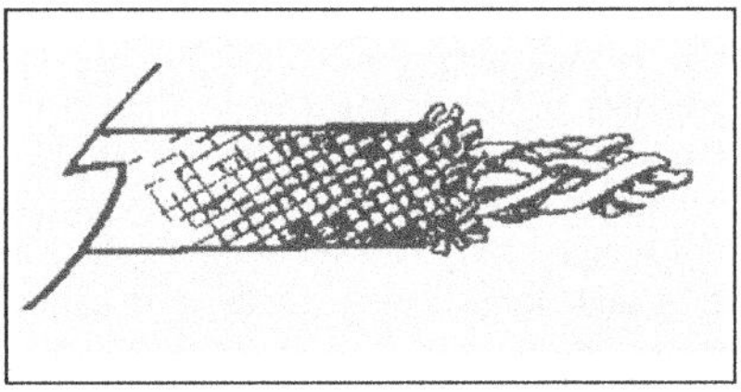

Figure 6-2. Kernmantle (core and sheath) rope.

6-2. ROPE SELECTION

Rope selection for military operations requires ropes of various types, lengths, and diameters; and depends on its intended use (static or dynamic), environment (water-repellant), climate, mission (lightweight ropes), and other factors. Static ropes should have minimal stretch; be smooth, flexible, and strong; resist abrasion and cutting; and have a low-impact force.

a. *Laid rope* is easier to inspect (by twisting the strands apart to observe the interior surfaces), has greater friction in holding knots, and is less expensive than kernmantle ropes. Its disadvantages include: much lower abrasion resistance, extremely high stretch factor for both static and dynamic loads, greater friction on rock, readiness to kink, tendency to spin a free-hanging rappeller, and greater absorption of water and dirt.

b. *Kernmantle* rope is easy to work with, has less friction, does not kink easily, and has less susceptibility to abrasion. The outer sheath protects the internal fibers from dirt, abrasion, and ultraviolet radiation. Its disadvantages are that it is hard to determine damage to the core fibers, and it is more expensive than laid rope. Kernmantle rope is not readily identifiable as dynamic or static and must be marked accordingly.

c. Rope diameter and length vary with intended use. A standard diameter for many uses is 11 millimeters (7/16 inch). For stirrups, utility ropes, and hauling lines, 7 millimeters (5/16 inch), which can also be used in double strands, is adequate. Standards to remember are that the smaller the diameter, the less the tensile strength; but the larger the diameter, the heavier the rope. The standard lengths are 36 1/2 meters (120 feet), 40 meters (135 feet), 45 meters (150 feet), and 50 meters (165 feet). Normally, standard lengths should not be cut. If a nonstandard length is needed, the rope should be precut before the mission and clearly marked.

d. A rope should be selected for a specific use. There are different types of ropes for various uses. A dynamic rope is designed for climbing, while a static rope is designed for rappelling, rescue operations, load hauling, and rope installations.

WARNING

Using a rope for other than its intended use could result in personal injury or damage to equipment.

6-3. CARE OF ROPES

Because the rope is the rappeller's lifeline, it requires a great amount of care and maintenance.

a. Before and after use, the rope should be carefully inspected for excessive wear, cuts, mildew, or rotten spots. If any of these defects are found, the rope is unserviceable. An unserviceable rope is destroyed by cutting it into usable parts. If no sections are usable, the rope is cut into smaller parts and discarded.

b. Ropes should never be spliced together, stepped on, or dragged. This greatly reduces the tensile strength of the rope.

c. The rope must not come in contact with sharp edges. If this is likely, the areas must be padded or taped before rappelling.

d. The rope must be kept as dry as possible. If it becomes wet, it must be dried out to prevent rotting. (Avoid excessive heat/heating during the drying process.)

6-4. COILING

Two methods of coiling are the *mountaineer's coil* and the *farmer's coil* (Figure 6-3).

a. **Mountaineer's Coil.**

(1) To use the mountaineer's coil, grasp the rope about 3 feet from the end with the left hand. The right hand is then placed next to the left hand and runs along the rope until both arms are outstretched. Grasping the rope firmly, bring the hands together to form a loop, which is laid in the left hand. Repeat, forming uniform loops that run in a clockwise direction until the rope is completely coiled. If the rope tends to twist or form figure eights, give it a slight twist with the right hand when each loop is formed. Coil the rope in a clockwise direction to conform with the lay.

(2) To complete the coil, form a bight about 30 centimeters (12 inches) long with the starting end of the rope and lay it along the top of the coil. Uncoil the last loop and use it to wrap around the coil and the bight. Wrap toward the closed end of the bight, binding the final wrap across itself to lock it in place. Make six to eight wraps to secure the coil. Push the end of the rope down through the closed end of the bight. Carry the running end of the coil either in the pack (by forming a figure eight, doubling it, and placing it under the flap) or by placing it over the shoulder and under the opposite arm. If the rope to be coiled is anchored as in coiling a belay or rappel lane, start the coil near the end closest to the anchor so that the kinks work themselves out of the free end.

b. **Farmer's Coil.**

(1) The farmer's coil is easier and faster to coil than the mountaineer's coil. Establish the center of the rope either by locating its center mark or by grasping both ends of the rope and feeding them out until a bight comes up that is the center of the rope. Begin coiling with the doubled rope the same as the mountaineer's coil method. Leave about 4.6 meters (15 feet) of rope uncoiled. Then, squeeze the coils together and make four to six wraps around the middle of the coil, ensuring that the first wrap is held in place by the other wraps.

(2) Form a bight with the two running ends and place it through the bight formed by the top of the coil. Then, run the two running ends over the top of the coil and through the bight that is formed. Dress them in the center of the back of the carrier. Run the two ends over the shoulders to form shoulder straps. Then, bring them under the arms, cross them in the back over the coil, bring them around the body of the carrier, and tie them off with a square knot at the stomach.

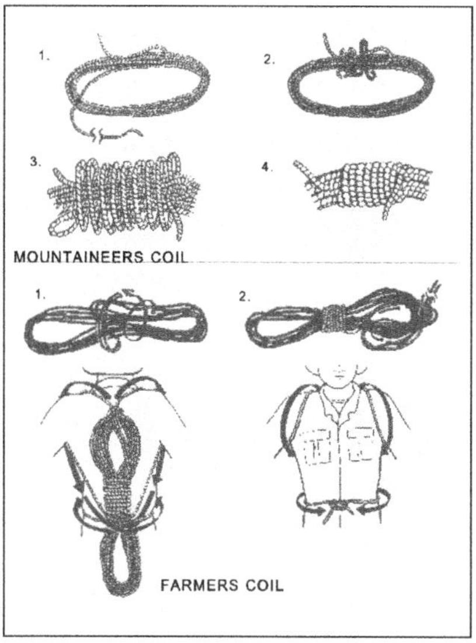

Figure 6-3. Two methods of tying a coil.

Section II. SNAPLINKS

Snaplinks are used to join equipment, rope, and people into a functioning system. When used properly, they are strong, versatile items. Snaplinks are metal (steel, aluminum, or alloys) loops with a hinged, spring-loaded gate on one side. They come in many sizes, shapes, and strengths (with and without locking gates) and in many types of metal. Steel is the strongest snaplink, but it is heavier.

6-5. DESCRIPTION

Snaplinks are available in many sizes and shapes to suit different needs. A locking snaplink is best for rappelling. Nonlocking snaplinks are easy to operate, but should be used only where they cannot be accidentally opened. Hollow snaplinks should be avoided since their use is limited. The following information applies to all snaplinks.

a. The weakest part of a snaplink is the gate. The gate must be closed before applying a load. When the gate is open, snaplinks should have little or no lateral movement of the gate.

b. Locking pins should be checked to ensure that they are not loose, worn, or corroded.

c. The metal should be checked for any cracks, grooves, burrs, flaws, or rust.

d. The spring-loaded gate should automatically close securely from an open to a closed position with no gap between the locking pin and notch.

e. If an engraver is used to mark snaplinks, it should be applied only to the gate, never to the load-bearing side.

f. A snaplink should never be side-loaded (across the gate) since this reduces the overall strength to the point of gate failure.

6-6. TYPES OF SNAPLINKS

The two types of snaplinks used during military operations are the standard snaplink (Figure 6-4) and the locking gate snaplink (Figure 6-5). Both snaplinks are available in a D-shaped variation or an oval variation.

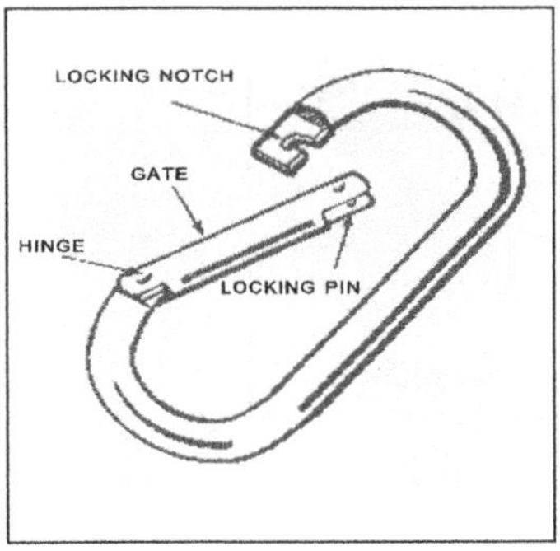

Figure 6-4. Standard snaplink (oval).

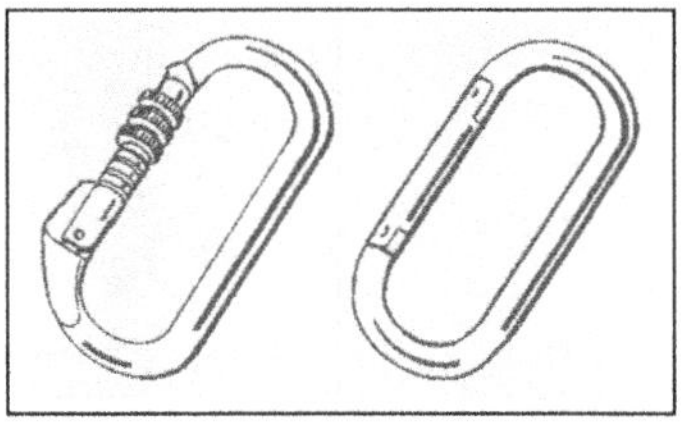

Figure 6-5. D-shaped snaplinks.

a. **D-Shaped Snaplinks.** D-shaped snaplinks are stronger than the oval type because the shape directs the largest part of the load to be applied to the longer, stronger side opposite the gate (Figure 6-5). D-shaped snaplinks are made of steel or aluminum alloys, and are available in many sizes and thicknesses, all of which can be with or without locking gates. Their strength and durability vary, which must be considered before use.

b. **Oval Snaplinks.** Oval snaplinks are versatile and have different applications. They are made of steel or aluminum alloys and are available in many sizes and thicknesses. Both sides of an oval snaplink bear the strain equally under load weight. Many modified ovals are available and most can be with or without the locking gate. All must meet military standards or UIAA specifications.

c. **Locking Snaplinks.** Locking snaplinks have a locking mechanism with a threaded sleeve on the gate. The sleeve screws tightly over the gate opening end or hinge end to hold the gate closed (Figure 6-6). A reverse locking gate is needed to prevent a moving rope from unscrewing the sleeve. The locking sleeve and threads should be kept free of dirt and grit. If the sleeve is forced to close, it may strip the threads. The locking mechanism ages and weakens after repeated use and should be routinely inspected for wear. The characteristics of locking snaplinks are as follows:

- Material: steel.
- Approximate strength: 2,000 pounds with the gate closed.
- Gate: spring-loaded without locking sleeve.
- Weight: 4.3 ounces.

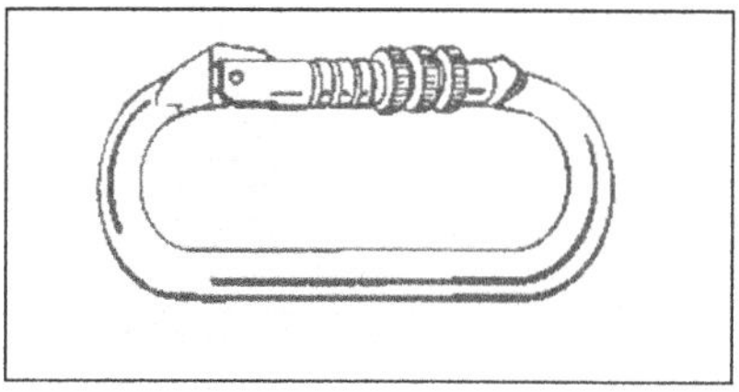

Figure 6-6. Locking snaplink.

6-7. INSPECTION

All rappelling material and individual rappelling equipment must be inspected before conducting a rappelling operation. The rappel master is responsible for ensuring the serviceability of rappelling material and individual rappelling equipment.

a. **Snaplinks.** Snaplinks should be inspected daily before, during, and after use.

(1) The metal should be checked for cracks, grooves, burrs, rust, and flaws. The gate should open and close freely without binding. There should be no lateral movement when the gate is open. The gate spring action should snap shut when released. The locking notch should have a slant or slot so that the gate remains shut under the impact of a rappeller's fall. The gate pins should not work their way out of their holes and should not be shorter than their holes. If there is a locking mechanism, it should be inspected to ensure that threads are not stripped and that the sleeve tightly locks the gate.

(2) If burrs, grooves, or rough areas are identified, the snaplink should not be used. Rust should be removed with steel wool and the spot rubbed with oil or solvent. The spring should be lubricated as needed. The snaplink must be boiled in water for 20 to 30 seconds to remove the cleaning agents since solvents and oils cause dirt to cling to the snaplink and to rub off on the ropes. It is better to use a dry graphite-based lubricant on snaplinks since such lubricants do not attract dirt.

b. **Ropes.** The ropes are inspected for serviceability. A rappel rope is unserviceable if it is saturated with petroleum products, is mildewed, shows excessive wear, has soft spots (internal damage), or if one strand is cut more than one-half of its diameter. Ropes should be inspected daily—before, during, and after use. An applicable annotation should be made in a corresponding rope log (DA Form 5752-R) (Figure 6-7).

ROPE LOG (USAGE AND HISTORY)

For use of this form, see TC 90-6-1; the proponent agency is TRADOC

UNIT ID MARKING: D-11-86-5			
NSN: 8455-73560366	DOCUMENT NUMBER: W330SV 61890007	SERIAL NUMBER: 6-6-6933	MFR LOT NUMBER: G620135115103
DATE OF MFR: 08 FEb 87	ISSUE DATE: 01 OCt 87	DATE IN SERVICE: 17 NOV 87	LENGTH: 45M (150 FT)
DIAMETER: 11mm (7/16 inch)	FIBER: TYPE 6.6 NYLON	COLOR: OD Green	CONSTRUCTION: DYNAMIC KERMANTLE

INSPECT ROBE FOR DAMAGE OR EXCESSIVE WEAR EACH TIME IT IS DEPLOYED AND AGAIN AFTER EACH USE. IMMEDIATELY RETIRE ALL SUSPECT ROPES.

DATE USED	LOCATION	TYPE OF USE	ROPE EXPOSURE	INSPECTOR'S INITIAL/DATE	ROPE CONDITION AND COMMENTS
20 NOV 87	Yonah Mtn GA	Direct Aid Climb 5.7 A2	65°F clear & dry Smoothe Granite	DLE 20 NOV 87	
21 NOV 87	Yonah Mtn GA	Top Rope Belayed Balance Climb	68°F Clear & dry Smoothe Granite	ACY 21 NOV 87	
15-18 Jan 88	Gulcana Glacier ALASKA	ROPED GLACIER Travel	2°F snow storm	JEY 15-18 Jan 88	
27 Jan 88	Smuggler's Notch Vt	Front Point Ice Climb	35°F Cloudy No Precip wet rope Due to Ice thaw	ATS 27 Jan 88	SUSTAINED ONE 15 FT Leader FALL
28 Jan 88	Smuggler's Notch Vt	Front Point Ice Climb	20°F Cloudy & Snowy	JMC 28 Jan 88	
8 Feb 88	Mt Shasta CAL	FLAT Foot ICE Climb	8°F Lightly Snowing	DCY 8 Feb 88	Approx ONE inch of Sheath Slippage on End of rope
22-24 Feb 88	DECOS MtN NM	TWO MAN PARTY Rock climb 5.7+	45°F Clear & DRY GRANITE	LLY 22 Feb 88	
10-22 Jan 88	Mt McKinley ALASKA	ROPED PARTY Climb CLASS VI	20-53°F ONE DAY of RAIN	JMK 10-22 Jan 88	Abrasion to Rope AT Points of TIE-IN (Ends & Middle)

DA FORM 5752-R, MAY 89

Figure 6-7. Example of completed DA Form 5752-R.

c. **Gloves.** The gloves are inspected for serviceability. They are unserviceable if holes are found in the friction-bearing surfaces or along the seams.

d. **Helmet.** The helmet is inspected for correct assembly and serviceability. It is unserviceable if it cannot be properly assembled or if any straps are cut or torn one-quarter of its width.

(1) Ensure that the chin strap attaching tabs are secured to the helmet by the locking nuts.

(2) Ensure that the pull-the-dot tab has four plies of nylon and that the chin strap is routed from the inside of the buckle around the horizontal bar and back toward the inside of the buckle.

Section III. ALTERNATE METHODS OF DESCENT

Snaplinks or carabiners that are used for rappelling are placed through the sewn loops of the seat harness or the sewn loops of the seat-chest combination.

6-8. FIGURE-EIGHT DESCENDER

The rappeller attaches a locking carabiner to the harness (Figure 6-8). He routes the rappel rope up through the large hole and places the bight over the collar. He inserts the figure eight with the rappel rope attached into the locking carabiner, and locks it down. The rappeller brakes to the rear and descends as in a seat hip rappel. If heavy loads are to be descended, he uses a double wrap around the collar.

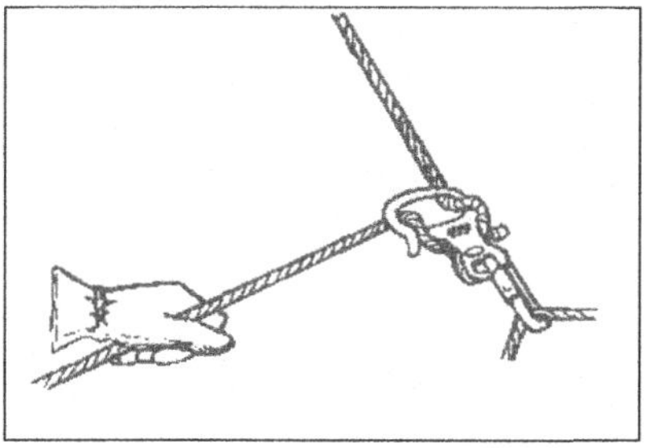

Figure 6-8. Figure-eight descender.

6-9. MUNTER HITCH

The rappeller attaches a large radius snaplink to the harness or rappel seat. He ties a Munter hitch and clip into the snaplink (Figure 6-9, page 6-9). He brakes to the front and descends with an L-shape body position.

Notes: 1. The guide hand should remain on the standing end of the rope to keep the rappeller's head and face away from the hardware.

2. The Munter hitch creates significant rope-to-rope friction and may cause premature wear on nylon ropes.

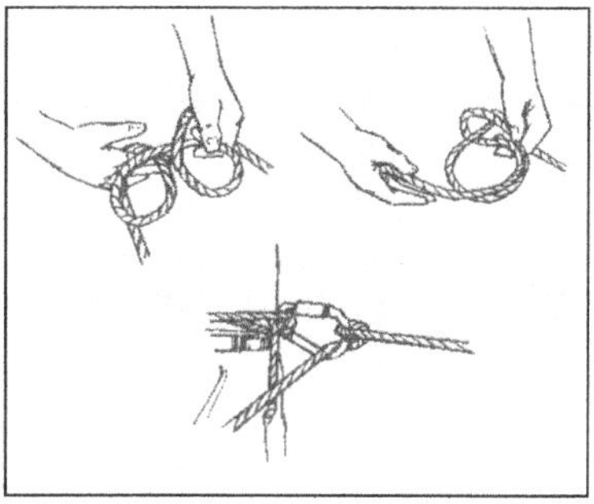

Figure 6-9. Munter hitch.

CHAPTER 7

KNOTS

A knot is a fastening made by intertwining or tying pieces of string, cord, rope, or webbing. All knots used by a rappeller are divided into four classes: Class I—joining knots, Class II—anchor knots, Class III—middle rope knots, and Class IV—special knots. These classes of knots are intended only as a general guide since some of the knots discussed herein may be appropriate in more than one class. The skill of knot tying may be lost if it is not used and practiced. With experience and practice, knot tying becomes instinctive and helps the soldier in many situations.

7-1. MILITARY MOUNTAINEERING TERMINOLOGY

The terms explained below are the most commonly used in military mountaineering (Figure 7-1).

a. A *bight of rope* is a simple bend of rope in which the rope does not cross itself.

b. A *loop* is a bend of rope in which the rope does cross itself.

c. A *half hitch* is a loop that runs around an object and locks itself.

d. The *running end* (working end) of the rope is the free end of the rope that can be used.

e. The *standing part* of the rope is the part that is static (anchored, coiled); the remaining part of the rope not being used (also called static end).

f. The *lay* of the rope is the same as the twist of the rope.

g. A *round turn* is a single complete wrap of the rope around an object providing 360-degree contact. The running end leaves the completed circle in the same direction as the standing part. In a round turn, the rope is wrapped around an object 1 1/2 times.

h. The *pigtail* is the short length of rope remaining at the end after tying a knot or coiling a rope.

i. *Back feeding* (or stacking) is taking off one wrap at a time from a coil, and letting it fall naturally to the ground.

j. *Dress down the knot* is tightening down a knot to its functioning form.

k. An *overhand* is a loop with the running end pulled through the loop.

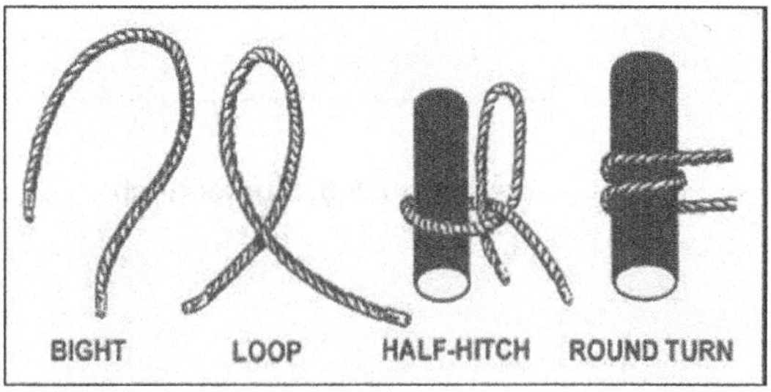

Figure 7-1. Examples of rope terminology.

Section I. JOINING KNOTS

7-2. SQUARE KNOT

The square knot is used to tie the ends of two ropes of equal diameter (Figure 7-2).

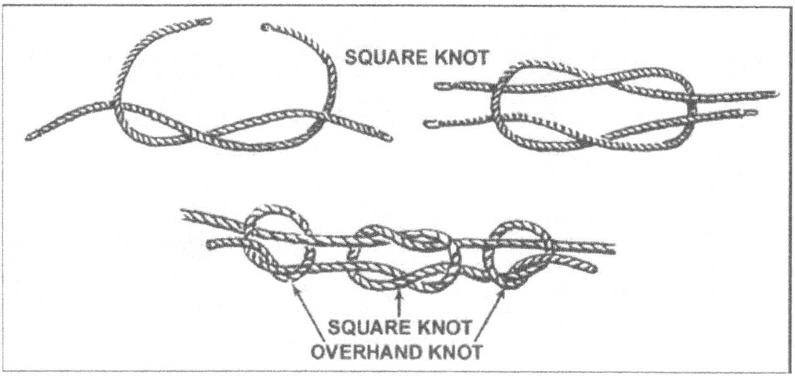

Figure 7-2. Square knot.

a. **Tying the Knot.**

STEP 1. Holding one working end in each hand, place the working end in the right hand over the one in the left hand.
STEP 2. Pull it under and back over the top of the rope in the left hand.
STEP 3. Place the working end in the left hand over the one in the right hand and repeat STEP 2.
STEP 4. Dress the knot down and secure it with an overhand knot on each side of the square knot.

b. **Checkpoints.**

(1) There are two interlocking bights.

(2) The standing parts are on the same side and properly secured with overhand knots.

7-3. DOUBLE SHEET BEND KNOT

The double sheet bend knot is used to tie the ends of two or more ropes of equal or unequal diameter (Figure 7-3). When a single rope is tied to a multiple of ropes, the bight is formed with the multiple of ropes.

Figure 7-3. Double sheet bend knot.

a. **Tying the Knot.**

STEP 1. Form a 30-centimeter (12-inch) bight in the left hand with the rope or ropes. Ensure that the short end of the bight is facing to the inside.

STEP 2. Place the index finger of the left hand on top of the bight. Bring the rope in the right hand up through the bight and over the index finger of the left hand, so the working end is away from the body.

STEP 3. Reach through the loop and grasp the working end of the rope and bring it back towards the body. Place it between the index finger and the bight, forming a round turn. Repeat this one more time, wrapping toward the tip of the index finger.

STEP 4. Remove the finger and dress the knot down.

b. **Checkpoints.**

(1) The two wraps around the bight are held in place by a locking bar.

(2) The two standing parts of the ropes exit parallel and from opposite ends of the knot. The two working ends form an L and exit the knot at a 90-degree angle and on opposite sides of the knot.

7-4. FISHERMAN'S KNOT

The fisherman's knot is used to tie two ropes of similar or dissimilar materials (Figure 7-4).

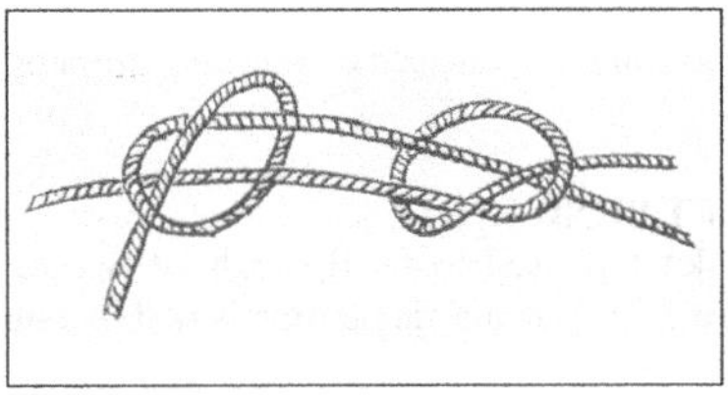

Figure 7-4. Fisherman's knot.

a. **Tying the Knot.**

STEP 1. Tie an overhand knot in one end of the rope.
STEP 2. Pass the working end of the other rope through the first overhand knot. Tie an overhand knot around the standing part of the first rope with the working end of the second rope.
STEP 3. Tightly dress down each overhand knot and tightly draw the knots together.

b. **Checkpoints.**

(1) The two separate overhand knots are tied tightly around the long standing part of the opposing rope.

(2) The two overhand knots are drawn snug.

7-5. DOUBLE FISHERMAN'S KNOT

The double fisherman's knot (also called double English or grapevine) is used to tie two ropes of similar or dissimilar materials (Figure 7-5).

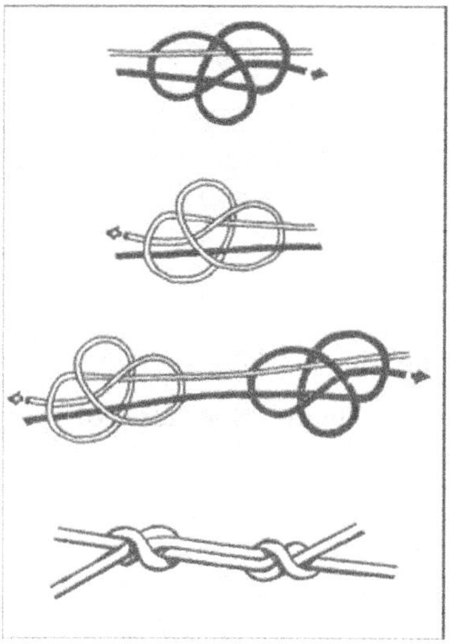

Figure 7-5. Double fisherman's knot.

a. **Tying the Knot.**

STEP 1. With the working end of one rope, tie two wraps around the standing part of another rope.
STEP 2. Insert the working end (STEP 1) back through the two wraps and draw it tight.

STEP 3. With the working end of the other rope, which contains the standing part (STEPS 1 and 2), tie two wraps around the standing part of the other rope (the working end in STEP 1). Insert the working end back through the two wraps and draw tight.

STEP 4. Pull on the opposing ends to bring the two knots together.

b. **Checkpoints.**

(1) Two wraps with the working end running through and drawn tight.

(2) Both wraps are drawn together with all the slack out of the knot.

7-6. WATER KNOT

The water knot is used to attach two webbing ends (Figure 7-6). It is also called a ring bend, overhand retrace, or tape knot. It is used in runners and harnesses.

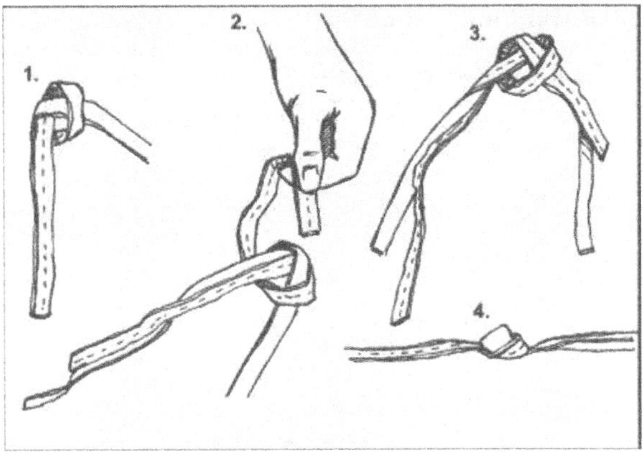

Figure 7-6. Water knot

a. **Tying the Knot.**

STEP 1. Tie an overhand knot in one of the ends.

STEP 2. Feed the other end back through the knot, following the path of the first rope in reverse.

b. **Checkpoints.**

(1) There are two overhand knots, one retracing the other.

(2) There is no slack in the knot, with the working ends coming out of the knot in opposite directions (at least 5 centimeters [2 inches] long).

Section II. ANCHOR KNOTS

7-7. BOWLINE KNOT

The bowline knot is used to tie a single fixed loop in the end of a rope (Figure 7-7). It does not slip under strain and is easily untied. This knot is always used when there is alternating tension. It can also be used to tie the end of a rope to an anchor.

Figure 7-7. Bowline knot.

a. **Tying the Knot.**

STEP 1. Bring the working end of the rope around the anchor, from right to left (as the rappeller faces the anchor).
STEP 2. Form an overhand loop in the standing part of the rope (on the rappeller's right) toward the anchor.
STEP 3. Reach through the loop and pull up a bight.
STEP 4. Place the working end of the rope (on the rappeller's left) through the bight, and bring it back onto itself. Dress the knot down.
STEP 5. Form an overhand knot with the tail from the bight.

b. **Checkpoints.**

(1) The loop is locked into place by a bight.

(2) The short portion of the bight is on the inside and on the loop around the anchor (or inside the fixed loop).

7-8. ROUND TURN AND TWO HALF HITCHES KNOT The round turn and two half hitches knot is used to tie the end of a rope to an anchor and must have constant tension (Figure 7-8).

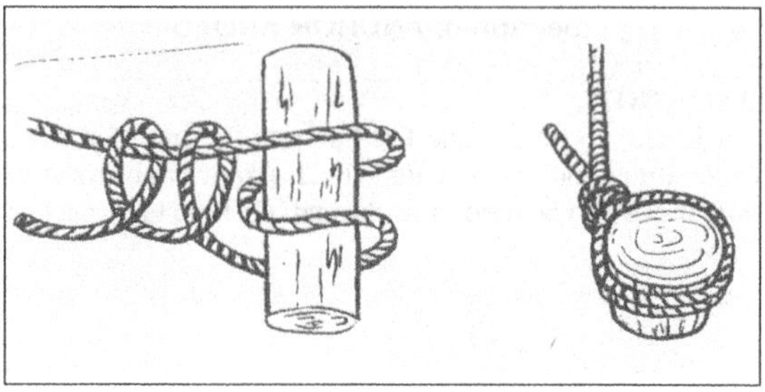

Figure 7-8. Round turn and two half hitches knot.

a. **Tying the Knot.**

STEP 1. Route the rope around the anchor from right to left and wrap down (must have two wraps in the rear of the anchor and one in the front). Run the loop around the object to provide 360-degree contact, distributing the load over the anchor.

STEP 2. Bring the working end of the rope left to right and over the standing part, forming a half hitch (first half hitch).

STEP 3. Repeat STEP 2 (last half hitch has a 15-centimeter [6-inch] pigtail).

STEP 4. Dress the knot down.

b. **Checkpoints.**

(1) A complete round turn should exist around the anchor with no crosses.

(2) Two half hitches should be held in place by a locking bar with no less than a 15-centimeter (6-inch) tail remaining.

(3) When the two half hitches are twisted to the left or the right, they will appear to for a figure eight.

(4) Looking at the top of the half hitches, the rope should appear to be one continuous rope.

7-9. FIGURE-EIGHT RETRACE KNOT

The figure-eight retrace knot is also called a rerouted figure eight. Although it produces the same result as a figure-eight loop, by tying the knot in a retrace it can be used to fasten the rope to trees or to places where the loop cannot be used (Figure 7-9).

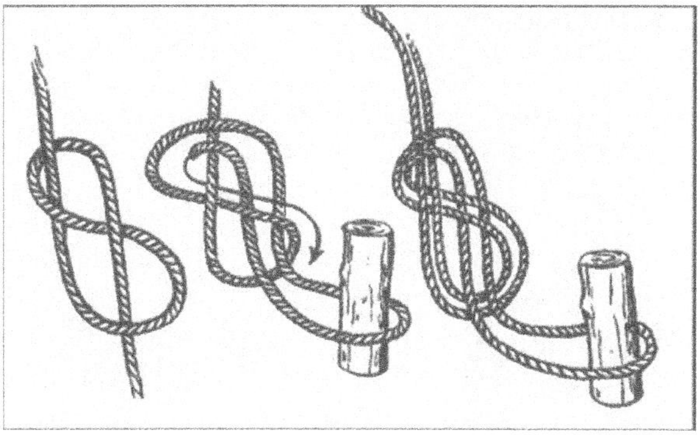

Figure 7-9. Figure-eight retrace.

a. **Tying the Knot.**

STEP 1. Use a length of rope long enough to go around the anchor, leaving enough rope to work with.
STEP 2. Tie a figure-eight knot in the standing part of the rope, leaving enough rope to go around the anchor. To tie a figure-eight knot:

- Form a loop in the rope.
- Wrap the working end around the standing part.
- Route the working end through the loop.
- The finished knot is dressed loosely.

STEP 3. Place the working end around the anchor point.
STEP 4. With the working end, insert the rope back through the loop of the knot in reverse.
STEP 5. Keep the original figure eight as the outside rope and retrace the knot around the wrap and back to the long sttanding part.
STEP 6. Safety off with an overhand knot.

b. **Checkpoints.**

(1) At least 15 centimeters (6 inches) of rope extends past the figure-eight knot.

(2) The original figure eight, tied with the standing end, remains as the outermost rope throughout the knot.

(3) The knot is dressed tightly.

7-10. CLOVE HITCH KNOT

The clove hitch knot can be used in the middle of the rope as well as at the end of the rope (Figure 7-10). The knot must have constant tension on it once tied to prevent slipping. It can be used as either an anchor or middle knot, depending on how it is tied. (Do not use a clove hitch knot on metal except on a picket-hold-fast.)

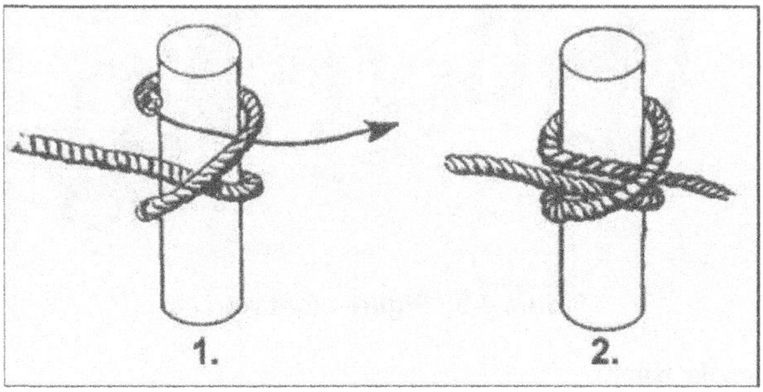

Figure 7-10. Clove hitch knot.

a. **Tying the Knot.**

(1) If there is access over the top of the anchor, the knot is tied as follows:

STEP 1. Hold rope in both hands, palms down with hands together. Slide the left hand to the left from 20 to 25 centimeters (8 to 10 inches).

STEP 2. Form a loop away from and back toward the right.

STEP 3. Slide the right hand from 20 to 25 centimeters (8 to 10 inches) to the right. Form a loop inward and back to the left hand.

STEP 4. Place the left loop on top of the right loop. Place both loops over the anchor and pull both ends of the rope in opposite directions to complete the knot.

(2) If the clove hitch knot must go around the anchor, tie the knot as follows (for instructional purposes, assume that the anchor is horizontal):

STEP 1. Place 76 centimeters (30 inches) of rope over the top of the anchor. Hold the standing end in the left hand. With the right hand, reach under the horizontal anchor, grasp the working end, and bring it inward.

STEP 2. Place the working end of the rope over the standing end (to form a loop). Hold the loop in the left hand. Place the working end over the anchor from 20 to 25 centimeters (8 to 10 inches) to the left of the loop.

STEP 3. With the right hand, reach down to the left-hand side of the loop under the anchor. Grasp the working end of the rope and bring it up and outward.

STEP 4. Dress down the knot.

b. **Checkpoints.**

(1) The knot has two round turns around the anchor with locking bar.
(2) The locking bar is facing 90 degrees from the direction of pull.
(3) The ends exit 180 degrees from each other.
(4) The knot has more than a 15-centimeter (6-inch) pigtail remaining.

SECTION III. MIDDLE ROPE KNOTS

7-11. WIREMAN'S KNOT

The wireman's knot forms a single, fixed loop in the middle of the rope (Figure 7-11).

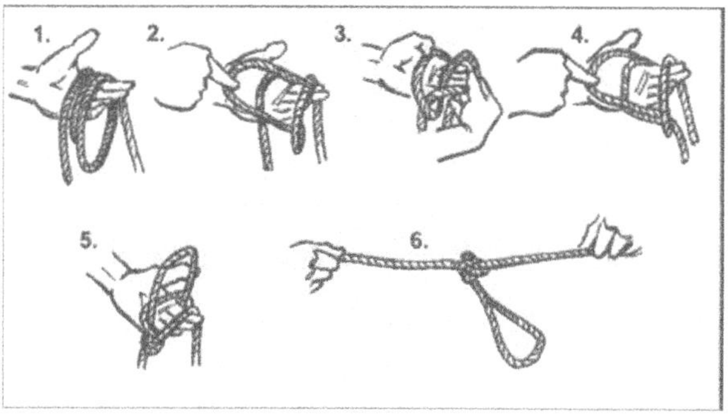

Figure 7-11. Wireman's knot.

a. **Tying the Knot.**

STEP 1. When tying this knot, face the anchor that the tie-off system will be tied to. Take up the slack from the anchor, and wrap two turns around the left hand (palm up) from left to right.
STEP 2. Take up a loop of 30 centimeters (12 inches) in the second round turn to create the fixed loop of the knot.
STEP 3. Name the wraps from the palm to the fingertips: heel, palm, and fingertip.
STEP 4. Grasp the palm wrap with the right thumb and forefinger, and place it over the heel wrap.
STEP 5. Grasp the heel wrap and place it over the fingertip wrap.
STEP 6. Grasp the fingertip wrap and place it over the palm wrap.
STEP 7. Grasp the palm wrap and pull up to form a fixed loop.
STEP 8. Dress the knot down by pulling on the fixed loop and the two working ends.
STEP 9. Pull the working ends apart to finish the knot.

b. **Checkpoints.**

(1) The completed knot should have four separate bights locking down on themselves with the fixed loop exiting from the top of the knot and laying toward the near side anchor point.

(2) Both ends should exit opposite each other without any bends.

7-12. DIRECTIONAL FIGURE-EIGHT KNOT

The directional figure-eight knot forms a single, fixed loop in the middle of the rope that lays back along the standing part of the rope (Figure 7-12).

Figure 7-12. Directional figure-eight knot.

a. **Tying the Knot.**

STEP 1. Face the far side anchor so that when the knot is tied, it lays inward.
STEP 2. Lay the rope from the far side anchor over the left palm. Make one wrap around the palm.
STEP 3. With the wrap thus formed, tie a figure-eight knot around the standing part that leads to the far side anchor.
STEP 4. Dress the knot down with the tail and the bight together.

b. **Checkpoints.**

(1) The loop should be large enough to accept a snaplink but no larger than a helmet-size loop.

(2) The tail and bight must be together.

(3) The figure eight is tied tightly.

(4) The bight in the knot faces back toward the near side.

7-13. BOWLINE-ON-A-BIGHT KNOT

The bowline-on-a-bight knot is used to form two fixed loops in the middle of a rope (Figure 7-13).

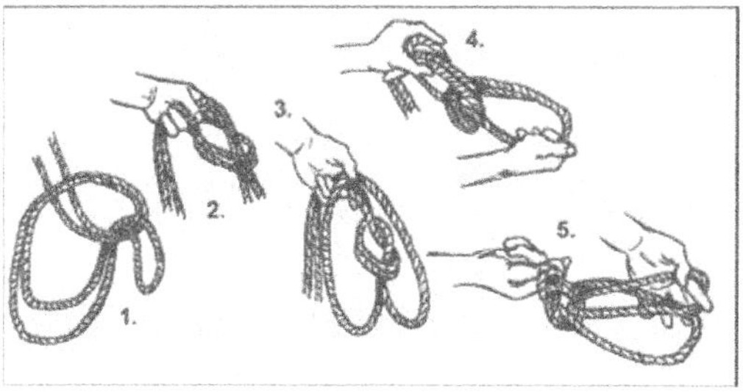

Figure 7-13. Bowline-on-a-bight knot.

a. **Tying the Knot.**

STEP 1. Form a bight in the rope about twice as long as the finished loops will be.
STEP 2. Tie an overhand knot on a bight.
STEP 3. Hold the overhand knot in the left hand so that the bight is running down and outward.
STEP 4. Grasp the bight with the right hand, fold it back over the overhand knot so that the overhand knot goes through the bight.
STEP 5. From the end (apex) of the bight, follow the bight back to where it forms the cross in the overhand knot. Grasp the two ropes that run down and outward and pull up, forming two loops.
STEP 6. Pull the two ropes out of the overhand knot and dress the knot down.
STEP 7. A final dress is required: grasp the ends of the two fixed loops and pull, spreading them apart to ensure the loops do not slip.

b. **Checkpoints.**

(1) There are two fixed loops that will not slip.
(2) There are no twists in the knot.
(3) A double loop is held in place by a bight.

7-14. FIGURE-EIGHT-ON-A-BIGHT KNOT

The figure-eight-on-a-bight knot is used to form two fixed loops in the middle of a rope . (Figure 7-14).

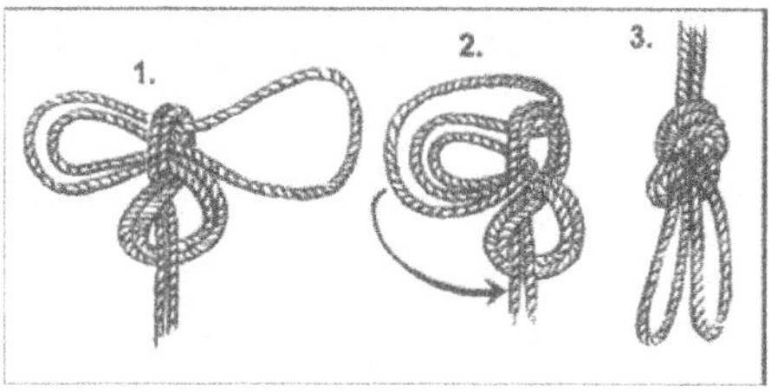

Figure 7-14. Figure-eight-on-a-bight knot.

a. **Tying the Knot.**

STEP 1. Using the doubled rope, form a 46-centimeter (18-inch) bight in the left hand with the running end facing to the left.
STEP 2. Grasp the bight with the right hand and make a 360-degree turn around the standing end in a counterclockwise direction.
STEP 3. With the working end, form another bight and place that bight through the loop just formed in the left hand.
STEP 4. Hold the bight with the left hand, and place the original bight (moving toward the left hand) over the knot.
STEP 5. Dress the knot down.

b. **Checkpoints.**
(1) There is a double figure eight with two loops.
(2) The knot will not slip.

7-15. OVERHAND LOOP KNOT

The overhand loop knot forms a single loop in the middle of a rope. It should not be used in a transport tightening system (Figure 7-15).

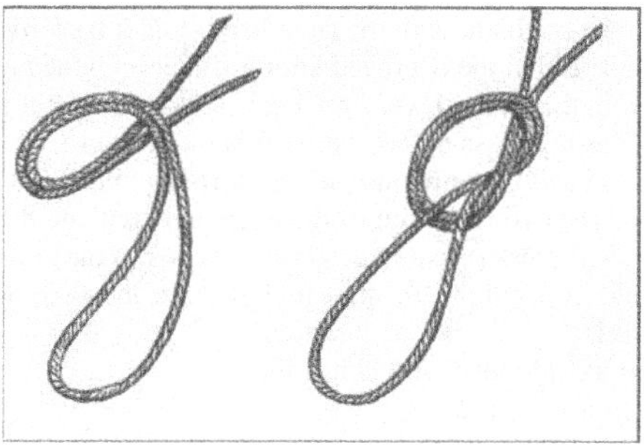

Figure 7-15. Overhand loop knot

a. **Tying the Knot.**
STEP 1. Form a bight in the rope about as long as the loop needed.
STEP 2. With the bight, form an overhand knot using the bight as the working end. Dress the knot tightly.

b. **Checkpoints.**
(1) The loop is the desired size.
(2) The ropes in the overhand knot are parallel and do not cross over each other.
(3) The knot is tightly dressed.

7-16. FIGURE-EIGHT LOOP KNOT

The figure-eight loop knot is a variation of the overhand loop and is easy to untie after it has held weight (Figure 7-16).

Figure 7-16. Figure eight loop knot.

a. **Tying the Knot.**

STEP 1. Form a bight in the rope about as large as the diameter of the desired loop.
STEP 2. With the bight as the working end, form a loop in rope (standing part).
STEP 3. Wrap the working end around the standing part 360 degrees and feed the working end through the loop. Dress the knot tightly.

b. **Checkpoints.**

(1) The loop is the desired size.
(2) The ropes in the loop are parallel and do not cross over each other.
(3) The knot is tightly dressed.

SECTION IV. SPECIAL KNOTS

7-17. SINGLE BUTTERFLY KNOT

The single butterfly knot is used to form a single, fixed loop in the middle of the rope without using the ends (Figure 7-17). The butterfly can be used for the middle man in a rope party as well as in a transport tightening system. The knot can be hard to untie when heavy weight has been placed on it for extended periods.

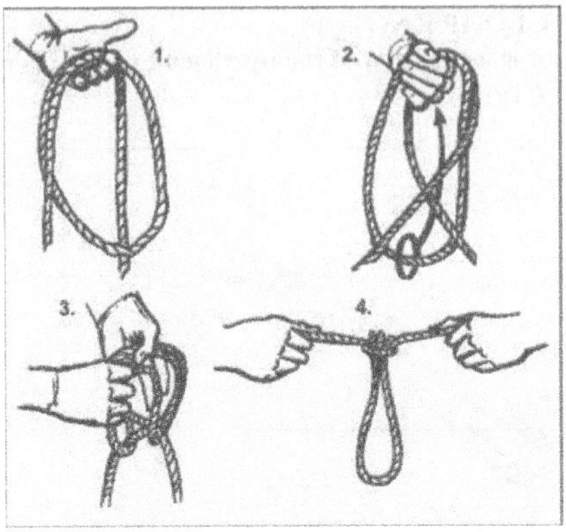

Figure 7-17. Single butterfly knot.

a. **Tying the Knot.**

STEP 1. Form a loop over the palm of the left hand.
STEP 2. With the right hand, cross the two long-standing ends; reach under and grasp the bottom of the loop formed in STEP 1.
STEP 3. Route the bottom of the loop over the crossed standing ends and back through the top of the loop.
STEP 4. Hold the newly formed loop and dress the knot down, removing all of the slack out of the knot.

b. **Checkpoints.**
(1) The wings of the knot are dressed down tightly and as close together as possible.
(2) The ropes between the wings are parallel with no crossovers.
(3) The loop is no larger than a helmet but large enough to accept a snaplink.
(4) All ropes in the knot are tightly dressed.

7-18. PRUSIK KNOT

The prusik knot is used to put a moveable rope on a fixed rope such as a prusik ascent or a tightening system. This knot can be tied as a middle, finger, or end knot.

a. **Tying the Knot.**

(1) *First method.* The prusik knot is tied at the middle or end of a rope; there are many ways to do this (Figure 7-18). The middle-of-the-rope prusik knot can be tied with a short rope to a long rope as follows:

STEP 1. Double the short rope, forming a bight, with the working ends even. Lay it over the long rope so that the closed end of the bight is 30 centimeters (12 inches) below the long rope and the remaining part of the rope (working ends) is the closest to the rappeller; spread the working end apart.

STEP 2. Reach down through the 30-centimeter (12-inch) bight. Pull up both of the working ends and lay them over the long rope. Repeat this process making sure that the working ends pass in the middle of the first two wraps. Now there are four wraps and a locking bar working across them on the long rope.

STEP 3. Dress the wraps and locking bar down to ensure they are tight and not twisted. Tie an overhand knot with both ropes to prevent the knot from slipping during periods of variable tension.

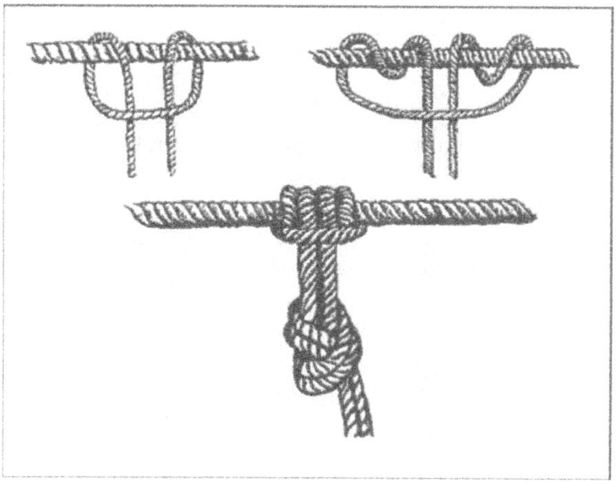

Figure 7-18. Middle-of-the-rope prusik knot.

(2) *Second method.* Another way of tying the prusik knot is the finger prusik (Figure 7-19).

STEP 1. Lay the rope across the thumb and index finger of the left hand.

STEP 2. Wrap the rope around the index finger twice (wrapping away from the rappeller). Wrap the rope around the thumb twice (wrapping towards the rappeller).

STEP 3. Place the thumb and index finger tip to tip. Push the wraps together so that the prusik is on the left thumb. Four wraps and a locking bar should be across them.

STEP 4. Pass the running end of the rope through the wraps of the prusik knot.

STEP 5. Dress the wraps and locking bar down to ensure they are tight and not twisted.

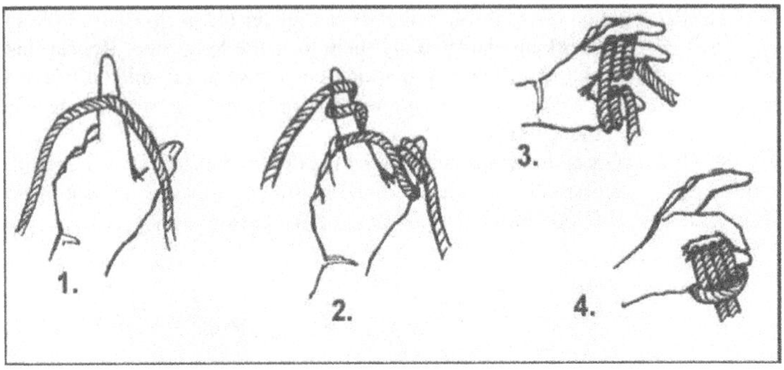

Figure 7-19. Finger prusik knot.

(3) *Third method.* Another way of tying a prusik knot is the end-of-the-rope knot (Figure 7-20).

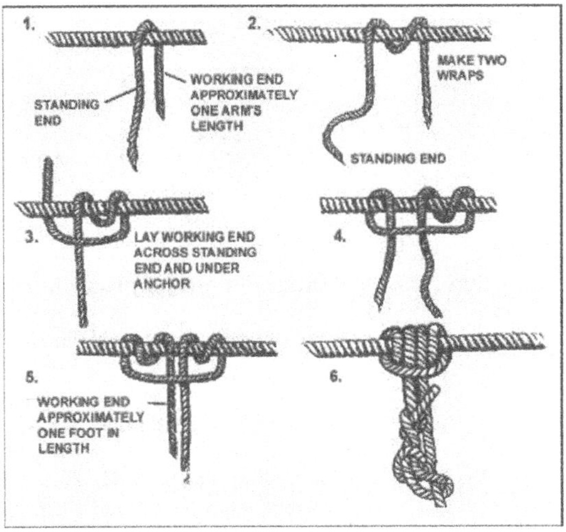

Figure 7-20. End-of-the-rope prusik knot.

STEP 1. Using an arm's length of rope, place it over the long rope.
STEP 2. Form a complete round turn in the rope.
STEP 3. Cross over the standing part of the short rope with the working end of the short rope.
STEP 4. Lay the working end under the long rope.
STEP 5. Form a complete round turn in the rope, working back toward the middle of the knot.

STEP 6. Four wraps and a locking bar should run across them on the long rope. Dress the wraps and locking bar down. Ensure they are tight, parallel, and not twisted.

STEP 7. Finish the knot with a bowline to ensure that the prusik knot will not slip out during periods of varying tension.

b. **Checkpoints.**

(1) Four wraps with a locking bar.

(2) The locking bar faces the rappeller.

(3) The knot is tight and dressed down with no ropes twisted or crossed.

(4) Other than a finger prusik, the knot should contain an overhand or bowline to prevent slipping.

7-19. BACHMAN KNOT

The bachman knot provides a means of using a makeshift mechanized ascender (Figure 7-21).

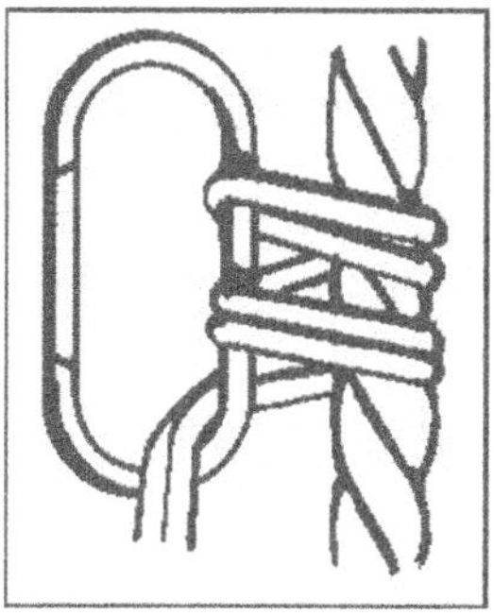

Figure 7-21. Bachman knot.

a. **Tying the Knot.**

STEP 1. Find the middle of a utility rope and insert it into a snaplink.

STEP 2. Place the snaplink and utility rope next to a long climbing rope.

STEP 3. With the two ropes parallel fromthe snaplink, make two or more wraps wraps around the climbing rope and through the inside portion of the snaplink.

Note: The rope can be tied into an etrier (stirrup) and used as a prusik-friction principal ascender.

b. **Checkpoints.**

(1) The bight of the climbing rope is at the top of the snaplink.

(2) The two ropes run parallel without twisting or crossing.

(3) Two or more wraps are made around the long climbing rope and through the inside portion of the snaplink.

7-20. BOWLINE-ON-A-COIL KNOT

The bowline-on-a-coil knot is an expedient tie-in used by rappellers in two- and three-man party climbs when a climbing harness is not available (Figure 7-22).

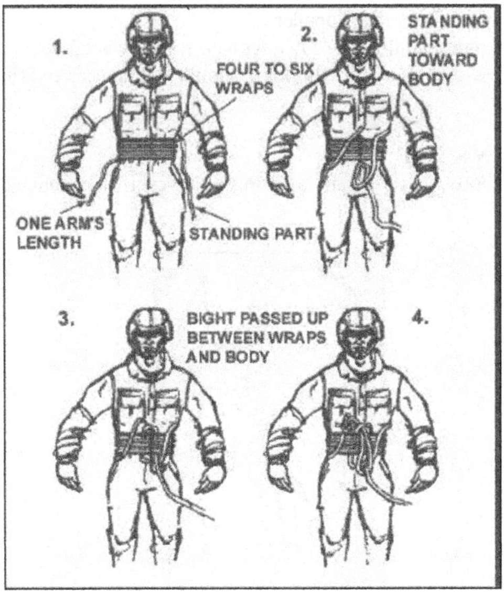

Figure 7-22. Bowline-on-a-coil knot.

Note: A safer tie in, when possible, is a swiss seat and figure eight around the shoulder, to avoid damaging the ribs if the climber falls.

a. **Tying the Knot.**

STEP 1. To tie a bowline-on-a-coil knot, start the same as tying a bowline around the waist. (The waist of the rappeller is the anchor point. The rappeller has about 40 centimeters [16 inches] of rope in the right hand. The standing end is coming from the remainder of the rope in his left hand.)

STEP 2. Draw slack from the standing end of the rope in the left hand, wrapping enough coils around the body to complete a minimum of four (but no more than six) coils.

STEP 3. With the rope in the left hand, make a loop inward with the standing end.

STEP 4. Slide the loop up between the horizontal wraps and the body, ensuring that the loop does not come uncrossed.

STEP 5. Bring a bight up through the loop. Bring the working end of the rope in the right hand through the bight and back onto itself.

OR

Insert the short working end to the right through the loop (just passed through the wraps). Pass the working end around the long standing end and back on itself, forming a bight.

STEP 6. Dress the knot down.

STEP 7. Safety off with an overhand knot against the knot formed on the top single coil. From 10 to 15 centimeters (4 to 6 inches) of rope should be left (standing part).

b. **Checkpoints.**

(1) There are a minimum of four (maximum of six) parallel wraps; the top and bottom ropes cross forward of the hips.

(2) The loop must be underneath all wraps.

(3) Other checkpoints are the same as the bowline.

7-21. THREE-LOOP BOWLINE KNOT

The three-loop bowline knot is used to form three fixed loops in the middle of a rope (Figure 7-23). It is used in a self-equalizing anchor system.

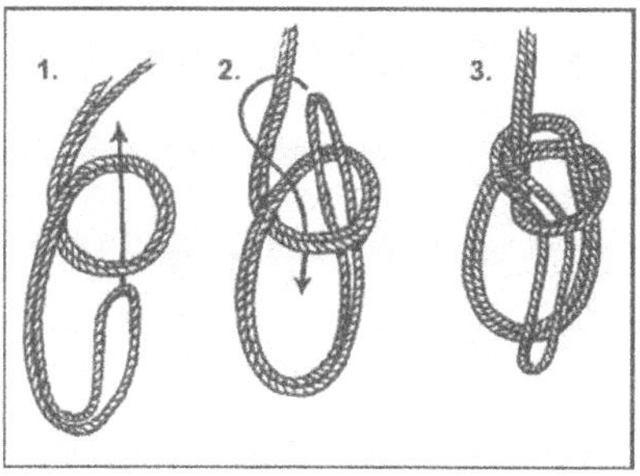

Figure 7-23. Three-loop bowline knot.

a. **Tying the Knot.**

STEP 1. Form about a 60 centimeter (24-inch) bight in the rope, laying it across the left hand, palm up, with the bight to the front.

STEP 2. With the right hand, form a loop with both ropes. Hold it in the left hand.

STEP 3. With the right hand, bring the long bight through the loop(formed in STEP 2) and around and around the two standing ends.

STEP 4. Follow the bight around the standing ends and back down into the loop.

STEP 5. Hold all three loops in the left hand. With the right hand, grasp both standing ropes and pull while dressing up the bowline.

b. **Checkpoints.**

(1) Checkpoints are the same as a regular bowline, except that it is tied with two ropes.

(2) An overhand knot must be tied if the third loop is not to be used as a secondary anchor.

7-22. FIGURE-EIGHT SLIP KNOT

The figure-eight slip knot forms an adjustable bight in a rope (Figure 7-24).

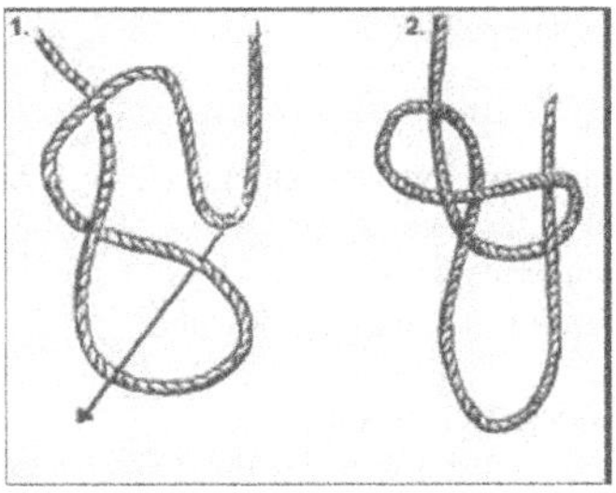

Figure 7-24. Figure-eight slip knot.

a. **Tying the Knot.**

STEP 1. Form a 30-centimeter (12-inch) bight in the end of the rope.

STEP 2. Hold the center of the bight in the right hand. Hold the two parallel ropes from the bight in the left hand about 30 centimeters (12 inches) up the rope.

STEP 3. With the center of the bight in the right hand, twist two complete turns clockwise.

STEP 4. Reach through the bight and grasp the long standing end of the rope. Pull another bight (from the long standing end) back through the original bight.

STEP 5. Pull down on the short working end of the rope and dress the knot down.

STEP 6. If the knot is to be used in a transport tightening system, take the working end of the rope and form a half hitch around the loop of the figure-eight knot.

b. **Checkpoints.**

(1) The knot is in the shape of a figure eight.

(2) The sliding portion of the rope is the long working end of the rope.

Section V. SPECIAL-PURPOSE KNOTS

7-23. KLEIMHIEST KNOT

The kleimhiest knot provides a moveable, easily adjustable, high-tension knot capable of holding extremely heavy loads while being pulled tight (Figure 7-25). This system requires a six-man pulling team when used in a transport tightening system.

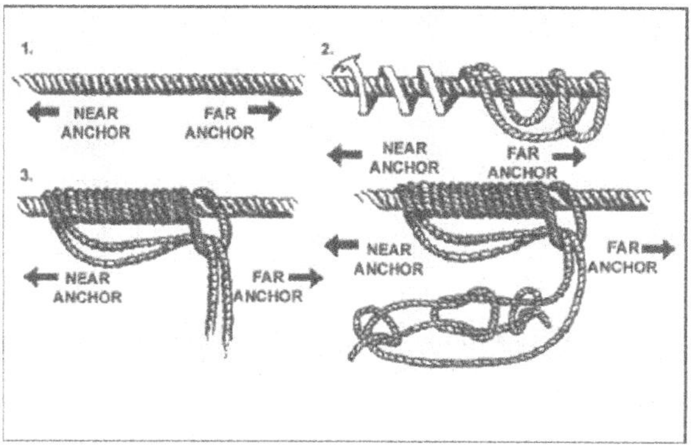

Figure 7-25. Kleimhiest knot.

a. **Tying the Knot.**

STEP 1. Using a utility rope, offset the ends by 30 centimeters (12inches). With the ends offset, find the center of the rope and form a bight. Lay the bight over a horizontal rope near the far side anchor.

STEP 2. Wrap the tails of the utility rope around the horizontal rope back toward the near side anchor. Wrap at least four complete turns.

STEP 3. Pass the remaining tails of the utility rope through the bight (see STEP 1).

STEP 4. Join the two ends of the tail with a joining knot (square knot with overhand safety knots).

STEP 5. Dress the knot down tightly so that all wraps are touching. Insert two snaplinks (opposing gates) into the large loop tied with the tails.

b. **Checkpoints.**

(1) The bight is closest to the far side anchor.

(2) All wraps are tight and touching.

(3) The ends of the utility rope are properly secured with the adjoining knot.

7-24. OVERHAND KNOT

The overhand knot (Figure 7-26) is used to make a knotted rope for a handline, to secure the ends of other knots, and to make stirrups in direct-aid climbing. This knot can also be used to temporarily whip the end of a rope. It is used with other knots.

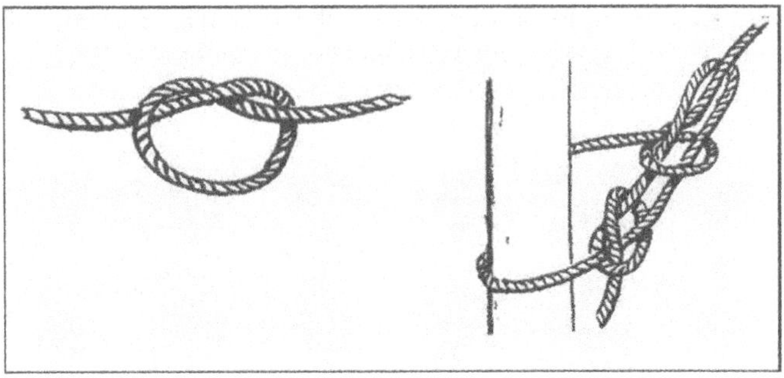

Figure 7-26. Overhand knot.

a. **Tying the Knot.**

STEP 1. Make a loop in the rope.
STEP 2. Pass the working end of the rope through the loop.
STEP 3. Dress down the knot.

b. **Checkpoints.**

(1) Running ends exit the loop in opposite directions.
(2) The knot is dressed down tightly.

7-25. FROST KNOT

The frost knot is used when working with flat webbing (Figure 7-27). It holds better than a standard overhand knot or square knot.

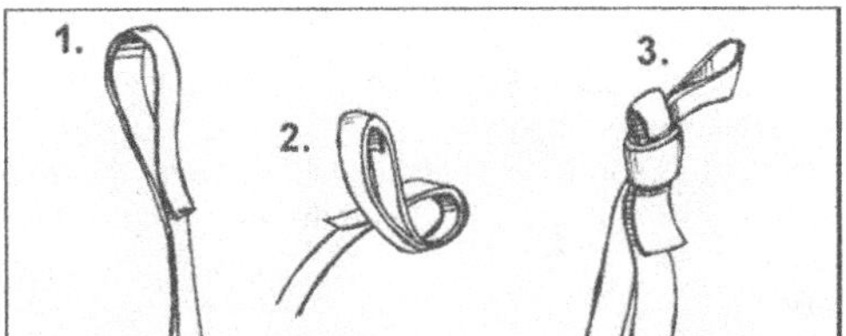

Figure 7-27. Frost knot.

a. **Tying the Knot.**

STEP 1. Lap one end (a bight) of webbing over the other (about 25 to 30 centimeters [10 to 12 inches]).

STEP 2. Tie an overhand knot with the newly formed triple-strand webbing; dress tightly.

b. **Checkpoints.**

(1) The tails of the webbing run in opposite directions.

(2) Three strands of webbing are formed into a tight overhand knot.

(3) A bight and tail exit the top of the overhand knot.

7-26. GIRTH HITCH KNOT

The girth hitch knot is used in tying a runner to a partly driven piton (Figure 7-28).

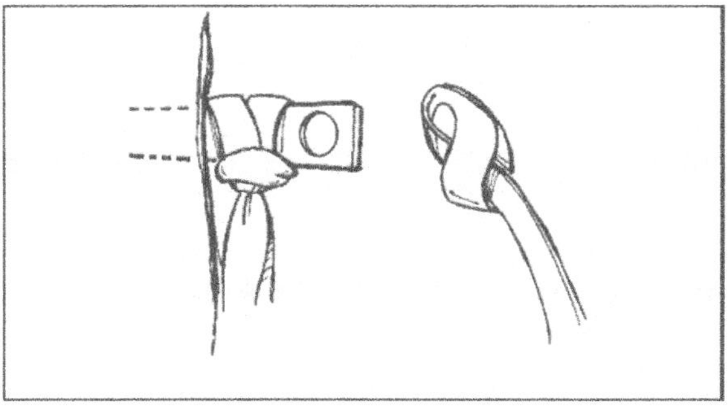

Figure 7-28. Girth hitch knot.

a. **Tying the Knot.**

STEP 1. Bring the standing ends back through the bight of a length of rope or webbing.

STEP 2. Dress the knot tightly.

b. **Checkpoints.**

(1) Two wraps exist with a locking bar running across the wraps.

(2) The knot is dressed tightly.

APPENDIX A

RISK ASSESSMENT

This appendix explains risk assessment for rappel operations. A risk assessement matrix is shown in Table A-1.

A-1. EFFECT

a. **Catastrophic.** Death or permanent total disability, system loss, major property damage.

b. **Critical.** Permanent partial disability, temporary total disability in excess of three months, major system damage, significant property damage.

c. **Moderate.** Minor injury, lost workday accident, compensable injury or illness, minor system damage, minor property damage.

d. **Negligible.** First aid or minor supportive medical treatment, minor system impairment.

A-2. PROBABILITY

a. **Frequent.**

(1) Individual soldier or item—Occurs often in career or equipment service life.

(2) All soldiers exposed or item inventory—Continuously experienced.

b. **Likely.**

(1) Individual soldier or item—Occurs several times in career or equipment service life.

(2) All soldier exposed or item inventory—Occurs frequently.

c. **Occasional.**

(1) Individual soldier or item—Occurs sometime in career or equipment service life.

(2) All soldiers exposed or item inventory—Occurs sporadically, or several times in inventory service life.

d. **Seldom.**

(1) Individual soldier or item—Possible to occur in career or equipment service life.

(2) All soldiers exposed or item inventory—Remote chance of occurrence; expected to occur sometime in inventory service life.

e. **Unlikely.**

(1) Individual soldier or item—Can assume will not occur in career or equipment service life.

(2) All soldiers exposed or item inventory—Possible, but improbable; occurs only very rarely.

A-3. RISK LEVELS

a. **Extremely High.** Loss of ability to accomplish mission.

b. **High.** Significantly degrades mission capabilities in terms of required mission standards.

c. **Medium.** Degrades mission capabilities in terms of required mission standards.

d. **Low.** Little or no impact on mission accomplishment.

RISK ASSESSMENT			HAZARD PROBABILITY				
			Frequent	Likely	Occasional	Seldom	Unlikely
			A	B	C	D	E
E F F E C T	Catastrophic	I	***Extremely High***				
	Critical	II			***High***		
	Moderate	III		***Medium***			
	Negligible	IV			***Low***		

Table A-1. Risk assessment matrix.

APPENDIX B

BASIC EQUIPMENT FOR RAPPEL OPERATIONS

The equipment listed in this appendix is the standard basic items required to conduct ground and air rappelling operations.

B-1. GROUND OPERATIONS

For ground operations, the following equipment is needed:

a. Nylon rope, 120 feet (NSN 4020-00-931-8793).
b. Snaplinks (NSN 8465-00-360-0228).
c. Gloves (NSN 8415-00-268-7868).
d. Rope coiling log or standard deployment bag.
e. Rappel seat.
f. Rappel rings (used as expedient anchor points).

B-2. AIR OPERATIONS

For air operations, the following equipment is needed:

a. All items in paragraph B-1.
b. Donut ring (UH-1H operations only) (locally produced TSC item).
c. Safety floating ring.
d. Two 6-foot safety ropes (with snaplinks).
e. Eight 120-foot nylon ropes.
f. Monkey harness for static rappel master.
g. Sixteen snaplinks (UH-60 operations).
h. Twenty-two snaplinks (UH-1H operations).
i. FM communications.
j. Aircraft communications headset.

GLOSSARY

ACRONYMS AND ABBREVIATIONS

AFRM	assistant fast-rope master
AGL	above ground level
AMC	air mission commander
AR	Army regulation
ATP	aircrew training program
attn	attention
BDU	battle dress uniform
CG	commanding general
CGU	modified helicopter safety strap
chemlight	chemical light
cm	centimeter
DA	Department of the Army
D-bag	deployment bag
F	Fahrenheit
FM	frequency modulated; field manual
FRIES	fast-rope insertion/extraction system
FRM	fast-rope master
GSO	ground safety officer
HQ	headquarters
IAW	in accordance with
ICS	interagency communications system
intercomm	intercommunication
LBE	load-bearing equipment
LZ	landing zone
MEDEVAC	medical evacuation
mm	millimeter
mph	miles per hour
MSO	mountaineering safety officer
NCO	noncommissioned officer
NCOIC	noncommissioned officer in charge
NSN	national stock number
NVG	night vision goggles

NWTC	Northern Warfare Training Center
OIC	officer in charge
PIC	pilot in command
POI	program of instruction
POL	petroleum, oils, and lubricants
RATELO	radiotelephone operator
RSO	rappel safety officer
SFC	sergeant first class
SOAR	Special Operations Aviation Regiment
SOP	standing operating procedures
SPIES	special patrol infiltration/exfiltration system
STABO	a system for extracting personnel by helicopter (the combined first letters of the surnames of the five persons who designed the system)
SSO	STABO system operator
TAPO	Technology Applications Program Office
TC	training circular
TM	technical manual
TRADOC	Training Doctrine Command
TSC	training support center
UIAA	Union of International Alpine Association
UH	utility helicopter
US	United States
USAF	United States Air Force
USAIS	United States Army Infantry School
USASOC	United States Army Special Operations Command
VIP	very important person

REFERENCES

DOCUMENTS NEEDED

These documents must be available to the intended users of this publication.

DA Form 2028	Recommended Changes to Publications and Blank Forms. 1 February 1974.
DA Form 5752-R	Rope History and Usage. May 1989.
FM 57-38*	Pathfinder Operations. 9 April 1993.
FM 7-93*	Long Range Surveillance Unit Operations. 3 October 1995.
TC 90-6-1*	Military Mountaineering. 26 April 1989.

READINGS RECOMMENDED

These sources contain relevant supplemental information.

AR 95-3	Aviation: General Provisions, Training, and Standardization, A Resource Management. 27 September 1990.
FM 8-35	Evacuation of the Sick and Wounded. 22 December 1983
FM 57-220	Basic Parachuting Techniques and Training. 1 June 1990.
TM 10-1670-262-12	Operator's and Organizational Maintenance Manual (Including Repair Parts and Special Tools List) for Personnel STABO Extraction System and Cargo Personnel Lowering Anchor Device. 7 April 1988.
USASOC Reg 350-2	Training Airborne Operations. 1 June 1995.
USASOC Reg 350-6	Training Special Operations Forces Infiltration/Exfiltration Techniques. 12 May 1995.
	The Air Assault School Handbook.

*This source was also used to develop this publication.

INDEX

TC 21-24
10 SEPTEMBER 1997

By Order of the Secretary of the Army:

DENNIS J. REIMER
General, United States Army
Chief of Staff

Official:

JOEL B. HUDSON
Administrative Assistant to the
Secretary of the Army
03807

DISTRIBUTION:

Active Army, Army National Guard, and U.S. Army Reserve: To be distributed in accordance with the initial distribution number 115059, requirements for TC 21-24.

Made in the USA
Monee, IL
05 November 2026

67807967R00203